RESOURCE CONSERVATION AND FOOD SECURITY

RESOURCE CONSERVATION AND FOOD SECURITY

An Indian Experience

Volume I

Edited by

TAPESHWAR SINGH

CONCEPT PUBLISHING COMPANY, NEW DELHI-110059

ISBN : 81-8069-100-4 (Set)
81-8069-101-2 (Vol. I)

First Published 2004

Published and Printed by
Ashok Kumar Mittal
Concept Publishing Company
A/15-16, Commercial Block, Mohan Garden
NEW DELHI-110059 (India)

Phones : 25351460, 25351794
Fax : 091-11-25357103
Email : publishing@conceptpub.com

CONTENTS

SECTION-II
SUSTAINABLE AGRICULTURAL DEVELOPMENT AND FOOD SECURITY

SECTION-III
WEATHER AND CLIMATE

SECTION-IV
GEO-SPATIAL TECHNOLOGY APPLICATION

VOLUME-II

SECTION-V
ENVIRONMENTAL CHANGE AND IMPACT

SECTION-VI
SOCIO-CULTURAL ASPECTS

INTRODUCTION

Tapeshwar Singh

In India nearly 70 per cent of population directly or indirectly depends on land for their livelihood. Sustained and equitable growth in agriculture is, therefore, crucial to the food security and sustainable development. Economic growth and evironmental preservation are no any more opposing objectives. Environment friendly economic growth has become a necessity. The day of frontier economy when abundant natural resources were available to propel economic growth and raise standard of living are over. We have reached a stage when the concern is with the sustainable development meaning thereby not only the development of the present but also of the future. Thus the focus has moved from sastained growth' to 'sustainable development'. Sustainable implies a notion of equilibrium, an equilibrium that simultaneously satisfy the needs of development as well as resource conservation. Tcchnology has helped us to harness the forces of nature thereby enhancing our capacity to improve living standard and quality of life. However, our past achievements in raising resources efficiency and productivity should not delude us into a thinking that any constraint can be overcome. It is known that technology has proved to be double edged sword in many cases. Green revolution technology for example, has been very successful in achieving spectacular results in food grain production during the last three decades. However, signs of fatigue in the natural resources have already emerged and have unleashed various agro-ecological problems. It has badly damaged the natural resource base of the country. Great apprehensions are, therefore, voiced about the stability of enhanced productivity and environmental safety as also the grave limitation to feed the teeming millions in 21st century. Realizing the dangerous implications of the emerging scenario particularly rapid depletion and degradation of natural resources and considering the paramount importance of the inevitable prudent use of natural resources an International Symposium on Land Degradation : New Trends Towards Sustainable Agriculture simultaneously with the Commonwealth Geographical Bureau Food Security Workshop was organized during April 7-12, 2002 at Ghaziabad to take stock of the work done in the past and to formulate future strategies for efficient integrated management of the natural resource for sustainable agriculture. The workshop and symposium were organized by the Department of Geography, MMH College (Ch. Charan Singh University), Ghaziabad. It was cosponsored by State Land Use Board, Planning Department, Government of Uttar Pradesh; Council for Scientific and Industrial Research; Indian Council of Social Science Reasearch; National Informatics Centre (Minisry of Communication and Information Technology, Govt. of India); Department of Science and Technology, Government of India; Bhoovigyan Vikas Foundation and Commonwealth Geographical Bureau. It was inaugurated by Dr. Panjab Singh, the then Secretary, Department of Agricultural Research and Education and Director General of Indian Council of Agricultural Research with Shri Som Pal, Member, Planning Commission in the Chair. Around 225 scientists and other interested in judicious use of natural resources representing different universities and organizations from India and abroad participated. In total 65 oral and 15 poster presentations were made. Some of the participants could not attend but sent their contributions.

Land degradation under increasing abiotic and biotic stresses has been increasing sharply in the last two-three decades and has become central issue to our developmental strategies. The five-day long proceedings and deliberations were particularly concerned with the contentious issues like (i) monitoring land quality and sustainable agriculture; (ii) Land use ethics and social and evironmental implications of land degradation—social relevance of soils function and natural resource endowment and biodiversity, climate change etc.; (iii) Land reclamation strategies and tchniques—land use, planning and prevention of degradation, sustainable land management : soil and water conservation and management; (iv) Public policy to achieve sustainable land use; (v) land degradation assessment—biotic and abiotic indices of land degradation, early warning indicators, tools (GIS, DEM, Simulation Models, and Decision Support Systems), rehabilitation of degraded land, use of traditional knowledge, early warning system and database, GIS application and techniques (application of tools and methods to assess the impact of various land use scenarios) and increasing vulnerability of land degradation risks, simulation and inerdisciplinary cooperaition and assistance to get from different components of planning and implementation and (vi) *The* land resource base : vulnerability; degradation risks to land submitted to different land usè pressure, resilience-potential of misused/degraded land to naturally recovered to its orginal condition. The symposium also focused on the impacts of natural disaster such as drought and floods and the impact of environmental changes on land use and cropping pattern in food security context. Learned speakers expressed serious concerns on surfacing noticeable stagnancy in food output in some cropping systems. In the contemporary state of science and development it is very important to improve mutual understanding of natural and social scientists, policy-makers and implementing agencies as well as better feeling of scientific understanding of land resources for addressing these issues.

There is no space to present excerpts of all the papers presented, therefore, I would like to concentrate only on some introductory presentations given by Dr. Panjab Singh, Seceretary, Dapartment of Agricultural Research and Education and Director General of Indian Council of Agricultural Research, Shri Som Pal, Member, Planning Commission; Dr. Victor Roger Savage, President of Commonwealth Geographical Bureau and Head Department of Geography, National University of Singapore; Dr. Tony Binns and Dr. Don Funnell of Sessex University, London; Dr. D.R. Sikka, Former Director, Indian Institute of Tropical Meteorology, Pune; Dr. Jagdish Singh, former Head, Department of Geography, Gorakhpur University; Dr. S.N. Das, Chief Soil Scientist, All Indian Soil and Land Use Survey; Dr. B.B. Singh, Ex-Head, Department of Geography, M.M.H. College; Dr. M. Moni, Deputy Director General, National Informatics Centre, Ministry of Communication and Information Technology, Government of India; and Dr. Tapeshwar Singh, Convener, CGB Worshop and Symposium.

Shri Som Pal, Member, Planning Commission stressed the need for solving the food security problems at both macro as well as micro-levels. At national level we have become food self-sufficient but at households level, yet not successful in solving the problems of poverty, hunger and malnutrition. It is a big challenge before the country and has to be solved for attaining universal food security. Thus, food security should be assured for those vulnerable groups who are food-insecure and resource-poor, and are unable to transact in the market. While talking about sustainable agricultural development Shri Som Pal lamented that land degradation is very serious problem and said that about 57 per cent area in the country is potentially exposed to various degradation forces such as water and wind erosion, floods, acidification, waterlogging, sodicity, shifting cultivation, ravines and gullies, riverine and torrent land etc. The Indian agriculture is running with net negative balance of plant nutrient at the rate of 8-10 million tonnes a year. The health of Indian soil is becoming poorer day by day and second generation problems of green revolution such as imbalance in micronutrient are surfacing. Depletion of groundwater and side by side waterlogging, alarming loss of biodiversities, and desertification are threatening the very sustainability of the agroecosystem. He called upon geographers to come forward with feasible solutions to overcome these challenges.

Tony Binns speaking on 'food security: changing the focus' presented various definitions of food security and elaborated extensively the different approaches in historical perspective—modern and post-modern approaches in development, post-modernism and food security, case study: issues concerning food security in Southern Africa, understanding indigenous livelihood systems: ensuring food security, future food security—an action plan etc. He particularly focused on the need of ACCESS (Food entitlement) and CONTROL over resources. Dr. Binns discussed in detail various facets of the coping strategies of food and livelihood on various scales—global, national, community, household and individual. The discussion on a 'food first' perspective to a 'livelihood security' perspective since 1985 drew good attention and appreciation. Lamenting on the action plan Dr. Binns suggested various strategies for attaining sustainable food security such as focus on supplying vulnerable people and households with secure access to food. Individual and household needs should take precedence over national food self-sufficiency and self-reliance. need for poverty-reducing economic growth to generate secured livelihoods, (with buffers against destitution) and buoyant national economies to provide jobs and, where necessary purchase of food, in agricultural sector—need for generating labour-intensive strategies which emphasise job creation and better incomes for poorest groups, balance between food and cash crops, emphasis on long-term comparative advantage rather than self-sufficiency for its own sake, efficient food marketing to store and distribute food at reasonable prices to all parts of the country in all seasons and in all years, improved famine preparedness and response, especially at local level, locally-based stock for relief distribution, programmes for vulnerable group feeding and rehabilitation etc. Dr. Binns also argued to recognize causes of food-insecurity, situation and strategies, providing households and individuals with choices, which contribute to self-determination and autonomy in developing livelihood strategies. Food security planning should follow a 'process' rather than a blueprint approach of decentralization, bias to action over planning, encourages risk taking and innovation, multi-disciplinary and multi-sectoral planning teams and strong international support in crisis management.

Don Funnell explained the significance of Geographic Information System in scientific studies, while presenting a paper on 'Food Security and Geospatial Analysis'. He presented comprehensive analysis of GIS application for resource appraisal and solving the food problem at different spatial levels (national, regional and local). He stressed the need to study the link between food availability and nutritional well-being which is not automatic, the impact of new agenda such as globalisation, biotechnology etc., whether food security remains a useful policy concept and the role of food aid in providing food security. Dr. Funnell observed that there has been a rapid growth in the development of affordable technologies associated with the provision of data which are flagged off to assist in the decision making process. Easily available computing power, cheap and relatively accurate GPS equipment, access to the Internet by more and more people and agencies and the development of PPGIS (sometimes known as GIS2) have widened the scope of accuracy. Information and Communication Technology (ICT) have been developed which promised not just improved methods of achieving existing aims but, arguably more important, providing the capacity for isolated and disenfranchised communities to participate in the whole process of managing food security in a manner that offers new ways of thinking about the process. The food security debate is becoming more nuanced in the light of a better understanding of the food and mutritional system but that ICT, especially geospatial technologies have become increasingly significant in tackling these problems.

Geospatial analysis must be involved in the following processes :

(i) *Generating data* : In this category falls the wide range of information gathering technologies ranging from 'high tech' use of satellite imagery and the use of GPS, through structured data collection by formal agencies.

(ii) *Handling data* : The particular forms in which data are categorised and stored.

(iii) *Interpreting data* : Food security policy usually requires the monitoring of the flows of food through an agricultural year. Depending upon the focus of attention concerning vulnerabilities. In some cases food vulnerability is not a function of gross supply or even of price *per se* but of weaknesses in other aspects of infrastructure such as transport, especially in the face of natural disasters such as earthquakes or floods. Data interpretation, therefore, is the process of using geospatial information.

(iv) *Developing policy* : National, regional, and local scales of policy formulation only really make sense when it is possible to obtain and analyse data which itself is available at these scales, too often, shortage of spatially referenced data at each of these scales.

(v) *Monitoring* : ICT can play a significant role in this area. One of the problems with most conventional data gathering procedures concerns the infrequency of data availability. Traditional data collection procedures is also time consuming hence using rapid appraisal techniques combined with input from satellite monitoring may help to update information continuously meanwhile checking spot anomalies.

Delivering the Valedictory Address Dr. D.R. Sikka, a renowned meteorologist underlined the vulnerability of the Indian agriculture as well as farmers, in view of global warming and changing climate and ecology. He laid stress on evolving adaptation strategies and resource-conserving techniques for sustainable farm development and improved productivity in the light of the current climate models prediction of a global warming of about 1.4 to 5.8 degree centigrade over the present century which is likely to have an impact on Indian agriculture resulting in up to 30 per cent drop in crop production by 2050. This is likely to have adverse impact particularly on rice and wheat yields. Eventhough there are some uncertainties in model prediction, it is important to keep these projections in view while formulating policies. South Asia, with its large population, low incomes, fragile ecosystems and dependence on agriculture and allied activities is particularly vulnerable to climate change. Dr. Sikka observed that projection of reduction in crop yields in our country were of grave concern, taking into consideration the growing food demands on account of rapidly increasing population. He also underlined the need to understand the social, environmental and ethical implications of land degradation and also warned against the danger of corporatisation of land in India and stressed the need to safeguard the interest of the small farmers from 'big sharks'. He also observed 'if we have to sustain agriculture we must have a proper database and information on land quality and all other determinants of agriculture'.

Jagdish Singh stressed the need of adopting the organic farming system (slowly and cautiously) that holds promise for the 'right of nourishment' of the masses and sustainable food security as the fatigues of Green Revolution have already emerged and chemical input-based farming is petering out without fully covering all crops and regions of the country. Under such a backdrop we must evolve a new paradigm of sustainable agricultural development based on job-led economic growth rooted in the principles of ecology and equity.

S.N. Das considered watershed management as the most advance integrated approach for sustainable development and management of our natural resources. Developmental planning on watershed basis evolves eco-friendly land management system to meet the needs of ever-increasing human and animal population and mounting biotic and abiotic stresses on the environment and calls for realistic planning of soil and water conservation measures and land use management in the sense of economic feasibility and people's acceptability. Watershed management involves detailed inventory of database of key natural resources—soil and water and their interactive analysis with socio-economic condition and environment. Application of modern technologies like space technology, bio-technology and information technology suitably integrated with the environment to develop a cost-effective, energy-efficient and environmentally benign management system which can make development sustainable over time. Sustainability has many aspects—ecological, socio-economic and cultural and all have to be integrated.

B.B. Singh in his presentation on 'Food Security Dilemma' observed that food security is not the function of mere food production alone but it is more a question of economic access to food by all sections of society, particularly the most vulnerable one. Food security without livelihood security is insurmountable. He suggested vigorous employment generation in the farm and off-farm sectors to enable the resource-poor to have direct access to income as an integral strategy for ensuring food security, promotion of employment through rural industrialization and giving immediate priority for creating opportunities for value-addition integrating production with processing and marketing, and full utilization of territorial resources and production potential. As per government claims the rate of poverty in India has come down to 26 per cent over 5 decades of Independence and planned development. However, the absolute number of the poor remains almost equal to the size of total population of the country at the time of independence. Dr. Singh, therefore, emphasized the need of re-defining 'food security' and arguably suggested to establish a separate Ministry at the Centre and the States to deal exclusively with the persistent problem of food insecurity, unemployment and poverty.

M. Moni observed that at the end of 20th century, the agricultural enterprises stand at the technological crossroads. The existence (or absence) of favourable natural resources can facilitate (or retard) the process of economic development. India is endowed with regional diversities for its uneven economic and agricultural development. Out of total geographical area of 329 Mha of the country nearly 265 Mha represent varying degrees of potential for biological production. More than 57 per cent of land suffers from different types of degradation and it is a very serious environmental problem in India in a acceptable technology management framework in the prevailing socio-economic infrastructure. Land degradation inversely affects the land productivity, food security, climate and environmental security, and eventually the quality of life.

Emergence of Information Technology on the national agenda in India and announcement of IT policies by the respective State Governments have recognized the convergence of core technologies and E-Governance as the tool for sustainable development and globalization of economy. The fusion of Information and Communication Technology, Biotechnology, Environment Technology, GIS and Remote Sensing Technology are expected to bring in the desired S & T inputs for sustainable development. As a step towards reaching technology to the Small Holders (Resource-poor-farmers) and also to usher in agricultural reforms through ICT in the country, the paper discussed an *Informatics led development Strategy—from digital divide to digital opportunities* for sustainable agricultural and rural development, the need to develop a comprehensive Agricultural Resources Informatics using Geomatics Technology in districts with public funding, and also suggests need to develop metadata and application of Open GIS model for optimal utilization of agricultural resources in India.

While introducing the themes of the Workshop and Symposium Dr. Tapeshwar Singh observed that economic growth and environmental preservation are no more opposing objectives. Environment friendly economic growth has become a necessity. The days of frontier economy when abundant resources were available to propel economic growth and raise living standards are over. Action is needed to prevent degradation and regenerate degraded resources. We urgently need positive steps to extend the forest cover, rejuvenate degraded lands and other natural resources especially two key resources such as land and water. Ecological improvement can certainly condition sustainable development. The use of geomatics (GIS, Remote Sensing, GPS) is very significant in geographical studies, hence be incorporated at all levels from primary to higher education. A renaissance in geographical thinking is necessary to maintain the Domain of Geography as a science. We must develop a vision for achieving the ecologically sustainable agriculture through income enhancing, productivity growth promoting, resource conserving and environment friendly practices, policies and technologies that will conform to the emerging paradigms of efficiency, sustainability and environmental security for reducing poverty and achieving sustainable food security.

Recommendations

The three-day long discussions and exchange of ideas on up-to-date information by the delegates on the main theme and sub-themes of the Symposium across the spectrum of the Earth Sciences, Environment, Agriculture, Information Technology, Meteorology and Social Sciences concluded with the following recommendations :

- Declining soil productivity in India is threatening the food grain production. The challenge before the scientific community to combat poverty and hunger can be achieved through combating land degradation and at large soil and water conservation and management. Excessive use of land resources by rapidly growing population and diversion of prime agricultural land for other uses should be prevented and brought to a minimum.
- The land use policy should be optimizing returns on a long-term basis rather than meeting short-term requirement. The emphasis should be on the management of land resources on watershed basis. Integrated and holistic development of rainfed area will have to be promoted by conservation of land and water with involvement of watershed community.
- The potential ability of the system to meet firture food need must be measured against constraints being imposed by limited resources, mounting degradation and unfavourable policies since these factors will hamper the future growth and development.
- Farming will have to be knowledge-intensive, using information from Remote Sensing, Geographical Information System (GIS), Global Positioning System (GPS) and information and computer techniques.
- With necessary caution to prevent unforeseen problems associated with biotechnology the adoption of this technology will play an important role in strengthening food security in 21st century.
- The gene revolution, the ecotechnology revolution and the communication revoluton have to be included as the main components of resource management and the national agricultural policy for sustainable development of agriculture in future.
- Soil, water, biodiversity and climate-basic natural resources for agricultural growth and development, are subjected to

various forms of deteriorating influences and indiscriminate exploitation, hence needs conservation and management, particularly under the threat of anticipated climate change scenarios.

- In fact, the complex and intrinsic relationship between environment and agricultural development has received insufficient attention, therefore, improvement of environmental health for the goal of achieving sustainable agricultural development and food security, should be central to planning.
- A second green revolution is needed particularly in those areas, which have already been benefited but have also suffered environmental problem and stagnancy in food output under the prevailing biophysical conditions, and for environmentally fragile areas and people. But without repeating the disadvantages of the first revolution. However, green revolution should also cover all the areas left out from the benefit so far. It needs to aim both at increasing yield and at preserving and developing the environmental base. Therefore we must strive for promotion of sustainable green revolution and curbing of on-going geed revolution.
- Value addition to the agricultural products should be made one of the important components of sustainable agriculture.
- The components of climate change and variability will have to be incorporated in formulation of development policy and planning, particularly as water may become a scarce commodity. Since modern agriculture is water-intensive activity, there should be proper and judicious exploitation and management of water for sustainable agriculture.
- The Symposium strongly recommended to have a proper database on land quality and all other determinants of agriculture for sustainable agricultural development, particularly on subregional basis where the adverse impacts are felt.
- We must evolve a mechanism to ensure food security to every citizen. The persistence of hunger, despite enough food production can be combated only through economic access to food and recognition of entitlement right to food security

particularly to the most vulnerable sections. To achieve these objectives, a Hunger-Free Area Programme and National Food Security Act should be initiated for universal food security. It is also recommended that a separate 'Food Security Ministry' be established at Centre and State level to achieve the objectives.

- Finally, our long-term agricultural policy should be ecologically sound, economically viable, socially relevant and sustainable for different agroclimatic systems keeping in view the needs of the present as well as the future generation.

In view of the concerns expressed by the experts on the significance of the issue and need of update information, this volume is undertaken to be published. This volume contains a series of commissioned articles by experts in various fields, ranging from natural scientists to socio-economists and remote sensing and geographic information system, in consonance with the themes of the workshop and symposium. This comprehensive scientific information on wide ranging issues—soil and water resources and their depletion and degradation, weather and climate, agro-bio-diversity, cropping system, integrated resource management, sustainable agriculture and food security, demography, remote sensing and GIS, organic farming and agribusiness, science and technology for food security, land use and land cover, drought and flood management, bio-technology, impact of environmental changes on plants, global environment, sami-arid and arid ecosystem, hill ecology etc.

The 51 contributions published in this document pertain to leading policy makers, planners, technologists and scientists, economists and geographers. They have been brought together in this book to help define shared perceptions on the evironmental issue and to further sustain and re-examine resource conservation polices needed for a long-term agenda for action in the era of knowledge. It is necessary for a more determined and drastic measures towards achieving sustainabitlity. The aim of the publication is not only to discuss and help set out the facts clearly, but also to encourage public awareness and to achieve a desired political effect.

The paper included in this volume are divided into six major sections :

Section I :	Land and Water Resource Management
Section II :	Sustainable Agricultural Development and Food Security
Section III :	Weather and Climate
Section IV :	Geo-spatial Technology Application
Section V :	Environmental change and Impact
Section VI :	Socio-cultural Aspects

Nineteen papers included in Section I discuss the 'Land and Water Resource Management'. Majority of the papers have expressed great concern on the ecosystem sustainability. Over-exploitation beyond the resource capability or an imbalanced use or/and misuse not in harmony with the local condition in an attempt to augment rapid production for meeting the immediate mounting demands of the burgeoning human and livestock population without long-term perspective leads to unabated degradation of resource base, loss of biodiversity, decline in total factor productivity in the farming systems and environmental degradation. With shrinking per capita availability of land and water, coupled with pollution, the prevalent increasing demographic trend is putting further enormous strain on the already overburdened resource base.

While expressing deep concern on the deteriorating environment and some papers made references to the policy issues at the national, regional and local levels and called for reorientation in the policies. Some discussed the importance of land and water resources and suggested a long-term perspective planning for these resources to achieve the goal of economic prosperity and environmental security to meet various competing demands.

Section II deals with 'the Sustainable Agricultural Development and Food Security.' In all 15 papers are included in this section covering different aspects—sustainable livelihood, population growth and food security, food security and geo-spatial analysis, agribusiness, science and technology for food security, concept and techniques for measuring food security, sustainability of the agro-ecosystem etc. and have pointed out the deficiency in programme and policy and neglect of environmental sustainability, and called for a halt to the indiscriminate exploitation of natural resources. Most of the papers stressed the need for ensuring sustainable availability of food, sustaining the productivity of the natural resources, ensuring adequate household incomes, providing entitlement to food, introducting national food security, minimizing economic and environmental costs, diversification of farming system, promotion of rural employment, provision of eco-technology for fighting the fatigue of the green revolution etc.

Section III of the book deals with 'Weather and Climate' and contains five papers. These papers identify the problems of the areas needing priority attention for promoting conservation of climate and weather resources for sustainable agricultural productivity and food security. The issues dealt with in this part include monsoon droughts, drought and environmental security, drought management, climate change, ENSO and drought, also cover aspects of ecological security for sustainable agriculture. In the absolute terms agricultural productivity is directly related to prevailing climatic conditions and any aberration in climate and weather results in low food grain production followed by famines. The threat of climate change and its impact on environment and food security is one of the major environmental issues of the present time. Considerable uncertainty still surrounds the magnitude and spatial pattern of expected climate change and the consequences on crop yields. It is also suggested to take on the weather and climate on priority basis as the main research agenda for natural resource management in our country.

Section IV deals with the aspects of 'Geo-spatial Technology Application' which consists of five papers. The papers facuses on the use of remote sensing, geopgraphic information systems (GIS), globlal positioning systems (GPS) etc. in the natural resource assessment and expected to bring in the desired S & T inputs for sustainable resource management, and also suggested to have a comprehensive agricultural resources informatics using geomatics technology at district level. An urgent need to develop metadata and application of open GIS model for optimal utilization of agricultural resource bases of India is also stressed.

Section V of this book deals with aspects of 'Environmental Change and Impacts'. This part consists of five papers pertaining to global environmental change, environmental degradation and land use changes, impact of environmental change on ornamental and medicinal plants, implication of biotechnology on vegetable crops etc. The paper etitled the 'Global Environment : Crossing National Boundaries' indicated that the drive to make more money, to consume more and own more material goods is laudable from a national economic perspective but certainly undesirable in trying to ensure a holistic balance between society and environments such societies live in. One wonders whether the senseless accumulation of meaterial wealth is only measure for the quality of life and human satisfaction.

Section VI deals with the 'Socio-cultural Aspects' which consists of only two papers. Paper one on 'population growth, economic development and food security in India' analyses trend of population growth in the world with special foucus on India and concludes that the problem of food security is the result of rapid population growth and underdevelopment. Even developing economies like ours quality of life is offset by rapid population growth and poor purchasing capacity of a big section of our population. Paper second on 'Ecology & Encounters' has discussed in detail, how land use and land cover change during the different political regimes from the medieval period have changed and consequently induced changes in cropping pattern which leads to degradation of agro-ecosystem.

The papers included in this volume are multidisciplinary in character and cover a wide range of spectrum, seeking to provide a better understanding of the complex nature of problems related to resource conservation, sustainable development and food security. Some papers appear to be out of place, however if looked into in detail these are found to be quite relevant. All the papers are written by different experts with differing perception, conviction and different levels of experiences. A recurring idea running through all the papers is the need for the developing countries to adopt developmental policies and programmes that are socially relevant, environmentally compatible and economically viable. The papers have also drawn attention of politician, policy makers and policy implementers to be highly careful in choice and adoption of policy and use and ensure food security.

The book will serve as a reference material to the researchers, students, planners, decision-makers and others interested in judicious integrated management of the natural resources for sustainable development and food security. I want to express my hearty thanks to the Concept Publishing Company, New Delhi for undertaking the publication of this volume. The list of contributors is given at the fag end of the book.

SECTION-I

LAND AND WATER RESOURCE MANAGEMENT

1

Soil Information System for Watershed Management

S.N. Das

The water sustains life where soil acts as medium. These natural resources are too precious to mankind as they meet all the needs and protect the environment as well as civilization. The growth of population resulting in increasing demand, over exploitation of ground water and inefficiency, in tapping the surface runoff water crested imbalance which ultimately affects sustained food production and eco-development.

It is estimated that global withdrawal of water resources will increase from 3750 km^3 to 5100 km^3 during 1995 to 2025, a rise of withdrawal of water resources from 8.4 per cent to 12.2 per cent. It is also predicted that in 2025, most of the population on earth will be living under low water supply and global warming could worsen the situation further (Shiklomanov, 2000). The increase of annual air temperature in the arid region of Asia may cause reduction of 60 per cent river runoff (Anon, 1998). All these points to enact water use policy comprising efficient water use, soil and water conservation, reduction in deforestation and diversion of major rivers. The scarcity of water in one region, flood havoc in another needs a holistic view for sustainable utilization and development of water resources.

The utilization of water resources effectively *vis-a-vis* development of water resources are intimately related to management of soil resources. Soil loss is prevalent in both water scarcity and water surplus zone due to wind and water erosion. Besides other forms of degradation of soil resources occur due to overexploitation and abuse of water resources. It has been reported that the annual rate of global loss of land is 5 to 7 mha which might increase to 10 mha by end of 2000 (Anon, 1996). It has also been assessed that 430 mha land is subjected to severe land degradation in the world and the impact of degradation in Asia will be around 18 per cent (Anon, 1991, 1998)

In India, water resource potential has been estimated to be 186 m ha m out of total annual rainfall of 400 m ha m of which utilizable water resource potential is 110 m ha m comprising 70 m ha m surface water resource potential and 40 m ha m ground water resource potential (Anon, 1997). It thus calls for tremendous challenge to harness the water resource potential and to combat the droughts and floods in the country. India being a riverine country, water induced soil erosion is very acute which emphasized the need of conservation of soil and water resources to arrest huge amount of soil loss.

Strategy for Soil and Water Conservation

Natural Resource Management Division, Department of Agriculture and Cooperation, Ministry of Agriculture, Government of India introduced watershed as basic unit of treatment with a view to develop the land resources under natural system in the catchments of River Valley Projects and Flood Prone River. The strategy adopted by the Department comprises (1) delineation and codification of watersheds; (2) Prioritization of watersheds through rapid inventorying of catchment area for micro level planning; (3) Detailed soil survey of the vulnerable priority watershed for macro level planning; (4) Release of funds for treatment of the vulnerable watershed in the catchment; (5) Awareness campaign and peoples participation in watershed management and (6) Evaluation and monitoring of the impact of a treatment. The soil and water conservation programmes are implemented by various State Governments covering 45 catchments in the country (Anon, 1992).

Generation of Soil and Land Information

Commensurate with the need of database for formulation of the soil and water conservation planning at macro level and implementation planning at micro level. All India Soil and Land Use Survey Organization (AISLUS) generates the database by employing the following working models:

(a) Rapid reconnaissance survey for identification and demarcation of priority watersheds in the catchment area using 1:50,000 scale toposheet aerial photographs/satellite imagery; and
(b) Detailed soil survey in the selected very high and high priority watersheds using large scale aerial photographs or Cadestral map for derailed acquisition soil and land characteristics.

Watershed Delineation and Codification

Planning to phase out the watershed management programme in the catchment area and to formulate the action progamme needs micro level delineation and codification of watershed. The task has been completed by superimposing the watershed boundary from Watershed Atlas of India on to a drainage map of 1:50,000 scale. The micro level delineation and codification enable to obtain a viable size of watershed ranging from 500 to 1000 ha.

Watershed Prioritization

Subsequent to the delineation and codification, watershed prioritization is carried out based on sediment yield index model/runoff index model developed by AISLUS (Anon, 1991). It allows the planners and policy makers to chalk out the watershed management programme in the catchment in a phased manner considering the constraints of funds and manpower besides, the local condition and political system.

The prioritization of watershed is a general phenomenon and any watershed could be prioritized based on the objectives of the various schemes of watershed development programme. An integrated model for prioritization of watershed could be developed to achieve the goal. The basic framework of watersheds as well as the basic soil and land information will remain same in the integrated model. Only interpretation of soil and land information according to the objectives of the watershed development scheme is essential for prioritization purposes (Anon, 2001).

Soil Resource Database

The prioritization of watershed is followed by acquisition of detailed soil information, which serves as primary inputs to the watershed management programme. It allows deriving information relating to land, soil, water, crop and nutrient management. The detailed soil information is generated by conducting detailed soil survey (Anon, 1970) that helps to develop soil resource database. Scientific land use planning, diagnosis of soil for optimal utilization, maintenance of soil health, suitability of land for various crops, fertility management and dissemination of soil information to the farming community etc. could be dealt with the soil resource data base (Jhonson, 1998; Sanchez *et al.* 1982 and Olson, 1981) which must have the following data for the benefit of users.

1. Statistics with spatial distribution of various types of soils at detailed level.
2. Morphological, physical and chemical properties of soils
3. Soil classification.
4. Interpretative groupings of soil data for various applications in the field of agriculture, forestry, horticulture, soil and water resource development and non-agricultural applications.
5. Land capability map.

Soil Information System

The management of voluminous map and statistical data becomes faster and easier due to the technological advancement in the form of Geographic Information System (GIS). It paves the way for not only storing of spatial data but it allows editing, analysis and manipulation of data besides updating and retrieval in desired format that enables dissemination of information to the users in a faster mode. The development of soil information system using GIS requires soil resource database. Soil information system will enable to resolve many issues related to the sustainability in agriculture development in the country. AISLUS has contemplated to develop spatial database using GIS and RDBMS (Anon, 1993). Some important aspects of Soil Information System would be as below :

- Dissemination of database to the users accurately, quickly and in desired formats.

- Optimal use of potential area and management of problem area.
- Scientific Land Use Planning.
- Proper Diagnosis of Soil.
- Monitoring soil health.
- Precision Agriculture.
- Generation of soil information for farmers in the forth of soil health card to establish linkage between the farmers and policymakers, scientists and others.
- e-Govemance for Natural Resources Management.
- Open avenues for e-Business.

Conclusion

Development of soil information system should be given priority not only to support the decision making process in a subtle manner for strategic planning but also to provide a platform for monitoring and implementation of the soil and water conservation programme for sustainable agriculture production and eco-development in the country.

REFERENCES

Anon (1970) *Soil Survey Manual,* All India Soil and Land Use Survey, New Delhi.

Anon (1983) *Manual of Remote Sensing* published by American Society.

Anon (1990)Watershed Atlas of India published by All India Soil and Land Use Survey, New Delhi.

Anon, (1991) *Methodology for Priority Delineation Survey*, All India Soil and Land Use Survey, New Delhi.

Anon (1992) *Guidelines: Soil Conservation in the Catchments of River Valley Projects*, Ministry of Agriculture, New Delhi.

Anon (1993) *Global and National Soils and Terrain Digital Databases* (SOTER), World Soil Resources Reports No.74, FAO, Rome.

Anon (1996) *Soil Management in relation to land degradation and environment,* Indian Society of Soil Science, IARI Campus, New Delhi.

Anon (1997) *Survey of Indian Agriculture*, 155-157; published by the Hindu, New Delhi.

Anon (1998) *International Year of Ocean*, http: //www. Zoto 98.noaa.gov

Anon (1998) *Global Assessment of the Status of Human induced Soil Degradation* (GLASOD), ISRIC, http//www. Isric.nl/GLASOD.htm

Anon (2001) *Consultancy Project Report on Monitoring and Evaluation of the Impact of Watershed Management under NWDPRA Project using Remote Sensing and GIS*, IARI Buildings, New Delhi-110012.

Sabins, Jr. F.Floyd (1977) *Remote Sensing—Principles and Interpretation*, published by W.H. Freeman and Co., NY.

Sancher, P.A., Cozo, Wut and Boul, S.W. (1982) *The fertility capability soil classification system: interpretation, applicability and modification*. Geoderma 27 (4): 283-309.

Jhonson Paul (1998) *Evaluation of Natural Resource's Management Programme*. Proceedings of the 8th International Conference on Watershed Management held at New Delhi during Dec.7-9, 1998.

Olson, G.W. (1981) *Soil and the Environment : A Guide to Soil Surveys and their Applications*, Chapman and Hall, USA.

Shiklomanoy, Igor A. (2000) *Water International*, Volume - 25, No. 1, pp. 11 to 31.

Rossiter, D.G. (2001) *A compendium of online soil survey information*, ITC, Netherlands.

—

2

Status of Groundwater Development and Management in India

D.K. Chadha

Introduction

In the past five decades since independence, the development of water resources in the country has increased rapidly in order to meet the demands for diverse uses. As a result, the country has achieved self-sufficiency in food grains besides providing assured drinking water throughout the country except for a few problem areas. However, due to rapid growth of population, urbanisation and industrialization, and improvements in living standards, India's finite water resources are now getting stressed with declining per capita availability and deteriorating quality.

Ground water has played a significant role in maintenance of India's economy, environment and standard of living. Besides being the primary source of water supply for domestic and many industrial uses, it is the single largest and most productive source of irrigation water. At present about 80 per cent of water supplies for domestic use in rural areas and 50 per cent of water requirements for urban and industrial areas as well as irrigation are being met from ground water. Uncontrolled development of the resource for meeting the increased demands has resulted in declining ground water levels. Further, haphazard disposal of domestic and industrial wastes and excessive application of fertilizers, pesticides/insecticides in agricultural fields have led to deterioration of ground water quality thereby further limiting the availability of fresh ground water resources, On the other hand, rejected recharge due to seepage from irrigation canals has led to problem of water logging and soil salinity. Keeping in view the constraints being faced in the development of ground water, appropriate management strategies need to be evolved to meet the growing demands of various use sectors.

Groundwater Availability

The repenishable ground water resources are estimated on the basis of long term annual water level fluctuations. The void space created due to withdrawal of ground water gets re-filled during monsoon season and ground water storage is built up. The annually replenishable ground water resources of the country have been assessed as 432 billion cubic metres (BCM). After keeping a provision of 71 BCM of ground water per year for domestic and industrial uses, 361 BCM/ year is available for irrigation. The Ganga basin has the highest potential (171 BCM/year) followed by Godavari basin (41 BCM/year) and Brahmaputra basin (27 BCM/year). Based on crop water requirement and availability of cultivable land, utilisable irrigation potential that can be created from ground water has been estimated as 64.05 million hectares.

In addition to replenishable resources, the country is blessed with enormous ground water resources below the lowest level of ground water fluctuation termed as instorage ground water resource. The in-storage fresh ground water resources up to the depth to 450 metres in alluvium and 100 metres in hard rock terrain, have been estimated as 10812 BCM. However, in alluvial aquifer systems such as in Ganga basin, there is scope for further deep ground water development. As per National Water Policy, the development of ground water resources at present is restricted to replenishable component only. Keeping in view the vast in-storage ground water resources in Indus, Ganga and Brahmaputra basins, there is ample scope for development of ground water in periods of scarcity. Statewise replenishable and in-storage ground water resources are given in Table 2.1.

Groundwater Development

The groundwater development programme in the country is generally

Table 2.1 : State-wise ground water resources availability (BCM)

S .No.	*State*	*Replenishable Ground Water Resource*	*In-storage ground water resources* Alluvium/ Unconsolidated	*Hard rocks*	*Total rocks*
1.	Andhra Pradesh	35.3	76	26	102
2.	Arunachal Pradesh	1.4	–	–	–
3.	Assam	24.7	92	–	92
4.	Bihar	33.5	2557	11	2568
5.	Goa	0.2	–	–	–
6.	Gujarat	20.4	92	12	104
7.	Haryana	8.5	42	1	421
8.	Himachal Pradesh	0.4	13	–	13
9.	Jammu & Kashmir	4.4	35	–	35
10.	Karnataka	16.2	–	17	17
11.	Kerala	7.9	5	6	11
12.	Madhya Pradesh	50.9	14	27	41
13.	Maharasotra	37.9	16	22	38
14.	Orissa	3.2	162	13	175
15.	Punjab	18.7	91	–	91
16.	Rajasthan	12.7	115	13	128
17.	Tamil Nadu	26.4	98	–	98
18.	Tripura	0.7	101	–	101
19.	Uttar Pradesh	83.8	3470	30	3500
20.	West Bengal	23.1	1625	1	1626
21.	Total U.Ts.	0.4	4	–	4
	Total	431.9	10633	179	10812

restricted to shallow zones within 50 metres depth and is primarily sustained by the investment by the farmers themselves or through finance obtained as loan from the institutional sources. The public sector outlay is limited to only such items as ground water surveys, construction of deep tube wells for community irrigation, services provided and grants extended to small and marginal farmers. The programme of ground water development, therefore, poses less burden on public exchequer. The details of investment in irrigation sector are given in Table 2.2.

The flow of institutional finance for ground water development is generally about 60 per cent of the total outlay for ground water development. Besides, share of public outlay in ground water development is 20 per cent and the balance is met from the private sources.

The past five decades have witnessed phenomenal increase in growth of ground water abstraction structures due to implementation

Table 2.2 : Investment in minor irrigation sector *(Rs. in crores)*

Period	*Investment on Minor irrigationI*		*Total investment on irrigation*		*Share of MI in expenditure on irrigation sector (per cent)*
	Plan	*Instnl.*	*Total*	*sector*	
1951-56	66	Negbl.	66	442	15
1956-61	142	19	161	542	30
1961-66	328	115	443	1019	44
1966-69	326	235	561	991	57
1969-74	512	661	1173	2416	49
1974-78	631	779	1410	4073	35
1978-80	502	480	982	3276	30
1980-85	1979	1438	3417	11529	30
1985-90	3118	3061	6179	18734	33
1990-91	836	676	1512	4456	34
1991-92	844	674	1518	4625	33
1992-97	3157	5331	8488	27480	31

of technically viable schemes backed by liberal funding and availability of power and diesel, good quality seeds, fertilizers, government subsidies etc. The green revolution in the post-independence era has also led to excessive development of ground water particularly in northern parts of the country.

Ground water development has, therefore, occupied an important place because of its role in stabilizing Indian agriculture and as means for drought management. The stage of ground water development in the country as estimated in 1991 was 32 per cent. Up to March 1998, the stage of development has reached approximately 40 per cent. This is evident from the growth of ground water abstraction structures from the pre-Plan period till date. The number of ground water abstraction structures has increased from merely 4 million in 1951 to nearly 17 million in 1997. With the growth of ground water abstraction structures, there has been increase to irrigation potential created from ground water from merely 6.5 million hectares to 45.73 million hectares by the end of VIII Plan. This rapid pace is likely to continue till the full irrigation potential estimated to be available from ground water is created. The growth of ground water abstraction structures and irrigation potential created from ground water since 1951 are given in Table 2.3.

Ground water is the main source of drinking water in rural areas contributing about 90 per cent of rural water supply. Under the Rajiv Gandhi National Drinking Water Mission launched in 1986 with the main objective of providing water supply to no problem villages/

Table 2.3 : Growth of ground water abstraction structures ('000) and irrigation potential (Mha)

Year	*Dugwells*	*Private tube wells*	*Public tube wells*	*Total*	*Cumulative irrig. pot. created from ground water*
March, 1951	3860	3	2.4	3865.4	6.50
Maroh, 1980	7786	2132	33.3	9951.3	22.00
March, 1985	8742	3359	46.2	12147.2	27.82
March, 1990	9407	4754	63.6	14224.6	35.62
March, 1992	10120	5379	67.6	15566.6	38.89
March, 1997	10501	6743	90.0	17334.0	45.73

habitations, a total of 1430543 habitations have neen covered. About 3626 lakh rural population has been served under the mission so far Statewise break up of habitations covered as on 1-4-1999 is given in Table 2.4. In areas having serious ground water quality problems like Fluoride, Iron, Arsenic and salinity, exclusive submission programmes are being implemented by the Mission. In such quality affected areas, where ground water is unfit for drinking, safe drinking water is supplied either through alternative sources, tapping surface water or by other means by taking remedial measures by defluoridation, iron removal and arsenic removal.

Role of Groundwater in Drought Management

Frequent drought conditions prevail in various states in the country. The States which normally experience below normal rainfall resulting in meteorological as well as agricultural drought are the States of Rajasthan, Gujarat, Madhya Pradesh and Orissa. Droughts have ripple effects on the Indian economy. Not Besides loss of agricultural production, there are several secondary effects. Vulnerable populations are at risk and are often forced to migrate in search of work. Public expenditure on drought mitigation and food distribution increases substantially. A total of 179 districts in various States have been categorised as drougnt prone.

Development of ground water has greatly led to increased drought proofing of Indian agricultural economy. During periods of droughts, ground water is the only source that can be tapped for meeting strategic water demands. Dependence is laid on this resource since the storage levels in surface reservoirs dwindle and the impact of vagaries of weather on ground water is not as pronounced and is detayed.

Table 2.4 : Status of coverage of habitations under Rajiv Gandhi National Drinking Water Mission

S.No.		Status as on 1-4-99			
		NC HABS	PC HABS	FC HABS	Total HABS
1.	Andhra Pradesh	0	24683	45049	69732
2.	Arunachal Pradesh	548	1227	2403	4173
3.	Assam	3901	26202	40566	70669
4.	Bihar	1069	564	203803	205436
5.	Goa	34	27	344	1405
6.	Gujarat	437	4639	25193	30269
7	Haryana	32	0	7513	7545
8.	Himachal Pradesh	3750	13592	28025	45367
9.	Jammu & Kashmir	2430	4047	9249	15726
10.	Karnataka	1522	10498	44662	56682
11.	Kerala	880	6719	2164	9763
12.	Madhya Pradesh	3563	17709	138596	159868
13.	Maharashtra	1515	31811	43798	77124
14.	Manipur	77	510	2204	2791
15.	Meghalaya	869	1276	6494	8639
16.	Mizoram	2	624	285	911
17.	Nagaland	428	703	394	1525
18.	Orissa	1978	4709	107412	114099
19.	Punjab	5845	3123	4481	13449
20.	Rajasthan	4715	38163	61188	104060
21.	Sikkim	0	732	947	1679
22.	Tamil Nadu	0	16971	49660	66631
23.	Tripura	726	1604	5082	7412
24.	Uttar Pradesh	0	0	274641	274641
25.	West Bengal	0	22547	57830	80377
26.	Andaman & Nicobar Islands	11	6	487	504
27.	Dadra & Nagar Haveli	128	190	198	516
28.	Daman & Diu	0	1	28	29
29.	Delhi	0	0	200	200
30.	Lakshadweep	0	10	0	10
31.	Pondicherry	0	0	276	276
32.	Chandigarh	0	0	24	24
	Total	34460	232887	1163196	1430543

Impact of Human Activities on Groundwater Regime

Ground water withdrawal for various uses coupled with haphazard disposal of untreated urban and industrial wastes and over-application of fertilizers, pesticides/insecticides in the agricultural fields has led to problems of ground water depletion and pollution. The problems discussed below require special attention of the water resources development planners and managers.

Over-extraction of groundwater

In many arid and hard rock areas, overdraft and associated quality problems are increasingly emerging. Situation of overdraft exists, i.e.

Table 2.5 : Name of the stateldistrict having pockets with fall in ground water level of more than four metres during pre-monsoon period (1980-99)

State	*Districts*
Haryana	Bhiwani, Rohtak, Jind, Yamunanagar, Kurukshetra, Kamal, Kaithal, Mahendergarh, Rewari, Gurgaon, Faridabad, Panipat
Punjab	Amritsar, Jalandhar, Firozpur, Fandkot, Ludhiana, Fatehgarh, Patiala, Sangrur, Bhatinda
Rajasthan	Alwar, Bikaner, Churu, Jhunjhunu Bharatpur, Dholpur, Dosa, Jaipur, Sikar, Nagaur, Jaisalmer, Ajmer, Tonk, Sawai Madhopur, Pali, Jalore, Sirohi, Barmer, Jodhpur, Bhilwara, Rajsamand, Udaipur, Kota, Mandsor, Chittorgarh, Bundi.
Gujarat	Bhul, Rajkot, Jamnagar, Junagarh, Amreli, Surendernagar, Banskantha.
Uttar Pradesh	Nanital, Moradabad, Bulandsahr, Bareily, Aligarh, Mathura, Agra, Gonda, Lucknow, Kanpur, Allahabad, Fatehpur, Banda, Hamirpur, Firozabad, Lalitpur, Gaziabad, Fatehgarh, Saharanpur, Meerut, Jounpur, Jhansi, Banaras, Balia,
West Bengal	Purulia
Assam	Golaghat, Sonitpur, Lakheempur, Kachchar
Madhya Pradesh	Guna, Panna, Satara, Sahjahpur, Sagar, Ambekapur, Ujjain, Dewas, Dhar, Indore, Khargone, Chhindwara, Narsinghpur, Hosangabad, Bilaspur, Raipur, Jagdalpur, Murena East Nimar, Rajnandgaon, Durg, Ratlam
Karnataka	Bangalore, Belgaum, Bellary, Bijapur, Dharwar, Kolar, Mandia, Uttar Kannada, Shimoga, Gulbarga, Mysore.
Tamil Nadu	South Arcot, Salem, Coimbatore, Puddukotai, Madurai, Tiruvarur, Kamarajar, Tirunelveli, Tiruchirapalli, Thanjavur
Bihar	East Singhbhum, Ranchi
Maharashtra	Jalgaon
Andhra Pradesh	Medak, Rangareddy, Mehboobnagar, Nalgonda, Cuddapah, Chittoor
Orissa	Koraput, Mayurbhanj

the ground water withdrawal has exceeded the annual replenishable resource in 231 blocks (out of 4272) in various states in the country, besides 6 mandals in Andhra Pradesh and 12 talukas in Gujarat. In addition, in 24 mandals in Andhra Pradesh, 14 talukas in Gujarat, 34 watersheds in Maharashtra and 107 blocks in rest of the states, the stage of ground water development has exceeded 85 per cent of the annual replenishable resource. The overdraft has led to decline in water levels resulting in failure of wells, shortage of water supplies, etc. necessitating deepening of existing wells thereby increasing pumping lifts and costs. Statewise details of areas showing more than 4 metres decline in ground water levels during pre-monsoon period (1980-1999) are given in Table 2.5. Rate of decline of ground water level is estimated to be 20 cm/year. In coastal areas, overdraft of ground water has led to problem of sea water ingress resulting in contamination of fresh water aquifers. Problem of sea water ingress has been manifested in Minjur area of Tamil Nadu and Mangrol-Chorvad-Porbandar belt along the Saurashtra coast and in Pondicherry region east of Neyveli Lignite mines.

Groundwater quality deterioration/pollution

Ground water quality is a major environmental concern. Factors affecting the quality of ground water include sea water intrusion, pollution due to human activities, aridity and water logging. Seepage from septic tanks; animal wastes and return flow from irrigation where fertilizers and pesticides/insecticides are applied has resulted in high levels of nitrate, potassium and phosphate in parts of Bihar, Haryana. Gujarat, Orissa, Uttar Pradesh and National Capital Territory of Delhi. Pollution of ground water with toxic chemicals in the vicinity of industrial areas and urban settlements has been observed in many parts of the country. In the industrial zone of Faridabad, high concentrations of chromium, copper, nickel and zinc have been reported from ground water. In Ludhiana, ground water is polluted with chromium. In Rajasthan, pollution of ground water has been reported from industrial areas of Udaipur, Pali, Khetri and Jodhpur. In Kanpur city in Uttar Pradesh, higher concentration of chromium and iron have been observed. In Mayurbhanj district of Orissa, the shallow ground water in a localised area is rendered acidic with pH ranging from 3.2 to 6.4. In Warangal city in Andhra Pradesh, effluents

Table 2.6 : Statewise details of habitations affected by water quality problems

S. No.	*State*	*No. of Habitations Affected with*				*Total*
		Flouride	*Arsenic*	*Salinity*	*Iron*	
1.	Andhra Pradesh	7548	–	14138	441	13127
2.	Arunactral Pradesh	–	–	–	–	–
3.	Assam	–	–	–	–	–
4.	Bihar	12	–	–	18669	18681
5.	Goa	–	–	2	2	4
6.	Gujarat	2378	–	677	–	3055
7.	Haryana	334	–	–	–	334
8.	Himachal Pradesh	488	–	106	450	1044
9.	Jammu & Kashmir	–	–	–	–	–
10.	Karnataka	860	–	769	274	1903
11.	Kerala	287	–	26	422	735
12.	Madhya Pradesh	201	–	87	1928	2216
13.	Maharashtra	39	–	–	–	39
14.	Manipur	–	–	–	157	157
15.	Meghalaya	–	–	–	315	315
16.	Mizoram				52	52
17.	Nagaland	–	–	–	–	–
18.	Orissa	1138	3	304	42830	44275
19.	Punjab	700	–	580	15	1285
20.	Rajasthan	16560	–	14415	440	31415
21.	Sikkim	–	–	–	–	–
22.	Tamil Nadu	527	–	300	173	1000
23.	Tripura	–	–	–	–	–
24.	Uttar Pradesh	1072	–	3426	3720	8218
25.	West Bengal	21	9000	7722	6516	23259
26.	Andaman & Nicobar Islands	–	–	–	–	–
27.	Daman & Diu	–	–	–	–	–
28.	Lakshdweep	–	–	–	–	–
29.	Pondicherry	–	–	–	–	–
30.	Delhi	46	–	–	–	46
31.	Dadra & Nagar Haveli	–	–	–	–	–
32.	Chandigarh	–	–	–	–	–
	Total	32211	9003	33552	76404	151170

from textile and tannery industries have affected ground water. In Tamil Nadu, discharges from tanneries in North Arcot district have affected the ground water quality.

Besides the pollution caused by human activities, instances of ground water pollution of grogenic origin have also been noticed. Fluoride concentration higher than permissible limit have been reported in vast areas of Andhra Pradesh, Haryana and Rajasthan and in some places in Punjab, Uttar Pradesh, Kartnataka, Tamil Nadu and Delhi. High iron content has been reported from eastern and north eastern parts of the country. Problem of ground water pollution due to excess concentration of arsenic has been reported from eight districts of West Bengal. Statewise details of habitations affected by water quality problems are given in Table 2.6.

Inland salinity in ground water

It has been estimated that over 1.93 lakh sq.km area in parts of Haryana, Punjab, Delhi, Rajasthan, Gujarat, Uttar Pradesh, Karnataka and Tamit Nadu is affected by inland salinity in ground water (EC > 4000) micro-mhos. Statewise details are given in Table 2.7.

Table 2.7 : Area affected by inland salinity in ground water—EC > 4000 micro-mhos

S.No.	*State*	*Total area of the State (Sq.km)*	*Area underlain by saline ground water (Sq.km)*	*per cent*
1.	Haryana	44212	11438	25.87
2.	Punjab	50353	3058	06.07
3.	Delhi	1485	140	09.43
4.	Rajasthan	342239	141036	04.12
5.	Gujarat	196024	24300	12.40
6.	Uttar Pradesh	294411	1362	00.46
7.	Karnataka	191791	8804	04.59
8.	Tamil Nadu	130058	3300	02.54
	Total	1250573	193438	15.47

Saline water can also be put to beneficial uses such as prawn culture and for growing salt tolerant crops. However, use of inland saline ground water calls for a multi-disciplinary approach because the use/application of saline water without proper scientific management techniques may render irrigated soil saline/sodic.

Waterlogging and soil salinity

Large areas particularly in the command areas of major and medium irrigation projects suffer from waterlogging due to rejected recharge. The waterlogging conditions have been created due to high intensity of irrigation without adequate drainage, which results in upward movement of water table. As per the assessment made by the Working Group on Problem Identification in Irrigated Area with suggested Remedial Measures (1991), about 2.46 mha of the area under surface water irrigation projects is water logged or threatened by water logging. As per the assessment made by CGWB in November 1998, about 31.12 mha area in the country was affected by water logging and 50.41 mha area was prone to water logging. In May 1999, an area of about 1.87 mha was logged and 10.02 mha area was prone to water logging. Statewise details of area water logged and prone to water logging are given in Table 2.8. Such areas offer scope for groundwater development by lowering the water table upto 6 metres or more.

Table 2.8 : Statewise details of area waterlogged, area affected by soil salinity and alkalinity ('000 sq.km)

S.No.	*State*	*Area water logged*	*Area affected by soil salinity*	*Area affected by soil alkalinity*
1	Andhra Pradesh	266.40	5.00	22.80
2.	Assam	-	-	-
3.	Bihar	619.70	224.30	-
4.	Gujarat	172.59	91100	-
5.	Haryana	249.00	125.20	72.00
6.	Himachal Pradesh	0.20	-	-
7.	Jammu & Kashmir	1.50	-	-
8.	Karnataka	24.54	34.23	17.12
9	Kerala	11.61	-	-
10.	Madhya Pradesh	73.12	-	35.79
11.	Maharashtra	15.35	5.35	
12.	Orissa	196.26	-	-
13.	Punjab	200.00	490.00	-
14.	Rajasthan	179.00	70.00	-
15.	Tamil Nadu	16.19	48.00	92.30
16.	Uttar Pradesh	430.00	1150.80	-
	Total	2455.96	3063.88	240.01

Projected Water Demands

While the availability of fresh water is decreasing on the one hand, the demand for water for various use sectors is increasing with the growth of population. India's population of over one billion is putting additional pressure on natural resources and complicating the development task. Census in 1991 estimated India's population at 847 million. The rural population in the country has grown from 213 million in 1901 (89.2 per cent of the total population) to 629 million (74.3 per cent of the total population). The short and long term Population Forecast have been made by Natrajan (1993), United Nations (1994 Revision), Registrar General of India (RGI,19961 and Visaria and Visaria (1998). With the growth of population, development and urbanisation, there shall be a shift from rural towards urban areas and the gap between rural and urban population shall shrink. The projected population figures in the 21st century as adopted by the National Commission for Integrated Water Resources Development Plan are as follows :

Table 2.9 : Projected population in the 21st century (*Population in millions*)

Year	*2010*		*2025*		*2050*	
Scenario	*High*	*Low*	*High*	*Low*	*High*	*Low*
Population	1146	1157	1333	1286	1581	1346
Rural	759	790	730	810	610	700
Urban	387	367	603	476	971	646

As urban and rural populations have different lifestyles, their needs for food and drinking water are quite different. However. with the access to information and demand for rural-urban equality the gap between urban and rural demand for food is likely to shrink gradually but differences are expected to persist for a long time. Though the rural population shall decrease from 759 million in high scenario in 2010 to 610 million in 2050 and 790 million in the low scenario in 2010 to 700 million in 2050, the demand for water would increase by 2050 on account of social upgradation and economic upliftment. Using the high and low population projections for the country and per capita per annum food requirement, the total food requirement for the country, has been estimated between 449 million tonnes (High demand scenario) and 382 million tonnes (Low demand scenario). Feed requirements, losses in storage and transportation seed requirement and carry over for years of monsoon failure in the country have been estimated at 12.5 per cent of the food grain production.

Depending upon the country's capacity to contain population growth, adoption of water saving techniques, rate of growth of gross domestic production, rate of growth of expenditure and a host of other factors, the total water requirement for various uses by the year 2050 as estimated by the National Commission for Integrated Water Resources Development Plan is expected to vary between 973 to 1180 BCM. Water requirements for various use sectors in the 21st century are given in Table 2.10.

Table 2.10 : Estimated sectoral water requirement (BCM)

S.No.	*Uses*	*Year 2010*		*Year 2025*		*Year 2050*	
		Low	*High*	*Low*	*High*	*Low*	*High*
1.	Irrigation	543	557	561	618	628	807
2.	Domestic	42	43	55	62	90	111
3.	Industries	37	37	67	67	81	81
4.	Energy	18	19	31	33	63	70
5.	Inland Navigation in addition to ecological need	7	7	10	10	15	15
6.	Flood Control	-	-	-	-	-	-
7.	Environment (1) Afforestation	-	-	-	-	-	-
8.	Environment (2) Ecology	5	5	10	10	20	20
9.	Evaporation	42	42	50	50	76	76
	Total	694	710	784	850	973	1180

The annual average water availability in terms of replenishable utilizable water resources of the country has been estimated at 1086 BCM (690 BCM from surface water considering that the available surface water resource shall be completely tapped +396 BCM from ground water). The per capita availability of utilizable water, which was 3000 cubic metres in the year 1951 is now about 1100 cubic metres and would reduce to 660 cubic metres by the year 2050. The water availability at national average, in the year 2010 shall be below 1000 cubic metres and shall start hampering health, economic development and human well being.

Current water resource constraints, therefore, are expected to manifest themselves even more rapidly in the coming years. While population growth rates have stabilized and are expected to gradually diminish over the next several decades, the composition of demand is significantly changing as a result of rising incomes and of rapid urbanization and industrialization. Competition nor water between urban and agricultural sectors will be a major challenge in the coming century. Further expansion in irrigation, industry and domestic water demands will have serious implications on competing non-consumptive uses, such as hydropower.

Management Needs

The new ethic of sustainable development of water resources reinforces and extends the main principles of water resources management. It has been agreed that the crisis today is more of water management than of water itself. Caused by bad management, bad institution, bad governance, bad incentives and bad allocation of resources, which are threatening the available water resources. The real solution, therefore, requires an integrated approach to water resources management.

Investment to capture additional surface runoff will become increasingly expensive in future. The massive public investment in irrigation development since the 1950s, which resulted in surface irrigation expansion from 16 million hectares in 1951 to 43.83 million hectares by the end of VIII Plan. is now reaching practical limits in many of the river basins.

Keeping in view the increased demands of water for various uses and decreased availability of fresh water resources, it would be a very difficult task at the national level to increase availability of water. By the middle of the 21st century, surface water resources shall be fully committed. Therefore, additional stress would need to be laid on ground water resources. Concerted efforts are, therefore, required at the national as well as regional levels to transform all types of waters into utilizable category. The measures that need to be adopted to meet the increased water demands in the coming decades would include:

Promoting development of ground water in eastern states

Even though ground water resources in some parts of the country are over-exploited, in certain areas particularly in the eastern states development of ground water has been low. Poor development of ground water in eastern states is attributed to fragmented land holdings, lack of efforts in public funding for construction of wells/tube wells, lack of consultancy services, non-availability of timely and assured power supply and lack of training and extension services.

Such areas need efforts for promoting development of ground water. Since small sized and fragmented land holdings do not qualify for assistance from the institutional sources because of the lack of economic feasibility, the farmers need to be pursuaded to form cooperatives and share water from the well. They may also be motivated to set up a Maintenance Fund in a local Financial Institution. The State Public Health Agencies should train a couple of village youth and women in maintenance of the structures. Availability of quality power at the right time is most essential for avoiding breakdown and enhancing the life of pumps.

Exploration of Deeper Aquifers

Till now the potential aquifers up to the depth of 400 metres in alluvial areas and 200 metres in hard rocks in the country have been explored. However, to increase the availability of water to meet the increasing demands and to retain it for future use, newer potential sources of ground water at greater depths need to be explored. Ganga basin, which extends up to an area of 2,50,000 sq.km, forms one of the largest ground water reservoirs of the world. The aquifers occurring at depths as great as 2000m hold a vast untapped source of fresh ground water in Siwalik sediments. Borehole electrical logs of ONGC have revealed occurrence of fresh ground water to a maximum depth of 2675 m at Madhubani in Gandak depression. CGWB has taken up exploration for delineation of deeper fresh water aquifers.

Groundwater Recharge

It has been estimated that about 865 BCM of surplus monsoon runoff is available in 20 river basins in the country. On the basis of available surplus monsoon runoff, which is not uniform in time and space, additional recharge to ground water storage that could be feasible has been estimated as 214 BCM out of which 160 BCM is considered to be retrievable. Details of basinwise ground water storage potential are given in Table 2.11. Therefore, rain water harvesting and recharge

Table 2.11: Basinwise ground water storage potential

S. No.	*Basin*	*Available water for recharge (MCM)*	*Water required to create sub-surface storage up to 3 mbgl, (MCM)*	*Feasible ground water storage based on water availability and sub-surface storage (MCM)*	*Utilisable ground water storage potential (MCM)*
1.	Indus	29836.80	143813.00	29836.85	20694.50
2.	(A) Gangs	248036.85	125076.00	125076.00	94930.00
	(B) Brahmaputra	339083.10	19177.00	19177.00	15883.00
	(C) Barak & Other	No Storage Potential	No Storage Potential	No Storage Potential	No Storage Potential
3.	Godavari	47437.20	8145.00	8145.00	5818.00
4.	Krishna	4648.50	8970.00	4648.50	3320.30
5.	Kavery	6151.05	4746.00	4746.00	3406.00
6.	Pennar	2152.35	1940.00	1940.00	1386.00
7.	East flowing rivers between Krishna & Pennar and between Mahanadi & Godavari	8549.10	832.00	832.00	594.00
8.	East flowing rivers between Pennar & Kanya Kumari	12409.65	5731.00	5731.00	4145.00
9.	Mahanadi	20051. 10	2055.00	2055.00	1468.00
10.	Brahmani & Baitarani	11615.65	483.00	483.00	345.00
11.	Subamarekna	3573.90	868.00	868.00	630.00
12.	Sabarmati	1136.25	8134 .00	1136.25	851.60
13.	Mahi	3464.10	1883.00	1883.00.	1345.00
14.	West flowing rivers of Kutch & Kai Including Luni	4349.70	150181. 00	4349.70	3270.10
15.	Namada	1851.30	77356. 00	1851.30	1389.20
16.	Tapi	395.55	4446.00	NIL	NIL
17.	West flowing rivers of Tapi & Tadri	74110.05	907.00	907.00	648.00
18.	West flowing river of Tandn to Kenya Kumari	53022.15	518.00	518.00	370.00
19.	Areas of Intand Drainage in Rajasthan	Nil	25310 .00	NIL	NIL
20.	Minor over basins draining to Bangladesh and Burma		No Storage Potential		
	Total	871874.55	590571.00	214184.00	160493.70

measures would need to be adopted so as to reduce the runoff to the sea. In urban agglomerates, where annual recharge to ground water has considerably reduced due to shrinkage in land consequent to urbanization and construction activities, roof top rain water harvesting would need to be made mandatory.

Model artificial recharge studies have been carried out in parts of Gujarat, Kerala, Punjab, Maharashtra, Karnataka, West Bengal, Tamil Nadu and Andhra Pradesh through different types of recharge structures such as construction of percolation tanks, check dams, subsurface dykes, recharge shafts, recharge wells and roof top rain water harvesting etc. suited to different hydrogeological environments. The studies have shown positive results which can be replicated in similar areas elsewhere in the country. Watershed management including construction of gully plugs, check dams, contour bunds, percolation tanks, ground water dams are very effective means to augment ground water storage in hard rock area. Recharge wells and basin spreading techniques can be utilised in alluvial areas. Further studies are thus being conducted in other parts of the country.

Development of aquifers in flood plains

Flood plains in the vicinity of rivers act as repositories for ground water. To meet the additional demands, ground water from the flood plains can be withdrawn during non-monsoon, which gets recharged during the monsoon every year. It has been estimated that in NCT of Delhi, about 200 MGD (0.91 MCM/day) of fresh ground water can be made available through construction of 1000 shallow tube wells in the Yamuna flood plains during the months of April, May and June. The dewatered aquifer zones would be recharged during the monsoon period through flood water adopting appropriate artificial recharge practices including construction of additional barrages. Similarly, it has been estimated that 400 MGD (1.82) MCM/day of fresh ground water can be made available through construction of 600 shallow/ deep tube wells in the Yamuna flood plains in the NCR. Therefore, the flood plains of rivers offer excellent scope for development of ground water to meet the additional demands to some extent.

Use of saline/brackish water

Yields of many crops. vegetables and fruit plants, e.g. barley, dates and pomegranate, when irrigated with saline or brackish water are not significantly affected saline/brackish water can be successfully used to irrigate such plants and fresh or good quality water can be saved for use by other sensitive crops or for other uses. Brackish water can also be utilized for pisciculture/aquaculture. Brackish water can also be used by blending with saline water for irrigating some crops. Blended water can also be used for domestic purposes other than drinking. Therefore, additional resource of 1164 BCM of saline/ brackish ground water resource would be available for use.

Recycling/reuse of water

Water used for domestic and industrial purposes is mostly non-consumptive and about 80 to 85 per cent of water is usually available for further utilization. Recycling and reuse are the two key actions in modern technology for waste water management. Reclaimed water after treatment can be used for irrigation, cooling, algae and pisciculture and other industrial uses. Recycling and reuse of waste water would help in minimizing the requirement of fresh water and reducing the quantity of water for final treatment before letting it off into the stream.

Development of ground water in waterlogged areas

Rejected recharge in water logged areas can serve the dual purpose of increasing ground water availability and mitigating problem of water logging in the command areas of major and medium irrigation projects. Conjunctive use studies have been completed in six irrigation commands, viz. Sarada Sahayak irrigation command in Uttar Pradesh, Hirakud command in Orissa, Indira Gandhi Nahar Pariyojana, Stage-I in Rajasthan, Mahi-Kadana in Gujarat, Ghatprabha in Karnataka and Tungabhadra in right bank canal command in Andhra Pradesh and Karnataka. These studies have suggested the optimal utilization of surface water and ground water to rectify water logging conditions, to mitigate water shortage in tail reaches and to fulfil the demand of water for increased irrigation intensity in these irrigation commands.

The best feasible scenario of conjunctive utilization of the two resources is given in Table 2.12.

Table 2.12 : The best feasible scenario of conjunctive utilization of groundwater and surface water

(i)	IGNP	Ground water development should be 18 per cent of canal water released at the head. There will be gradual decrease in the spread of the water logging conditions in next 15 years.
(ii)	Hirakud	To achieve 200 per cent irrigation intensity the contribution of surface water and ground water should be 90 per cent and 10 per cent respectively.
(iii)	Sarada Sahayak	The surface water utilization should be reduced from 1481.34 MCM to 1015.89 MCM, and that of ground water should be increased from 906.32 to 1356.23 MCM. This will result in increase in cropping intensity from 153 per cent to 200 per cent and decline in water level of 0.5 metres in one decade.
(iv)	Tungabhadra	The contribution of surface water and ground water should be 63 per cent and 37 per cent respectively. This will lead to irrigation of 85.8 per cent of the canal water shortage area.
(v)	Ghatprabha	The contribution of surface water ground water should be 84 per cent and 16 per cent respectively from 200 per cent cropping intensity in left bank canal, while 69 per cent and 31 per cent respectively from 140 per cent cropping intensity in right bank canal.
(vi)	Mahi Kadana	The present utilization of surface water and ground water in the proportion of 65:35 is optimal situation of conjunctive use without causing any adverse effect.

Creation of groundwater sanctuaries/protection zones

In many areas in the country, though the present sources are adequate enough to meet the requirements of drinking and domestic uses, they will not sustain the pressure of population after ten or fifteen years. In such cases, it will be desirable to locate alternate aquifers and declare them as reserved for drinking and domestic uses. These could be termed as Ground Water Sanctuaries. Similar zones could be set apart near urban agglomerates. Growth of any industry that may cause degradation of ground water quality near such areas may need to be prohibited. Such zones may be declared as "Protection Zones" and set apart for future drinking and domestic water uses.

Regulation of groundwater development

The Central Ground Water Authority, constituted under Environment (Protection) Act of 1986 is playing a key role in regulating the development for assuring sustainable management of ground water. The Authority is vested with the responsibility of regulation of ground water development and management including conservation, protection and augmentation of ground water through artificial recharge, recycling and reuse of water. The approach of the Authority is to place a major thrust on education and bringing in awareness among the people of the need to protect and conserve ground water and its proper utilization. The Authority is promoting adoption of artificial recharge in water deficit areas. With the objective of building up a data base on ground water utilization and draw up future policy the Authority has initiated registration of tubewells/ borewells in many areas, including NCT of Delhi, Faridabad and Ballabhgarh in Haryana, Ludhiana in Punjab and Ghaziabad in Uttar Pradesh.

Drought mitigation

For drought mitigation in drought prone areas, it is essential that exploration of ground water be accelerated. Further rain water harvesting and artificial recharge measures need to be implemented on large scale so that effect of drought in not much pronounced. Efforts made by various NGOs working in parts of Rajasthan, Gujarat and Madhya Pradesh in this field have proved to be successful. There is, therefore, need to promote augmentation of ground water in drought prone areas as a first priority.

Conclusion

The growth process, increase in population and expansion of economic activities would lead to increasing demands for water use for various purposes. By the year 2050, requirement of water will exceed the available utilizable water resources. The water resources would, therefore, have to be treated as a national asset and their development and management would have to be planned in an optimal manner by adopting different techniques of water conservation and appropriate water management techniques. Efforts would need to be made to increase the availability of fresh water by harnessing the surplus monsoon runoff and recharging it underground and/or recycling/reuse of water. The available water resources would need to be used in an optimal manner so as to avoid wastage of water. Stress would need to be laid on conjunctive use of water resources and use of saline/brackish water resources.

REFERENCES

Ministry of Water Resources. (1991) *Waterlogging, soil salinity and alkalinity*—Report of Working Group on Problem identification in irrigated areas with suggested remedial measures, New Delhi.

Natarajan, (1993) *Population Projections for India*—presented in 17th Conference of Indian Association for Study in Population. Annamalai University, Tamil Nadu.

Central Ground Water Board, (1995). *Ground water resources of India*. Central Ground Water Board, Ministry of Water Resources, Faridabad.

United Nations, (1995). *World Population Prospects—1994 Revision*. New York.

Government of India, (1996). *Population Projections of India and States 1996-2016*, Census of India. Registrar General of India. Ministry of Home Affairs. GOI, New Delhi.

Visaria, L. and Visaria, P. (1996). *Prospective Population Growth and Policy Options for India*—1991-2101. The Population Council, New York.

Central Ground Water Board, (1997). *Inland ground water salinity in India*, Central Ground Water Board, Ministry of Water Resources. Faridabad.

Central Ground Water Board. (1997). *Status of ground water quality including pollution aspects in India*, Central Ground Water Board, Ministry of Water Resources, Faridabad.

World Bank, 1(998). India water resources management sector review—*Ground water regulation and management report.*

Government of India, (1999). *Integrated Water Resources Development - A plan for Action*. Draft Report of National Commission for Integrated Water Resources Development Plan, Ministry of Water Resources, New Delhi.

3

Integrated Planning of Land and Water Resources with Special Reference to U.P.

Mridula Singh

Judicious management of the land resources is a major and complex challenge facing us today. On the other hand India's water resources are also under a state of deep crises. Traditional springs are dying up, largely on account of massive deforestation and erosion of catchment areas caused by *ad hoc* developmental programmes and infrastructural projects. Many models of resource management have been developed in the past.

India has 16 per cent of the world's population, 15 per cent of the livestock population, 1 per cent of the forest area, 0.5 per cent of pasture lands and only 2 per cent of the geographical area. The per capita availability of forests in India is only 0.08 per cent per ha as against the world average of 0.8 per cent leading to the pressure on land and forests.

Land is a non-renewable resource that supports all primary production system as well as provides essential social environment in terms of shelter roads and other facilities. It is scarce and inexpansible resource which has to be used very judiciously to meet various competing demands. Land which includes soil, water, associated plants and animals is a complex combination of factors like geology, topography, hydrology, soils, micro climates that are continuously interacting under the influence of climate and peoples' activities. The use to which land is put is determined by owners, farmers and government institutions etc. according to their own perception of needs.

Land has limited carrying capacity of population beyond which there will be degradation and loss in productivity due to excessive use. The fast pace of the technological and economic development and growing population on limited resources not only lead to environmental pollution but also their degradation. The concept of optimal land use is basic to land use planning which aims at balancing the competitive demand in a scientific manner for sustainable development of the economy. The purpose of land use planning is to regulate uses on a scientific ecological basis and to curb misuse, over use and abuse of land. Land use planning considers the eco-system holistically, i.e. an integrated system of land, water, plants and animals along with human beings. The management of one use is complementary to other.

An appropriate land use plan takes into account the sustainability and capability of land so that the superior quality of land is not put to use for a purpose suited for the land of inferior quality and *vice versa.* In this context effective control on the diversion of prime agricultural land for non-agricultural purposes is very important.

Optimum land use is one which is ideal for the geographical and socio-economic features of the region at a point of time. It is not necessarily one which gives the maximum yield per hectare of land but also gives best input-output ratio per unit of land. Land is not only inelastic but also heterogeneous in different parts and regions of the country with definite set of capabilities and suitability for different uses. India being a vast country and a land of ancient settlement; its present pattern of land use is the result of long and continued inter-action of environmental and cultural factors. Naturally the concept of optimum land use which was relevant for one region of the country in past may not hold good in future and for other regions also, i.e. the concept of optimum land use is dynamic and varies as the spatio-temporal framework is changed.

Shrinking resources of prime lands, deforestation and accelerated erosion, deterioration of soils—physical environment, increasing water-logging and salinity in canal-irrigated areas, declining water-table in well irrigated areas, poor management of rain-water, lower productivity of dry land areas, lower efficiency of inputs of water, fertilizer and agrochemicals, rapid industrialization coupled with pollution and environmental degradation and hazards have aggravated the problems related to this resource.

In this context as far as Uttar Pradesh is concerned the land use statistics of the last three decades are as follows:

Table 3.1 : Land Utilization Statistics of U.P.

(Thousand Hectares)

Sl.	*Item*	*1990-91*	*1998-99*	*1999-00*	*2000-01*
1	Reporting Area	24245	24200	24202	24201
2.	Forest	1698	1714	1689	1689
3.	Barren & Unculturable Land	2323	2390	2413	2436
4.	Area put to non-agricultural uses	720	635	625	617
5.	Culturable wasteland	709	571	550	535
6.	Permanent Pasture & Other Grazing Land	74	67	69	90
7.	Land under Msc. Trees & Groves	330	330	356	340
8.	Current Fallow	1152	1017	1020	1048
9.	Other Fallow	809	675	661	641
10.	Net Area Sown	16430	16801	16819	16825

In addition to the above it is also worthwhile to mention that approximately 46 per cent of total land in the State suffers from degradation of some form or the other.

Land Utilization in Uttar Pradesh : Growth Trends

The growth in different land uses in the State during 1980-81 and 1990-91, 1990-91 and 1999-2000 and 1980-81 and 1999-2000 had been negative except in area under other fallow land, non-agricultural uses and forest that showed a positive growth of 20 per cent, 6.92 per cent and 3.06 per cent respectively during 1980-81 and 1990-91. The net area sown also showed a negligible growth of 0.10 per cent during this period. During 1990-91 and 1999-2000, area put to various non agricultural uses, miscellaneous trees and net area sown had growth of 4.41 per cent, 9.54 per cent and 1.72 per cent respectively. All other uses of land experienced negative growth. The analysis of the period of 1980-81 and 1999-2000 revealed that area under forest showed a marginal increase of 0.27 per cent, net area sown by 1.82 per cent and a substantial growth of 11.63 per cent in area converted for various non-agricultural uses. Thus, during two decades of 1980-81 and 1999-2000, there was remarkable growth of around 12 per cent cent in the area shifted for non-agricultural uses in Uttar Pradesh.

The other uses of land showed negative growth except the forest and net area sown.

In western region of the State, the area put to various non-agricultural uses during 1980-81 and 1999-2000 showed fastest growth of 9.86 per cent followed by 1.12 per cent growth in net area sown and 9.44 per cent growth in area under miscellaneous trees. The other uses of land had negative growth. The different land uses in central region of the state showed typical growth pattern during 1980-81 and 1999-2000. There had been nominal growth of less than one per cent in forest and net area sown. The fallow land registered a substantial growth while all other uses of land had negative growth. In eastern region of the state, land area used for non agricultural purposes demonstrated maximum growth of 17.78 during 1980-81 and 1999-2000 followed by 14.26 per cent growth in pasture area, 1.14 per cent growth in net area sown and 0.29 per cent growth in the area of forest. All remaining uses of land experienced negative growth. The land under non-agricultural uses in Bundelkhand region had substantial growth of 25.58 per cent during 1980-81 and 1999-2000 followed by 10.36 per cent growth in area of forest and 9.08 per cent in net area sown. Other uses of land had negative growth.

A fast process of conversion of agricultural land to non-agricultural purposes particularly for the construction of buildings, sites and roads had remained operative in all parts of the state during last decades, resulting into substantial diversion of agricultural areas to such non-agricultural purposes. If the process continues unabated, it will have serious repercussion on food production, affecting the large section of our society, which is already vulnerable to poverty. Now the question arises whether the process of land conversion is to be taken as a natural phenomenon arising out of increasing demand of land by the growing population, which is difficult to be managed unless some controls are placed over the growth in population or what other measures are to be adopted? In the given circumstances, the best course of action would be to streamline the process, which could restrain such unmindful diversion of fertile agricultural land to non-agricultural uses and only genuine needs of land for other purposes could be met.

Today the need is to have a land use policy of the State which focuses on following important issues :

1. Expansion of effective forest cover to restore ecological balance.
2. Reclamation of vast area under waste lands.
3. Increase productivity of agricultural land and preserving prime agriculture land.
4. Conjunctive use of ground and surface water resources.
5. Moisture conservation through effective watershed management.

Food Security and Land Use

The agricultural strategy lays emphasis on the increasing food grains production in the country. The agricultural growth must be viewed not merely as means towards food security, but as a strategy to subserve the broader goal of poverty eradication. The extent of poverty in the country is unacceptably high. Here the factor which are critical to the issue are, land degradation and water depletion. The pace of employment generation is also affected in the process. The single most important issue in raising agricultural production to my mind is land and water management. Corrective actions for the various problem facing this resource need to be taken immediately.

As said earlier the statistics show that almost 50 per cent of the total land at the national level and similarly at the state level is degraded in one form or the other. Public investment in irrigation, rain water harvesting and watershed management must be given priority. People's participation in management of these precious resources should be ensured. The crucial issue today is long term sustainability of our agriculture. The need is to nurture nature and control the depletion of soils, diversion of agricultural land to non-agricultural uses and adopt suitable cropping patterns and proportionate use of water. Adoption of an integrated approach for planning of land resources is inevitable in terms of sustainability.

The FAO guidelines defined Land Use Planning as :

> "Land Use Planning is the systematic assessment of land and water potential, alternatives for land use and economic and social conditions in order to select and adopt the best land use options. The driving force in planning is the need for

> change, the need for improved management or the need for a quite different pattern of land use dictated by changing circumstances."

To ensure that future generations are secured as for as conservation and management of land resources are concerned the land use plan should aim not only at achieving the production targets but should also fulfil the conservation aims. This will be the real intergrated approach. The physical, social and political causes of the problems of land resources should be identified and keeping in mind the long term perspective, a suitable and appropriate plan of action should be evolved. In no way, the long term sustainability should be sacrificed for immediate needs.

In India most of the land available to meet current and future food requirements is already in production. The number of marginal holdings is much higher. Any ambitious, future expansion obviously targets the marginal land which is already fragile more so with high growth of population, increased poverty and weak institutions to deal with the situation.

Water which is another key resource for life is also under great stress. As population is increasing, aquifers are being exploited at faster rates and in turn giving rise to alarming situation.

The national annual average of per capita water availability in 2001 is about 1829 cu.m. With the existing water resources of the country the per capita availability of water varies in the range of 300 to 13754 cu.m. per year. It is estimated that by 2015 AD, the annual per capita availability of water will decrease to about 1557 cu.m. The conditions will be further worsening due to continued population growth and by 2050 A.D. the projected annual per capita availability of 1168 cu.m.would take the country to the threshhold of water scarce conditions. It is estimated that by 2050 AD 22 per cent of the geographical area and 17 per cent of population of the country will be under scarcity conditions having access to water availability of less than 500 cu.m. per year and 70 per cent of the area and 76 per cent of population will be on the verge of affected health and economic activities with access to water availability of less than 1000 cu.m. per year which is identified as stress on scarcity level.

As population increases more water would be required for agriculture, industry and domestic uses. A rough estimate indicates that nearly 1000 ton of water is consumed to produce one ton of wheat and double of this amount is required for one ton of rice. A substantial quantity of water is also consumed by the industrial sector. Presently nearly 56 ton of water is required to produce 1 ton of synthetic fibres. A thermal power station of 1000 megawatt capacity consumes 3.28-4.38 million metre cu. (Mm^3) water per day.

The water resources of the state of U.P. are as follows:

1.	Rain Water precipitation	
	(i) Average annual rainfall (depth)	1167 mm
	(ii) Average annual rainfall volume	34.37 lakh ha m
2.	Surface water resources	
	(i) River flow (gross)	13.77 lakh ha m
	(ii) Utilizable	13.11 lakh ha m
	(iii) Utilizable for irrigation	9.86 lakh ha m
	(iv) Utilized at present	4.38 lakh ha m
3.	Rain Water dispersion	
	(i) Evaporation	6.01 lakh ha m
	(ii) Ground water recharge	15.47 lakh ha m
	(iii) River flow	12.89 lakh ha m
4.	Ground Water resources (dynamic)	
	(i) Annual gross recharge	8.42 lakh ha m
	(ii) Utilizable for irrigation	7.15 lakh ha m
	(iii) Utilized at present	4.01 lakh ha m
5.	Ground water resources (static)	
	(i) Gross availability	8.80 lakh ha m
	(ii) Utilizable for irrigation	7.48 lakh ha m

Source : Pandey, R.N. and Rai, D. 1996. *Report of Subworking Group on Water Resources of UP.*, UPCAR, Lucknow and U.P. State Water Policy, 1999.

In the above scenario a well thought out action plan for integrated use of land and water is the need of the day. Managing the resources optimally should be the main concern. Plans should be formulated at state, district, watershed or village level. The process indeed involves the participation of land users and all efforts are needed to watch their interests. In short, the plan should aim to make best use of limited resources.

Issues for Consideration

1. A reliable, strong, continuous data management system should be evolved.

2. Effective institutional arrangements for planning, coordinating and subsequent management should there at the grassroots level.
3. Through proper land use plan and its management efforts should be made to reduce disparities and regional imbalances prevalent in the State.
4. *Sectoral programmes are generally spelt out in isolation. They should form an integral part of a well thought out and coordinated land use policy.*
5. Legal, institutional or organizational arrangements should be clearly spelt out so that land resources are used in integrated, sustainable and optimal way.
6. *Scientific generation of database for various land uses and degraded lands at appropriate scale need to be standardized for various category of users, national and regional Planners district and watershed level interpreters.*
7. The area under irrigation specially command areas is facing problems of salinity, water-logging and seepage. The priority should be to minimize the problem and ensure conjunctive use of surface and ground water.
8. *Another area of great concern is the per capita availability of land/arable land. A suitable population policy is the need of the day. Instead of coercion this should be implemented through education and awareness campaign.*
9. With legislation a system of appropriate price policies, incentives and other fiscal mechanism have to be evolved.
10. The sectoral inter-dependence and linkages in the process of economic development should be clearly spelt out in the Perspective Plans for conservation, management and development of land resources.
11. The success stories in this regard should be published and implemented.
12. Rain water conservation and ground water recharge should be given importance through development and conservation of water bodies and development of wetlands.

4

Land Degradation : New Trends Towards Sustainable Agriculture

Murari Lal

Introduction

Land has limited carrying capacity of population beyond which there will be degradation and loss in productivity due to excessive use. The Land use policy should aim at optimizing returns on long term basis rather their meeting short term requirements on *ad hoc* basis. An appropriate land use plan has to take into account the suitability and capability of land. The suitability for particular use depends upon agroclimatic conditions like—climate, rainfall, water requirement, soil slope etc. Whereas land capability would reflect the cap y of land to carry on a given activity on sustainable basis without erosion or damage to it productivity. Zero Soil Loss (ZSL) should be aimed at. It is at the micro-level that soil capability can be assessed and decision about land use can be taken. Therefore, the land use plan would be operationally meaningful if it is workable and feasible at micro level.

Optimal Landuse

The requirements of optimal land use plan are as under :

1. To meet the consumption needs of growing population by raising resource use effciency.
2. To restore and preserve ecological balance, ensure sustainable development and prevent future damage to land resource.

3. To restore productivity of degraded/marginal land.
4. To identify capability of land and put them under different uses based on capability, national goal and ecological balance.

Watershed Management

Where land is poor, people obviously remain poor. Watershed Management is, therefore, most effective way of removal of rural poverty. Objectives of watershed management are :

1. To increase infiltration into the soil,
2. To control damaging runoff, and
3. To manage and utilize runoff for useful purposes.

Thus, watershed management means harmonious development and management of land and water within natural boundaries of a watershed so as to produce on sustainable basis, abundance of plants for good of man and deliver clean and controlled flow of water to down stream. Watershed development and management projects. should have two major components :

1. Foundational component of land and water management and conservation.
2. Super structure component of conservation or sustainable production system (all aimed at sustainable production).

These systems include :

1. Conservation farming,
2. Conservation forestry/agro forestry,
3. Conservation of plantation crops (tea, coffee, coco),
4. Agriculture,
5. Animal Husbandry,

or any other as suited to physical/socio-economic conditions within watershed. The main threat of watershed management can be judged through the following :

1. Economics,
2. Ecology,
3. Employment generation,
4. Equitable distribution of benefits,
5. Enhancements of bio-energy specially biomass based energy.

Following are the salient points which must be kept in mind while examining the watershed management:

1. No two watersheds are alike,
2. No general rule can be laid down for all the watersheds,
3. Each watershed is to be studied in terms of its own problems/characteristics.

For natural resource investigation for effective watershed management following data/surveys are prerequisite:

1. *Land resource survey*—slope, erosion, texture, depth,
2. *Soil resource survey*—classification of physical/chemical characteristics of soil profile, location, length and course of drainage, fertility status of arable and non-arable land.
3. *Water resource*—Hydrological survey, Rainfall data, Groundwater level depth of water table in different seasons, Hydrochemistry of groundwater, Hydrological/Hydro-morphological data, surface flow, runoff.
4. *Vegetative resource survey*—forest type, flora, economic utility, biomass, fuel, fodder, species, wastelands, grazing grounds.

These statistics along with statistics available through socio-economic survey, demographic survey, area skill survey should be made use of to prepare integrated watershed development plan. For this data on three levels will be recognized, viz.

1. *Khasra level :* Khasm map of watershed presents land use, crops grown, cropping pattern, water bodies, drainage, irrigation, slope etc.
2. *Watershed level :* Soil characteristics, topography, natural vegetation, forest etc.

3. *Village level :* Infrastructure facilities, livestock, agricultural inputs, demography, education, medical/health facilities, socio-economic conditions of cultivators etc.

Sustainable Agriculture in Context of Uttar Pradesh

The targeted growth for agriculture in the current plan (2002-07) is pegged at 5.12 per cent per annum. The major problems in the development of agriculture are being listed below:

1. High degree of heterogeneity in agroclimatic zones and diversity in economic profiles of regions.
2. Land degradation problem in 74.48 lakh hectares.
3. Preponderance of marginal and small holdings.
4. Problem relating to transfer of technology.
5. Problem relating to rural infrastructure.
6. Credit facilities available.

With the vision to transform the state into granary of nation to ensure food security and through sufficiency in food grain production, to promote economic growth and prosperity of rural masses through diversification to cash crops and allied activities and to bring about qualitative improvement in village life by using clean and pollution free environment, and framework for progressive and prosperity, i.e. the Agriculture Policy of Uttar Pradesh has been chalked out as early as in 1999.

The main objectives laid down in the policy are :

1. To develop appropriate eco-friendly farming systems which would improve soil fertility.
2. To maintain ecological balance and appropriate land and water-management.
3. In order to substantially increase the income of farmers to diversify present agriculture towards cash crops and other complimentary sectors, i.e., horticulture, animal husbandry, sericulture, milk production etc.

Establishment of agro-based industries has also been listed in the priorities list. It has been envisaged, in the policy that present seed replacement rate of 12 per cent will be raised to 20 per cent by the end of 2007 AD.

Issues for Considerations

On the basis of foregoing analysis the following issues are for consideration :

1. Poverty alleviation programmes/schemes should be so tailored as to ensure the sustainable use of natural resources.
2. Productivity of farming systems should be measured by net income per unit of land, water and energy not in terms of yield per hectare.
3. Area skill surveys should be carried out for promotion of Rural Employment and diversification of agriculture.
4. The Research and Development Programmes/efforts relating to sustainable food/nutritional security should be attached due importance.
5. The symbiotic relationship between Man and Nature should not lose the sight.
6. The food security strategy should be ensured in terms of "Food Security Act". This act is the need of hour.
7. Area specific policy measures should be adopted for sustainable use of natural resources through efficient watershed management.
8. Land is the subject of State List. Hence State Land Use Boards should be suitably strengthened so that their apex bodies at state level may properly take care of land resources.
9. A close coordination amongst different departments/set ups may yield better results.
10. The land use on private lands cannot be dictated by legislation alone. This problem can be tackled through price policies, incentive and other fiscal mechanism too.
11. In place of agro-chemical agriculture the bio intensive cropping system should be adopted.

5

Neglect of Grazing Lands in India
Environmental and Economic Costs
(A Case Study of Pasture and Grazing Lands in Himachal Pradesh)

Surya Kant

Introduction

Grazing lands are one of the three important components of our land, other two being croplands and forestlands. Of them, grazing lands are the most neglected one. Large parts of former grazing lands have been brought under the cropland, leading to an overstocking of the remaining areas. Overgrazing prevents regeneration of forests and accelerates soil erosion, turning even foresters and environmentalists into enemies of the animal population. A variety of factors contribute to the neglect of grazing lands in our country, resulting to further deterioration of India's lands and impoverishment of our rural populace. This is all the more true of hill and desert areas, where environment is highly sensitive and animal husbandry form the main base of predominantly rural economy.

Himachal Pradesh makes a typical case of this kind. It is entirely a hill state, where the number of livestock population had been almost the same (5.11 million in 1992) as of human population (5.17 million in 1991); and animal husbandry contributes nearly 15 per cent of the total income of the state. As a source of manure and drought power, the cattle are worth of their weight in gold for the thousands of small farmers in the state. A number of village industries are also dependents on animal life for raw materials. But this huge mass of livestock in the state needs fodder, and there simply is not enough of it. Only about 1.2 million hectares of the total land area in the state is classified as 'permanent pasture and other grazing lands'. On one hand, this much of land is not enough for livestock population in the state, on the other these lands are highly degraded. Therefore, animals scrounge for whatever they can get on fallow and uncultivated lands. Grazing on the steep slopes of the high hills is now a common practice in the state. In fact, livestock graze all accessible vegetation in the state, snapping up everything except non-palatable weeds. Herds and flocks migrate to places where prospects are better, especially in tribal areas of Chamba, Lahul & Spiti and Kinnaur districts. Low growth rate of livestock population and decline in animal productivity are now well known features of animal wealth in the state.

The effects of haphazard grazing on the environment in the state are alarming. Land degradation due to overgrazing is resulting in desert conditions. Moreover, animal productivity has been reduced, putting enormous pressure on the socio-economic conditions of households, depending on animals for their livelihood. Notwithstanding, the grazing lands have and continue to be the most neglected lands. The planners and policy-makers in the state feel least concern with this sorry state of affairs.

Although the number of animal population has been declining in the recent years in the state yet such population is quite large especially in the context of dwindling size and degraded conditions of pasture and grazing lands. The size of land under permanent pastures and grazing has declined to 1.20 million hectares in 1997-98 from 1.58 million hectares in 1969-70. Against this, animal population increased to 5.26 million in 1992 from 4.70 million in 1972. A large proportion of grazing lands in the state belongs to the Panchayats and the powerful individuals have encroached upon them. The government has remained nearly the silent spectator to this. In several cases, the government itself has launched programme like irrigation schemes, turning earlier grazing lands into croplands. During this process, no genuine concern was shown for the rehabilitation of grazers. Besides, there has been a shift in the cropping pattern and a change in crop varieties. Additionally, absence of organised fodder production and pasture management has further intensified pressure on grazing lands. The growing population in the state means increased pressure on lands so as to produce food grains and raising cash crops. Forage crops do not find an important place in crop rotation because of small land holdings.

In the light of above statement, the present paper makes a geographical study of the grazing lands in Himachal Pradesh with the following research questions:

1. What has been the proportional share of and changes in land under grazing and pastures in the state?
2. What are the sub-regional variations in distributional pattern of such lands in the state?
3. What factors are responsible for decline in and degradation of grazing lands?
4. What are the socio-economic and environmental consequences of degradation in such lands especially in tribal areas of Chamba, Lahul & Spiti and Kinnuar? and
5. What are the remedial measures to deal with the situation?

For the purpose, a detailed analysis of district level data on land use with a focus on land under permanent pasture and grazing lands for the period 1970-71 to 1995-96 has been done. In addition, overall growth and inter-district distribution of livestock population in the state has been examined to see its impact on pasturelands in the state. Data has been analyzed with the help of suitable cartographic and statistical techniques.

The discussion in this paper has been organized in three sections. In the first, salient features of geography of Himachal Pradesh have been examined. This is followed by a brief study of livestock population in the state. Thirdly, land use in general and pasturelands in particular has been examined.

I
HIMACHAL PRADESH: GEOGRAPHIC CHARACTERISTICS

Situated in the heart of the Western Himalayas, Himachal Pradesh is one of the youngest and the most dynamic among the hill states in India. Its per capita income of Rs. 16144 in 1998-99 (at current prices) was the highest among the hill states. The state has achieved a remarkable success in the fields of horticulture, hydel-power generation, tourism and female education.

At the time of its inception in April 1948, it had an area of 27,454 km^2 and a population of 935 thousand persons. The whole territory was reorganized into four districts of Chamba, Mandi, Sirmaur and Mahasu. Passing through a number of territorial and administrative reorganizations, it attained to present shape and size in 1966, when under Punjab State Reorganisation Act, *Pahari* (a dialect of Hindi language) speaking areas of then Punjab were merged with Himachal territory. This enlarged not only its size but also provided with physical compactness and resource richness required for efficient administration and resource utilization.

The statehood granted on 25th January, 1971 amounted to giving a full freedom to state leadership in deciding upon the development priorities for the state. Also, this ushered in an era of renewed enthusiasm and vigour among the masses and State bureaucracy to dream for the new hopes and aspirations in future.

Currently, the State is divided into 12 districts, regrouped into three administrative divisions, 75 tehsils and 69 development blocks. It covers an area of 55,673 km^2 that makes it larger in size than both Punjab and Haryana. Among the hill States, the area of Himachal Pradesh is next only to Jammu & Kashmir and Arunachal Pradesh. Its population was 6.08 million persons at the 2001Census. Distributed in 20175 settlements (20,118 villages and 57 towns) of varying size and functions, population is highly concentrated in longitudinal river valleys.

The region is veiled from the Punjab plains by the Shiwalik hills. Elevation generally increases from South-West to North-East. The various landforms find typical longitudinal arrangement. The glaciated and snow-capped peaks are prominent features of the landscape. The Dhauladhar (the White Ranges) over look Kangra valley in supreme majesty. The Pir Panjal, the Great Himalayan and the Zaskar ranges stand guard over Chamba, Lahul & Spiti, Kullu and Kinnaur districts. The slopes of the mountains are covered with forest and meadows. The peaceful valleys below are interspersed with many streams, fields and quaint homesteads. The altitude ranges from 300 to 7000 metres above sea level. About one-half of the total area is higher than 3000 metres. More than one-third of the total area in the Trans-Himalayan zone is cold desert.

Himachal Pradesh may be divided into four physiographic regions:

(i) The Shiwalik Hills and the intervening *duns*, (ii) The Lesser Himalayas, (iii) The Great Himalayas, and (iv)Trans-Himalayas. These differ widely in terms of natural resources, human adaptations, development constrains and opportunities. Trans-Himalayan region, a zone of unfavourable climate, is a cold desert where crop productivity is low. Transhumance farming system is practiced. Since the resource base is poor and the area remains snow-bound for about six months in a year (November to April), the farmers shift their livestock (sheep and goats) to the lower hills for grazing. Even in the Shiwalik hill zone rainfall is both low and unreliable, making crop cultivation risky. Occasional high intensity rainfall in Shiwalik Hills, formed of loose materials, causes flesh floods leading to deep gully erosion. Droughts and floods, being the reoccurring problems here, agriculture is low productive. Farm families rely on non-farm income and remittances to stabilize their annual incomes and consumption. Moreover, livestock rearing is the only feasible activity where undulating topography suits the grazing of animals. In the Lesser and the Greater Himalayan zones, where stone and citrus fruits, apple cultivation and off-season vegetables are grown, soils are infertile; arable land is limited; and water resources are limited for irrigation.

Notwithstanding the sharp physical and socio-economic differentials among the four physiographic divisions in the state, all of them, on one hand, commonly suffer from an intense pressure of human and livestock populations on limited land resources and on the other possess highly fragile ecology.

Land available for cultivation is limited and the size of land holdings is small. Only about one-tenth or 583 thousand hectares of total geographical area of the state (5567 thousand hectares) is classified as net sown area. As per village papers, it comes to about 17 per cent, against the national average of about 47 per cent. Cultivated land is mainly confined to the Kangra Valley, the *duns*, valley of the Satlej and the terraced hill slopes. Only about 19 per cent of the net sown area in the state is irrigated against the national average of about 39 per cent. Average size of landholding is 1.2 hectares, and more four-fifths of holdings are marginal or small (of less than 2 hectares).

Low productive subsistent agriculture contributes only about 28 per cent to the state income nevertheless more than two-thirds of the total population is directly engaged in agriculture. To supplement meagre incomes from crop farming farm households go for livestock farming. Over the period, both human and livestock populations have grown. In last 30 years, between 1971 and 2001, human population has increased by 2.62 million, registering an annual growth of 1.89 per cent, whereas livestock population has increased to 5.2 million in 1997 from 4.7 million in 1972, registering 0.41 per cent annual increase. This speaks of ever intensifying pressure of human and livestock populations on limited land resources in the state.

The lack of transport and communication is one of the most serious impediments to the development of the State. Owing to undulating hill terrain and harsh climate, building and maintaining infrastructure is not only difficult but also requires greater investment. Despite a high priority accorded by the state to road development during the Five Year Plan, only 18160 km of road length was available by the end of 1992-93. Only 7521 or 44.2 per cent of the villages connected with roads. Inter-town and rural-urban transport linkages are very weak in most parts of the State.

Briefly, Himachal Pradesh is among the most dynamic hill states of India. Its most notable accomplishments have been in the fields of infrastructure development, horticulture and hydel-power generation. Nevertheless, state suffers from serious ecological and environmental problems. Some of such problems are imbibed in ecological setting of the state, while others have emerged with advances in development process in the state. More than one-third of the total area in the Trans-Himalayan zone is cold desert. About one-half of the total area is higher than 3000 metres. Cultivated land is restricted to only one-tenth of the total land area, and is mainly confined to the Kangra valley, the *duns*, valley of the Sutlej and the terraced hill slopes. Remarkable expansion has taken place in fruit cultivation. The state is now described as the 'apple belt' of India. The traditional household industry predominates the industrial scenario in the state. Majority of agricultural population is of the owner-cultivators, with small to marginal land holdings. Seeking or looking for a government job including the defence services is an integral part of the social ethos. Basic development problems of Himachal Pradesh relate to strengthening of social and development infrastructure, an optimal utilization of natural resources especially hydro-electric potential and scenic beauty and repair of any damage to environment and ecology caused by deforestation or intensification of cultivation on marginal lands or overgrazing of pasture lands.

II
LIVESTOCK POPULATION

In India, livestock population is recorded quinquennially, for example in 1992, 1997 and 2002. It does not coincide with human population census, conducted after every ten years, for example 1981, 1991 and 2001. The first livestock census in Himachal Pradesh was conducted in 1972. It has completed its seventh livestock census in 2002 but published data are not yet available.

A total of 5.2 million livestock was recorded in 1997 (Table 5.1), making one per cent of the total livestock population of the country. It was close to the number of human population in the state.

Table 5.1: Livestock Population in Himachal Pradesh, 1972-1997

Census Years	*Livestock (in million)*
1972	4.7
1977	4.8
1982	5.0
1992	5.1
1997	5.2

Source : *Integrated Sample Survey for Estimation of Animal Products, Milk, Eggs, Wool and Meat, 2001-2002*, Directorate of Animal Husbandry, Himachal Pradesh, Pasudhan Bhawan, Shimla, pp. 9-10.

Owing to small and ill-terraced fields, steep slopes, poor soils, inadequacy of irrigation and fertilizer use, severity of climate and lack of innovations, agriculture in the state is low productive. To supplement meagre income from crop farming, farm as well as other low-income rural households go for livestock farming. Not only the animal power is the dominant source of energy in farm operations and rural transportation but also provide essential food and non-food items like milk, meat, hides, skins, wools etc. This is the reason that nineteen out of each twenty households in the state keep at least one of the species of livestock (ISS, p. 15) to meet their domestic needs and or to earn livelihood. About 90 per cent of the households maintain bovines and more than 50 per cent rear goats and sheep. Poultry birds accounted for 5.54 per cent of the total households. In 1992, the state accounted for 1.1 per cent of India's livestock population, while its share in total human population of the country made only 0.6 per cent.

The composition of livestock population in the state has undergone change during 1972-97 (Table 5.2). While the share of cattle population declined to 41.6 from 46.3 per cent, that of buffaloes increased to 14.3 from 11.6 per cent. Similarly, while the share of sheep came down to 20.7 from 21.1 per cent that of goats experienced an increase to 22.4 from 19.3 per cent. The number of horses and ponies come down by one-fifth. However, poultry recorded an increase by 4.6 times: from 187 thousand in 1972 to 864 thousand in 1997.

Table 5.2 : Composition of Livestock, 1972 and 1997

Livestock	*Total*		*Percentage to total livestock*		*Trend*
	1972	1997	1972	1997	
Cattle	2175690	2173575	46.26	41.60	Decline
Buffaloes	543887	748246	11.56	14.32	Increase
Sheep	1039946	1080464	22.11	20.68	Decline
Goats	906415	1167992	19.27	22.35	Increase
Horses and ponies	16234	12884	0.34	0.24	Decline
Others	20283	40906	0.43	0.78	Increase
Total	4702455	5224067	100	100	

Source: Same as Table 5.1.

While there has been a general increase in livestock population during 1972-97, five out of twelve districts in the state registered absolute decline. These include Hamirpur, Kullu, Lahul & Spiti, Shimla and Una. Nearly twenty per cent of the total livestock were concentrated in a single district of Mandi, followed closely by Kangra district with about 19 per cent (Table 5.3).

Within livestock population, more than half of the total cattle population in the state is concentrated in three districts of Mandi (21.31), Kangra (18.07) and Shimla (14.92). Against this, Kinnaur and Lahul and Spiti have a small number. In spite of efforts towards improving the quality of cattle, only about one-sixth of them are crossbred and rest indigenous.

Table 5.3: Change in livestock population by districts, 1982-1997

Districts	*1982*		*1997*		*Trend*
	Total livestock	*% in state total*	*Total livestock*	*% in state total*	
Bilaspur	237582	4.76	238448	4.56	Decline
Chamba	727149	14.57	739036	14.14	Decline
Hamirpur	250236	5.01	232521	4.45	Decline
Kangra	809434	16.22	981483	18.78	Increase
Kinnaur	93869	1.88	123179	2.35	Increase
Kullu	332314	6.66	327628	6.27	Decline
Lahul & Spiti	68023	1.36	62172	1.19	Decline
Mandi	871093	17.46	996565	19.7	Increase
Shimla	630695	12.64	560647	10.7	Decline
Sirmaur	410276	8.22	430083	8.2	No change
Solan	317703	6.36	329953	6.3	No change
Una	240166	4.8	202352	3.87	Decline
Total	4988540	100	5224067	100	

Source: Statistical outline, Himachal Pradesh, 2002, p.85.

Buffaloes are highly concentrated in Kangra, Mandi, and Hamirpur districts (Table 5.4). These districts in combine share one-half of total buffaloes in the state. Lahul & Spiti, Kinnaur, Kullu, Shimla and Sirmaur have very few of them.

Table 5.4: Concentration of livestock population, 1997

Category	*Three top most districts in proportional shares*
Cattle	Mandi (21.31), Kangra (18.07), Shimla (14.92)
Buffaloes	Kangra (20.4), Mandi (15.77), Hamirpur (13.53)
Sheep	Chamba (23.81), Mandi (18.48), Kangra (17.56)
Goat	Kangra (20.24), Mandi (17.94), Chamba (16.79)
Horse	Kangra (28.67), Mandi (15.34), Shimla (14.84)
Yaks	Chamba (41.41), Lahul & Spiti (35.48)

* Figures in parentheses indicate to percentage in total animals in the state under respective category.

Source: As Table 5.1

In Chamba, Mandi and Kangra districts sheep population has high concentration. The number of sheep in Una, Solan, Bilaspur and Sirmaur districts is small; these together do not have even 6 per cent of the total sheep in the state. Geographical distribution of goat population finds a close association with that of sheep. Like sheep population, goats are also highly concentrated in Kangra, Mandi, and Chamba districts, having in combine more than half of the state's goat population. Kangra leads in poultry birds followed by Mandi & Solan districts. Poultry is more popular in districts that adjoin Punjab.

Livestock combinations reveal that low altitude gentle sloping terrain of districts adjoining Punjab such as Bilaspur, Hamirpur and Una have a dominance of Buffaloes in their livestock populations, whereas cattle population dominates in Kangra, Kullu, Mandi, Shimla, Sirmaur and Solan districts. In high altitude dry zone in districts of Chamba, Lahul & Spiti and Kinnaur sheep population dominate in total livestock.

III
PASTURE AND GRAZING LANDS

Largely subsistence farming on limited cultivated land in the state does not produce enough for large and growing population. Add to it the so large and generally inefficient livestock population compete with population beings for the limited agricultural resources. In this competition, food naturally gets preference over the fodder production. Firstly, there is hardly any important fodder crop grown in the state and secondly, not even one per cent of the cultivated land is devoted to such crops. Forest areas are consequently called to meet fodder requirements of vast population of livestock (Sharma *et al.*, 1988, p. 206).

The forest land is used for grazing and leaves and twigs are lopped for green fodder. Interestingly the members of farm community in the state measure their wealth by the number of animals owned by a household, leading overstocking and further pressuring the limited grassland resources. Due to low yield, farmers in the state keep larger herds to meet their requirements of milk, animal manure, and farm operations. Thus, livestock population increased by more than 11 per cent during 1972-1997, when it increased to 5.22 million in 1997 from 4.70 million in 1972 (Table 5.6). Buffaloes recorded as high as 37.6 per cent increase following by goats with 12.3 per cent. On the whole the pressure of animal population (number per hectare) has been growing on cropped land. This could happen in spite of the fact that cropped area record a marginal increase of 5.38 per cent during

1972-1997. However, in case of pasture and grazing lands pressure of livestock population intensified greatly. In fact, there has been a significant increase in number of livestock population per hectare of pasture and grazing lands in the state. While there had been 397 animals after 100 hectares of pasturelands in 1972, this ratio increased to 434 by 1997. Hence it may be inferred that the demand for fodder increased substantially. Since cultivators in the state devote very little land to grow fodder crops, the burden is passed on to pasture and forest lands.

Table 5.6 : Pressure of livestock population on Land Resources in Himachal Pradesh, 1972-1997

Item	*Cattle*	*Buffaloes*	*Sheep*	*Goats*	*Others*	*Total livestock*
Livestock Population						
(i) 1972	2175690	543887	1039946	906415	36517	4702455
(ii) 1997	2173575	748246	1080464	1167992	53790	5224067
(iii) Change in %	-0.1	37.6	3.9	12.3	47.3	11.1
Livestock Density (Number per100 hectares)						
Total Area						
(i) 1972	74	19	35	31	1	160
(ii) 1997	64	22	32	34	2	154
Total Cropped Area						
(i) 1972	241	60	115	101	4	522
(ii) 1997	229	79	114	123	6	550
Net Area Sown						
(i) 1972	397	99	190	165	7	858
(ii) 1997	390	134	194	209	10	937
Pasture and Grazing Lands						
(i) 1972	183	46	88	76	3	397
(ii) 1997	181	62	90	97	4	434

District level analysis of livestock population pressure on land resources is highly revealing. There is a higher pressure of livestock population on land resources in the districts of Una, Hamirpur and Kangra falling in the Shiwalik Hills and the Lesser Himalayan zones, whereas there is lesser pressure in districts of Lahul & Spiti, Kinnaur and Chamba falling in the Alpine zone. In this zone due to severity of climatic conditions and difficult terrain, human population is sparsely distributed; and undulating terrain provides favourable conditions for grazing. As far as pressure on forest area is concerned, there is a high pressure of livestock population in Bilaspur, Solan and Hamirpur districts (Table 5.7).

There has been a steep increase in goat population in the state during 1972-1997. The number increased to 11.68 lakh in 1997 from 9.06 lakh in 1972, registering an increase of more than 12 per cent. Goats when let into forests cause the maximum harm as they are voracious browsers and very fond of nibbling young seedlings and growing buds. The rapid growth of goats is often interpreted to severely disturb the eco-balance in the hill region. It is alleged that close browsing by this animal leads to denudation of forest cover, destruction of rangeland and eventually soil erosion (Sharma *et al.*, 1998, pp.207-208). In view of the importance attached to goats in socio-economic and religious life of the hill people and their negative role in preservation and conservation of hill environment and ecology, there is a need to create a proper balance between their population and feed quantity, available during the lean period.

Table 5.7: Livestock pressure on grazing and forestland by districts, 1997

District	*Livestock/100 persons*	*Livestock/100 hectare of grazing land*	*Livestock/100 hectare of forest land*
Bilaspur	70	529	1901
Una	54	2502	840
Hamirpur	61	25024	1251
Kangra	60	1787	99
Solan	64	403	1581
Shimla	436	314	550
Sirmaur	90	677	846
Mandi	97	900	501
Kullu	87	NA	NA
Chamba	158	218	268
Kinnaur	112	49	248
Lahul & Spiti	205	32	50

It emerges from above analysis that, if the official records are to be believed, the pasture and grazing lands in the state have not declined it rather registered increase during the study period (1972-1997). Nonetheless, it is true that the pressure on such lands has been growing over the period due to rapid growth of human and animal populations in the state. This is truer of the Shiwalik hills and the Trans-Himalayan zones. Both the zones are ecologically very sensitive for geological and climate reasons. In the Trans-Himalayan zone harsh arid climate puts severe limits on the growth and development of vegetation cover, whereas in the Lower Himalayan zone soils are formed of loose materials that are highly reliable to deep and intense soil erosion. Under pressure from rapid growth of population cultivation is now being extended even to ecologically sensitive marginal areas. Fuel and fodder problem has compounded over the period.

Moreover, forest and pasturelands are neither properly managed nor well protected in the state. Illegal falling of tree cover, unauthorized grazing over forestland and encroachment on pasture and grazing lands are well known and widely reported activities. Hence, notwithstanding the official records showing increased area under pasture and grazing lands ground realities are somewhat different. Not only the area under the pasture and grazing lands is much less than the official estimates but also they are overgrazed and neglected. Overgrazing coupled with carving of agricultural fields on sloping hillsides has exposed topsoil to intense erosional activities. In recent decades forestland has been cleared to feed the rapidly growing population. This is evidently clear from a substantial increase in cropped land over the period. With expansion and intensification of cultivation in the marginal areas, the soil erosion has intensified.

The soil erosion is so rampant throughout the state that two-fifths or 0.27 million hectares of total 0.62 million hectares cultivated area in the state suffers from high intensity of soil erosion. More than a half (.15 million hectares) of this is categorised as very high intensity erosional area. It is estimated that, as a result of enormous loss of soil nutrients. According to an estimate the nutrient loss comes to about 5.640 tons which cost about Rs. 22.4 million, annually (Govt. of Himachal Pradesh). Water conservation has been badly affected as a result. The perennial rivers also show signs of drying up during summers, or water flow is reduced to a minimum.

Soil erosion caused by running water is typical of the Outer and Lower Himalayas (Bilaspur, Una, Hamirpur, Solan, Sirmaur and Kangra districts). It is a result of deforestation—gradual in historic times and rapid during last one century. It has been greatly assisted by the intensity of rainfall and the unconsolidated nature of soils on

Outer Himalaya hills. Here xerophytic conditions exist despite an annual rainfall of more than 100 cms.

In the Greater Himalaya zone, high hill slopes are very steep. During the rainy season landslide is a frequent phenomenon. With increased grazing on hilltops, the soil erosion and landslides have intensified in recent decades. The frequency of and damages from the earthquakes and avalanches are also on increase.

Evidently, all this involves not only the enormous damage to ecology but also heavy economic cost. There is thus need to rethink on ongoing government policies and involve general masses in management and development of natural resources including the pasture and grazing lands. For example, government policy of raising income and assets of small and marginal farmers along with landless labourers by providing them livestock under various development programmes without giving a proper thought to fodder producing capacity of individual households and village commons in respective villages and their surroundings. This has led not only to increase in further pressure on village commons but also they are getting more and more degraded. Hence, there is an urgent need to incorporate quick-growing legumes in the existing farming system along with seeking cooperation from local bodies to ensure full protection and proper management of village commons. Fodder and pasture improvement programme will go a long way to improve livestock. Stalk-feeding for selected animals can become popular, if fodder is made available (Sharma *et al.*, 1988, p.208). This will require an effective programme for leaf fodder production. The sampling of quick growing and high-yielding fodder tree species suited to local ecology should be distributed among farmers. The production of grass in posture land required to be improved through rationalized grazing system and introduction of better quality grasses and legumes. To effect change in the existing land use, there is a need to close government and private lands to grazing and divert these into hay fields. Above all, there is an urgent need to combine the policies of economic development with that of ecology promotion at all spatial scales in the state.

REFERENCES

Bhatti, J.P. *et al.*(1992), "Diversity of Mountain Farming System in Himachal

Pradesh, India", In N.S. Jodha *et al.* (eds) *Sustainable Mountain Agriculture: Perspective and Issues,* Vol. II, pp.497-516.

Govt. of Himachal Pradesh (2003), *Integrated Sample Survey for Estimation of Animal Products, Milk, Eggs, Wool and Meat, 2001-2002*, Directorate of Animal Husbandry, Himachal Pradesh, Pasudhan Bhawan, Shimla, pp. 9-10.

Sharma, T.R. *et al.* (1988), "Livestock Pressure on Land: A Study in Himachal Pradesh", *Agriculture Situation in India*, June 28, New Delhi, pp. 205-08.

Surya Kant (1998), "Sustainable Development in Himachal Himalaya", In R.B. Singh (ed.) *Sustainable Development of Mountain Environment in India and Canada—CIDA-SICI Project Experience*, Oxford & IBH, New Delhi, pp.182-96.

6

Ethics, Ecology and Agriculture

M.S. Sekhon

Introduction

India's most prosperous state Punjab is staring at an impending ecological crisis. The economic prosperity brought about by Green Revolution has come at a price. Economic growth has enhanced the standards of living but severely eroded the ethics and damaged the ecology of the region.

In this paper, no quantitative techniques have been used. Some data has been used to highlight or emphasise the point being made. It is simply a qualitative narrative of economic development and its effects on the natural environment of Punjab. The primary economic activity of Punjab is agriculture, so impact of agriculture has been made the focus of study. Since the goverment has had a major role in implementation of Green Revolution along with the farmers, a comment on ethics of planning, policy implementation and practice has been added.

The State of the ecology has been highlighted with the help of a set of indicators. These are area under forests, wetlands, biodiversity, ground water quality, ground water availability and waterlogging.

Economic Development

Punjab is one of the most developed parts of India. This state has benefited the most from Green Revolution and enormous amount of economic development has taken placed in last four decades. At present, the economic development indicators of Punjab are much superior to the National averages (Table. 6.1).

Table 6.1 : Economic development indicators in Punjab

Indicator	*India*	*Punjab*
Population growth rate	2.134%	1.976%
Literacy	65.38%	69.95%
Urbanisation	27.78%	33.95%
Sex Ratio	933	874
Per capita food grain production	201 kg	957 kg
Road Length per 000 km	544	1010
Per capita bank deposit	3562	7084
Shares of Agriculture in SDP	34%	47%
Per Capita SDP	Rs. 5781	Rs. 9643
Growth in real per capita SDP	3% p.a.	3.4%

Punjab is perhaps one of the few states in India where 100 per cent electrification of villages has been long completed. All villages are accessible by metalled, all weather roads. Almost 97 per cent of the gross cultivated area is irrigated, per capita food production is almost five times as much as the national average and per capita bank deposits are almost double. All this has become possible in the last four decades as the fruits of Green Revolution. Agriculture was very well developed even at the time of Maharaja Ranjit Singh. When the British annexed Punjab, Lord Dalhousie observed that everybody ate well and there was abundance of food (Gazetteer of India, 1851).

The Punjabis, when made aware of high yielding varieties (HYV), chemical fertilizers, pest control and mechanical methods took to them like fish to water. The new techniques introduced in 1960s provided the fillip to an inherent trait. In two decades, Punjab was producing enough food grains for the country to stop importing. As the Green revolution spread to other parts, India became self-sufficient in food grains. Agriculture has become the mainstay of Punjab's economy and society.

Agriculture Development

Punjab has a very advanced agriculture practice. The methods, techniques and yields match international standards. Government policy and awareness of farmers is responsible for introduction of latest techniques, machines, seeds and other inputs. Punjab Agriculture University has been contributing a lot in the form of introduction of new seeds and techniques. The change and high standards achieved by Punjab in agriculture are evident from Table 6.2.

Table 6.2 : Agricultural standards achieved in Punjab

Indicators	*1961*	*2001*
Share of agriculture in SDP		47%
Area under cultivation	67%	84.18%
Net irrigated area	2.02 mill hect.	7.544 mill Hect.
Percentage of irrigated to net sown area		96.1
No. of Tube wells 1971	192,000	925,000
Area under wheat	1.4 mill hect.	3.388 mill hect.
Average wheat yield	1244 kg.	4696 k. hect
Area under rice	0.227 mill hect	2.604 mill hect.
Average rice yield	1009 kg.	3347 kg. hect.
Area under grams (chick pea)	0.838 mill hect	6,000 hect.
Area under *bajra* (millet)	0.123 mill beet	5,000 hect.
Total food grain production		19.63 mill tonnes

Impact of Agriculture on Ecology

All this development and changing agriculture practices have yielded huge economic benefits. At the same time, the ecology of the region has been mauled to such an extent that the present methods and rated of cultivation have become unsustainable.

FAO's definition of Sustainable Agriculture

Sustainable development is the management and conservation of the natural resource base and the orientation of technological and institutional change in such a manner as to ensure the attainment and continued satisfaction of human needs for the present and future generations. Such sustainable development (in the agriculture, forestry and fisheries sectors) conserves land, outer, plant and animal genetic resources, is environmentally non-degrading, technically appropriate, economically viable and socially acceptable.

The practice of agriculture in last four decades has taken a heavy toll of the physical environment and ecology of the region. The effects became evident in late seventies but the government policies and human greed took no heed. Despite all kinds of warnings, the policy and practice is not changing.

Wheat-rice monoculture is depleting the land of its fertility. It is also responsible for the depletion of groundwater reserves. Punjab falls in semi-arid agro-climatic zone while wheat and rice need wet agro-climatic regions to grow in. The deficiency of natural moisture is met by irrigation. On top of that, area under wheat has doubled since 1961 while area under rice has increased by as much as ten times. At the same time area under other crops like millets (*bajra*) and gram has decreased considerably (Table 6.2). These traditional crops were less water intensive. But their cultivation is not as profitable as wheat and rice, so less farmers are growing them. Once again it is profitability taking precedence over ecological ethics.

Forest Cover in Puniab

The forest cover in Punjab has increased from 3.72 per cent in 1961 to 6.07 per cent in 2001. It is a fair achievement as far as the statistics go but the truth is something else. The areas along the roads, railway tracks and canals have been handed over to the forest department and planted with trees. Most of the trees planted are Eucalyptus. These few feet wide strips of trees criss-cross the region and are termed as forests. They also add up to area under forests and look good as statistics. This is a government policy duly approved by legislation under the aforestation drive.

Planning of Eucalyptus is another of the policy decision which is ecologically unsound. In a region like Punjab, where the sub-soil water reserves are depleting at an alarming rate; we cannot have such trees whose evapo-transpiration rate is very high.

The main forested areas of Punjab lie in the Shivalik belt, where we still have uninterrupted stretches of forests. Even these are severely effected by human interference. The situation was further accentuated in 1998 when, the government of Punjab passed a new townships Act, and a master plan was prepared to build Anandgarh, a township to commemorate 300 years of Sikhism. This town ship was to come up in the Shivalik piedmont plain and 1500 hectares of forest land was being acquired for urbanisation. The Union Ministries of Environment and Urban Development along with Environmentalist Groups saved the day. Government of Punjab had to shelve the plan on Apex Courts' judgment. It was a politically correct policy in gross violation of environmental ethics.

Biodiversity

The ecological health of a region can be best measured by its bio-diversity. Since the onset of Green revolution many indigenous varieties of cereals have vanished and been replaced by HYV. Other domesticated plants like lentils, vegetables and fruits have also been replaced by higher yielding, disease resistant varieties. It has been a major loss to the biota.

With 84.18 per cent of Punjab's area having been brought under cultivation, lots of small groves and woods found near villages and referred to as *'Birs'* on *'Chirris'* have been axed. These small biomes were home to species of plants and trees whose fruit formed a part of the diet. Some of these are 'Kareer' (Capparis aphlla), 'Ban' *(Rubus fruticosus)*, 'Theu' *(Artocarpus lackoocha)*, 'Karonda'*(Carissa carandas)* and 'Ber' *(Zizyphus zuzuba)*. The fruits of these trees were pickled and supplemented the diet. With cutting down of these groves the species have almost vanished or are on the verge of vanishing. Along with the trees, these pickles, which are rich in minerals and vitamins, have also vanished from the diets of people.

The indiscriminate use of pesticides coupled with large scale habitat destruction has brought a large number of birds and animals on the verge of extinction. Vultures have almost vanished from Punjab. Common wrens or house sparrows have suffered alarming decrease in the populations. Grey and Black Partridge are on the red list. Peacocks, the pride of country side, are rarely sighted. Weaver birds, which nested in abundance near water bodies, are not seen any more. Most of the antelope and deer species, once common all over the region, are confined to forested pockets where their numbers are dwindling.

Wetlands

Most of the wetlands of Punjab are shrinking at a rapid rate, while some have vanished. There are a number of reasons to it.

- Multipurpose projects have checked the flow of water in Sutlej, Beas and Ravi, the three major rivers, thus shrinking their channels. The wetlands along their channels have vanished or shrunk. The area thus recovered has been put under cultivation. This has resulted in habitat loss for hundreds of species living in these biomes.

- Land reclamation for agriculture : Many wetland fringes have been drained and reclaimed for cultivation. Government has done that as a policy matter. One of the main wetlands of Malwa region, Bhupindra Sagar, a large lake surrounded by a swamp has vanished completely and the entire region is under cultivation now.

Ground water

The ground water of this region has suffered a double onslaught. The aquifers are getting depleted and contaminated by chemical agriculture inputs. The data of Ghaispura village in Ludhiana district is an indicator of the levels of contamination (Table 6.3)

Table 6.3 : Levels of Contamination in Ghaispura Village

Chemical	*Hand pump (85' deep)*	*Tube well (300' deep*
Nitrate	4.6	1.7
Chloride	82	27
Sulphate	83	9
Cadmium	0.012	0.012
Fluoride	0.61	0.27

All chemicals found in ground water are used in agriculture as fertilizers, pesticides or weedicides. The data shows us that contamination has even percolated up to the depth of 300 feet. The Government is not educating the farmers about the harmful effects. Instead, as a policy matter, there are heavy subsidies on all these chemical inputs.

The subsoil aquifers are getting depleted at an alarming rate. In 1971 there were 192,000 wells and tube wells. By 2001 the number had increased to 925,000. Over extraction of ground water in Punjab has led to a situation where the ground water table has been falling at a rate of 25-30 centimetres/annum. Out of 118 development blocks, ground water in 62 blocks is over-exploited and eight blocks are already dark areas indicating no ground water. *The State of the World*

Report 1998, published by Washington based World Watch Institute, says gap between water use and sustainable yield of the aquifer is so great as to exhaust it by the year 2025.

This has again resulted from the government sponsored wheat-rice rotation in preference to less water intensive traditional crops. Secondly, the government policy of indiscriminate grant of electricity connections for tube wells is leading to an increase in exploitation of the aquifers.

In last six years, successive governments have made power to tube wells free, thus leading to a wasteful pumping out of water. All these government policies have been formulated even after being aware of the situation of ground water in the region. The policy of annual water budget prepared by Government of India has not been adopted by Punjab Government.

Waterlogging

The drainage system of the Punjab is designed to efficiently drain monsoons. Traditionally, farmers have irrigated their crops along naturally formed canals, and in this way the farmers subsisted in a delicate balance with the cycles around them. However, the introduction of canals that cut across land features, and the construction of major dams such as the Bhakra Dam and Nangal Dam, upset this delicate balance. The crops of the Green Revolution need a large readily available water source. Traditionally legumes/oilseeds used to be grown as inter-crops or as break crops. In last few decades, acreage of HYV has increased substantially. A matter of serious concern is large area coverage under a single crop such as wheat or rice. Inter-cropping has been replaced by clean cultivation, reducing bio-diversity and increasing risks of build-up of new insect pest complexes as a consequence of continued cultivation. Traditional agriculture has gradually given way to monocultures and more or less clean cultivation with fewer trees in cropped area.

Increased cropping intensity as a consequence of irrigation expansion particularly in the low quality aquifer zones has led to water logging forcing farmers to grow rice in monsoon season in place of cotton, gram etc. Replacement of traditional systems without anticipating the impact may well endanger millions of people in this region.

> *"Twenty years of Green Revolution agriculture, have succeeded in destroying the fertility of Punjab soils which had been maintained over generations for centuries and could have been indefinitely maintained if international experts and their Indian followers had not mistakenly believed that their technologies could substitute land, and chemicals could replace the organic fertility of soils."* (Shiva, 1991)

On Ethics

> *Left to them, individuals pursuing their own private economic interests have little incentive to protect environments, except in cases where their own property values or conditions of life are concerned*—Hardin.

We live in a pluralistic world and do not share a belief, a god or a tradition with others. We do not share a set of religious or political beliefs either. This is evident from our difference of opinion over many issues. The only possible alternatives to collective decisions are ethics or reasoned judgement and force, both economic and violent. We are seeing more of economic force being put into play at global or regional scale.

Punjab is a classical example of economic development taking precedence over ethical planning leading to economic boom and ecological disaster. Green Revolution coupled with the tradition of hard work has put the state at the top or charts. But this great leap forward has resulted in irreversible damage to the ecology. The agricultural practice and productions have reached such levels that they have become unsustainable. The main reasons are unethical planning by the government agencies and equally unethical agricultural practices by the farmers. To conclude, policy formulation in Punjab seems to be influenced by politico-economic factors rather than any ecological ethics and it has taken a heavy toll on the ecology of the state. Some of the changes are irreversible and the pace of agricultural development is unsustainable.

REFERENCES

Director, Census Operations Punjab: *Census of India 2001, Provisional Population Totals, Paper* 1, N. Delhi.

Economic Adviser, Government of Punjab: *Statistical Abstract of Punjab - 2000,* Chandigarh.

FAO, (1993a) An overview of pollution of water by agriculture. J.A. Sagardoy. In: *Prevention of Water Pollution by Agriculture and Related Activities,* Proceedings of the FAO Expert Consultation, Santiago, Chile, 20-23 Oct. 1992. Water Report 1. FAO, Rome.

FAO, (1990b) *International Code of Conduct on the Distribution and Use of Pesticides. FAO,* Rome.

FAO/ECE. (1991) *Legislation and Measures for the Solving of Environmental Problems Resulting from Agricultural Practices (With Particular Reference to Soil, Air and Water), Their Economic Consequences and Impact on Agrarian Structures and Farm Rationalization.* United Nations Economic Commission for Europe (UNECE) and FAO, Agri/Agrarian Structures and Farm Rationalization Report No. 7. United Nations, Geneva.

Glaeser, Bernhard, The *Green Revolution Revisited,* (1987), Allen & Unwin: Boston.

Mugerauer, Bob. *Theories of Sustainability: Environmental Ethics, Mixed Communities and Compassion,* http://www.utexas.edu/depts/grg/eworks/eworks.html .

Parihar, S.S., Kheparar S.D., Raghbir Singh, Grewal, S.S. and Sondhi, S.K. (1993) *Water Resources of Punjab—A Critical Concern for the Future of its Agriculture,* Second Edition, Punjab Agriculture University, Ludhiana.

Shiva, Vandana, (1991) *The Violence of the Green Revolution,* Third World Network: Penang Malaysia.

World Watch Institute, (1998) *The State of the World Report,* Washington.

7

Challenges of Water Management in India

Paramita Majumdar

The holistic management of freshwater as a finite and vulnerable resource, and the integration of sectoral water plans and programmes within the framework of national economic and social policy, are of paramount importance for action...

Agenda 21, Chapter 18*

Introduction

Water comprises a most basic and critical component in all aspects of human life. Freshwater resources are an essential component of the earth's hydrosphere and an indispensable part of all terrestrial ecosystems. Whether in the wet or dry areas of the country, or in urban or rural settings, competition over the allocation and management of water resources is emerging as a top environmental issue. With wide biogeographic, socio-economic and cultural variation, it is difficult to generalize about the nature of these issues in India—each area of the country seems to have a unique set of water management challenges. The decentralized form of government also means that the management of watercourses shared by multiple states within India's boundaries may present some of the same challenges inherent in transboundary river basin management involving multiple nations. Water management in India is a rising crisis that already affects national and regional economic and political stability. The following

* Report of the United Nations Conference on Environment and Development, Rio de Janeiro, Vol. I, *Resolutions Adopted,* Resolution I, Annex I and II, 3-14 June, 1992.

paragraphs highlight some of the challenges that we face in resolving the crisis.

The Challenges

The current situation in water resources

As mentioned, freshwater is a finite and precious resource that is essential for sustaining life, as are the natural systems that provide and maintain its supply. Less than 3 per cent of the Earth's water is fresh and most of it is in the form of polar ice or too deep underground to reach. On the whole, the water environment is characterized by the hydrological cycle, including floods and droughts, which in some regions have become more extreme and dramatic in their consequences. The widespread scarcity, gradual destruction and aggravated pollution of water resources in many world regions, along with the progressive encroachment of incompatible activities have triggered a range of water crises. Global freshwater consumption rose six fold between 1900 and 2000—more than twice the rate of population growth[1] — and the rate of increase in consumption is still accelerating. More than 2.7 billion people will face severe water shortages by the year 2025 if the world continues consuming water at the same rate, the United Nations has warned. Demand for water resources is increasing both because of population growth (particularly in developing countries) and because of rising demand per person due to such causes as irrigation, development, industrialization and increasing use by individuals as incomes rise. A potential crisis is looming where available resources can no longer meet needs.

Table 7.1 presents the annual withdrawals within continents and across sectors. It illustrates the continental-scale variations that exist. The uneven distribution of the world's population exacerbates variations in the amount of water potentially available per person. In Asia for example, water availability per person is half the global average. Although annual withdrawals currently represent an apparently small proportion of the available freshwater resources—some 9 per cent in 2000—this is deceptive. In addition, the annual withdrawals are also forecast to increase globally to 12 per cent by 2025.

Table 7.1: Estimated annual water resources and withdrawals, by region, 1995

Continent	Annual renewable Water Resources (m^3per person)	Annual withdrawals in m^3 per person	Annual withdrawals as % of average renewable water resources	Sectoral withdrawals as % of annual withdrawals) Agri-culture	Indus-try	Muni-cipal	Evapora-tion losses from reser-voirs
Asia	4000	680	17	80	9	8	3
Europe	4200	760	18	38	45	14	3
Africa	5700	325	6	61	4	9	26
North America	17000	1530	9	47	38	10	5
South America	38000	1140	3	57	12	21	10
Australia/ Oceania	84000	840	1	50	24	11	15
The World	7600	680	9	66	19	10	5

Source : Comprehensive Assessment of the Freshwater Resources of the World, Stockholm, Environment Institute 1997.

In India the total water resources are estimated to be 1953 cubic km according to Central Water Commission (CWC). However, there are sharp disparities between different regions. For instance, the water resources of the Ganga-Brahmaputra basin is 60 per cent (1200 cubic km) of the nation's resource flows while the basin occupies only 33 per cent of its land area. On the other hand 64 per cent of the areas (west and north-west India) has a meagre 533 cubic km of water resources. Table 7.2 depicts the surface water potential of some of the important rivers of India. It is not only the uneven distribution of water resources that is of concern but the way the rivers are being polluted is a major challenge.

According to the CWC, the development of water resources has been given a place of pride in all the five-year plans, but the water quality in Indian rivers does not reveal this. Country's water resources bear the brunt of unplanned urbanization and rapid industrialization. Polluting a river is dangerous as it is the primary source of drinking water for towns and cities downstream of the point of pollution. Broadly the causes of water pollution can be ascribed to :

(i) *Urbanisation and Industrialization :* Many of the towns and cities that came up on the banks of a river did not give a

Table 7.2 : Basinwise Surface Water Potential of India

(Cubic Km/Year)

S. No.	*Name of River Basin*	*Average Annual Potential in River*
1.	Indus (up to Border)	73.31
2.	(a) Ganga	525.02
3.	(b) Brahmaputra Barak & Others	585.60
4.	Godavari	110.54
5.	Krishna	78.12
6.	Cauvery	21.36
7.	Pennar	6.32
8.	East Flowing Rivers Between Mahanadi & Pennar	22.52
9.	Mahanadi	66.88
10.	Brahmani & Baitarni	28.48
11.	Subernarekha	12.37
12.	Sabarmati	3.81
13.	Mahi	11.02

Source : Ministry of Water Resources, Government of India.

thought to the problem of urban sewage. Most of the sewage was conveniently allowed to flow into the rivers. For instance, prior to the Ganga Action Plan (GAP), over 95 per cent of the sewage entered the river without being treated. Added to this effluents from heavy industries containing toxics and heavy metals is drained into the rivers as municipal water treatment facilities in India mostly do not remove the traces of heavy metals. Along with this, the unplanned growth of towns and cities has directly affected the rivers.

(ii) *Agriculturdl Runoff and improper agricultural practices* : When pesticides and fertilizers are used more than the recommended doses, they remain in the soil. While some of this may become inert over a period of time, traces end up in the closest water body. This could be in the form of irrigation or monsoon runoff. As Indian rivers are the primary source of drinking water in most municipalities, their pollution means that the municipal drinking water is contaminated with pesticides and fertilizer residues.

(iii) *Religious and social practices* : Carcasses of cattle and other animals disposed of in the so-called 'holy' rivers add to the pollution load. Mass bathing in a river during religious festivals is another environmentally harmful practice. Studies

> have indicated that the biochemical oxygen demand (BOD) goes up drastically when tens of thousands of people simultaneously take a 'holy dip'. Religious practices also demand that offerings from a *puja* be immersed in a river, thus adding to the pollution load of a river.

Therefore Indian rivers, the major water resource of the country show case the price paid for mindless development; a classic example of pay-off between rapid population and economic growth and environmental degradation.

But it is not only the surface water that is threatened. According to the Indira Gandhi Institute of Development Research (IGIDR), Mumbai there are more than 4.79 million electric and 3.7 million diesel pumps drawing groundwater across the country. The use of ground water needs immediate attention to prevent any rise in aridity. The importance of this source is best highlighted by the fact that more than 50 per cent of the irrigation in the country is with ground water. With agriculture contributing roughly 27 per cent of India's GDP (Table 7.3) and production from irrigated land claiming the lion's share, a large percentage of the country's GDP is closely tied to the availability of groundwater. The decline in agriculture's share in recent years to India's GDP is mainly due to the scarcity of water which has marked a structural shift in the composition of GDP.

Table 7.3 : Sectoral share of India's GDP

(Per cent)

Year	*Agriculture & Allied Sector*	*Industry*	*Services*
1950-51	55.80	15.20	29.00
1980-81	38.10	25.90	36.00
1990-91	31.00	29.30	39.70
1999-2000	28.50	73.70	47.80
2000-01	27.30	24.60	48 20

Source : Statistical Outline of India 2002-2003, Table 6, p.14.

Besides, ground water is now the source of 4/5ths of the domestic water supply in rural areas and around half that of urban and industrial areas. Further, in drought years, ground water is the predominant source of irrigation. While ground water development has had important implications for the economy, the overuse of ground water is emerging as a major concern. The total replenishable ground water is estimated at 432 cubic km (Table 7.4), of which 396 cubic km is deemed utilizable.

Table 7.4 : Basinwise Ground Water Potential of India

(Cubic Km/Year)

S. No.	*Name of River Basin*	*Total Repleneshable Ground Water Resources*
1.	Brahmani with Baitarni	4.05
2.	Brahmaputra	26.55
3.	Cambai Composite	7.19
4.	Cauvery	12.30
5.	Ganga	170.99
6.	Godavari	40.65
7.	Indus	26.49
8.	Krishna	26.41
9.	Kutch & Saurashtra Composite	11.23
10.	Madras and South Tamil Nadu	18.22
11.	Mahanadi	16.46
12.	Meghna	8.52
13.	Narmada	10.83
14.	Northeast Composite	18.84
15.	Pennar	4.93
16.	Subarnarekha	1.82
17.	Tapi	8.27
I8.	Western Ghat	17.69
	Total	**431.42**

Source : Ministry of Water Resources, Government of India.

While the pressure on ground water resources has built up, the policy framework pursued over the last 50 years for the management has not promoted judicious and equitable use of a vital natural resource. The state's policy of subsidizing power and water has, in fact, led to a greater use of ground water sources for irrigation.

Most ground water structures are privately owned and have therefore been outside the purview of direct regulation by the state. Measures to regulate groundwater extraction chiefly through restrictions on credit or electricity have had limited impact. On the contrary, wherever the water table is high, affluent farmers use diesel pumps if electricity is in short supply. In the legal framework, for the management of groundwater in India, there are no *de jure* rights to groundwater; *de facto* all landowners have the right to the ground water in their land. Thus, groundwater is viewed essentially as an add-on to the land. As a consequence, there is no limit to the amount of ground water a landowner can extract from his land.

Table 7.5 : Name of the Districts (in pockets) Showing Fall in Water Level as per the Long term Water Level trends during the Post-Monsoon (1981-2000)

State	*Fall in Water Level >4 metres (@ > 20 cm/year)*
Andhra Pradesh	Adilabad, Ananthapur, Chittoor, Cuddapah, East Godavari, Guntur, Hyderabad, Karimnagar, Khammam, Krishna, Kurnool, Mahaboobnagar, Medak Nalgonda, Nellore, Nizamabad, Prakasam, Rangareddy, Srikakulam, Vijayanagaram, Visakhapatnam, Warangal, West Godavari.
Bihar (Including Jharkhand)	Giridih, Palamu.
Chattisgarh	Bastar, Durg, Mandla, Rajnandgaon, Rewa.
Gujarat	Ahmedabad, Amreli, Banaskantha, Bharuch, Bhavnagar, Jamnagar, Junagarh, Kheda, Kuchchh, Mehsana, Panchmahal, Rajkot, Sabarkantha.
Haryana	Bhiwani, Faridabad, Gurgaon, Hissar, Kaithal, Kurukshetra, Mahendergarh, Panipat, Rewari, Sirsa.
Karnataka	Bangalore, Bellary, Belgaum, Bidar, Bijapur, Chickmagalur, Chitradurga, Coorg, Dharwar, Gulbarga, Hassan, Kolar, Mandya, Mysore, Raichur, Shimoga, Tumkur, N. Kanara.
Madhya Pradesh	Bhind, Bhopal, Chhatarpur, Chhindwara, Datia, Dewas, Guna, Gwalior, Hards, Hoshangabad, Indore, Jabalpur, Jhabua, Khandawa, Khargone, Mandsaur, Morena, Narsinghpur, Neemuch, Punna, Raisen, Rajgarh, Ratlam, Sagar, Sehore, Shivpuri, Ujjain, Vidisha.
Maharashtra	Ahmadnagar, Amravati, Aurangabad, Bhandara, Buldhana, Chandrapur, Dhule, Gadchiroli, Jalgaon, Jalna, Kolahapur, Nagpur Osmanabad, Parbhani Pure, Satara, Solapur, Wardha, Yavatmal.
Orissa	Angul, Balasore, Bargarh, Bolangir, Boudh, Dhenkanal, Ganjam, Jaipur, Kalahandi, Keonjhar, Koraput, Malkangiri, Suvarnapur.
Punjab	Amritsar, Bathinda, Faridkot, Fatehgarh Sahib, Ferozpur, Jalandhar, Kapurthala, Ludhiana, Moga, Panda, Sangrur.
Rajasthan	Alwar, Barmer, Bhilwara, Bikaner, Chittorgarh, Churu, Dungarpur, Jaipur, Jalore, Jhalawar, Jhunjhunu, Jhodhpur, Kota, Nagaur, Pali, Rajasmand, Skar, Sirohi, Udaipur.
Tamil Nadu	Coimbatore, Cuddalore, Dharmapuri, Erode, Kancheepuram, Karaikal, Madras, Madurai, Namakkal, Perambalur, Pudukkottai, Ramanathapuram, Sivaganga, Trunelveli, Tiruvallur, Trichirapally, Tuticorin.
Uttar Pradesh (Including Uttaranchal)	Agra, Aligarh, Allahabad, Badaun, Bulandashahr, Etah, Etawah, Farrukhabad, Fatehpur, Ghaziabad, Hardoi, Kanpur, Lucknow,. Mathura, Meerut, Moradabad, Rai Bareli, Saharanpur, Unnao.

Source : Ministry of Water Resources, Government of India mentioned in http://www.water-mgmt.com/daen/tabase.htm

A hidden dimension to this *de facto* right on ground water by landowners is now beginning to unfold. Increasing demand for water in urban areas is prompting landowners sell their ground water. Not only is such a trend causing irreparable damage to the hydrology of the area, it is also creating a rural-urban divide in ground water utilization. Table 7.5 shows the fall in water level in different states during the period 1981-2000.

The above paragraphs therefore point out to the fact that although India is endowed with sufficient water, there are significant variations in the spatial and temporal availability of this resource. There are significant variations in water availability even within a river basin. Consequently, at any given time, there are areas of both water excess and water stress in the country.

The current situation in water supply and sanitation

The International Atomic Energy Agency (IAEA), estimated that 1.1 billion people have no access to safe drinking water, 2.5 billion lack proper sanitation and more than five million people die from waterborne diseases each year—10 times the number of casualties killed in wars around the globe. A substantial majority of these people live in Asia.

But over the past 15 years, good progress has been made in water supply, notably in Asia where the population with access to water has increased rapidly. In India, some greatest achievements have been made. Table 7.6 shows the status of drinking water and sanitation facilities in India. The proportion of the population with safe drinking water available at home or with reasonable access was 84.3 per cent in 1993 for urban areas and 82.4 per cent in 1995 for rural areas. The proportion of the population with adequate excreta disposal facilities was 49.9 per cent in 1993 in urban areas and 3.7 per cent in 1995 in rural areas.

Thus there has been increase in the number of habitations covered by safe drinking water supply and sanitation facilities over the years. The provision of drinking water to all villages has been identified as one of the priority tasks of the government. As a result number of habitations (72859) not covered during the Ninth Five Year Plan has come down to 14356 in January 2003. But the challenge still remains.

Table 7.6 : Population covered with Drinking water and Sanitation Facilities

Item/Area	*1985*	*1990*	*1995*	*1998*	*1999 (P)*
Drink, Ware, Supply					
Rural	56.3	73.9	82.8*	92.5*	98.0*
Urban	72.9	83.8	84.3#	90.2@	90.2@
Sanitation Facilities					
Rural	0.7	2.4	3.6*	8.1*	9.0*
Urban	28.4	45.9	49.9#	49.3@	49.3@

As on 31-3-1993

* With Government Initiative only under CRSP, MNP, JRY and IAY. Latest authentic data i.r.t. coverage through private initiative is not known.

@ As on 31-3-1997

Note : (i) Perecentage coverage in respect of Rural Water Supply and Sanitation are based on Popuplation covered in current year to corresponding 1991 census population.

(ii) Percentage coverage in respect of Urban Water Supply and Sanitation are based on Population revered in current year to corresponding current population.

Sources : Ministries of Rural Develonment & Urban Development.
Economic Survey, 2000-2001.

The 54th round of National Sample Survey (NSS, July 1999) on drinking water, sanitation and hygiene in India estimated that 50 per cent of rural households were served by a tubewell/hand pump, 26 per cent by a well, and 19 per cent by taps, and only 17.5 per cent were using latrines. But the area of concern is the re-emergence of uncovered villages in view of non-involvement of people in design/ operation/maintenance of assets and fast depletion of ground water levels leading to problems of arsenic and fluoride content being enhanced.

In urban areas on the other hand according to NSS estimates 70 per cent of the households had tap water and 21 per cent had access to tubewell or hand pump. Forty-one per cent had sole access to their principal source of drinking water, which means that 59 per cent were sharing a public source. In terms of toilet facilities, 26 per cent had no latrine, 35 per cent used the septic tank and 22 per cent, the sewerage system, i.e. 43 per cent of households in urban areas either had no latrines or no connection to a septic tank or sewerage. But the coverage of water and sanitation in urban areas varied according to the size class. Class V towns (population between 5000 and 10000) registered a high percentage of households not covered by toilets (61 per cent) and drinking water (28 per cent) in 1991.

The per capita water also varies according to the urban growth and population concentration, which is clearly revealed very high consumption in the major cities like Delhi, Mumbai, Chennai, Calcutta and Hyderabad. But even in such big cities there is lack of spatial and social equity in water supply. The following table shows the spatial inequity in water supply in various zones of Delhi.

Table 7.7 : Spatial inequity in Water Supply

Zone	*Water supply in litres/capita/day*	*Zone*	*Water supply in litres/capita/day*
Mehnuli	29	Narela	31
Najafgarh/Dwarka	74	Shahdara	130
New/South Delhi	148	Paharganj	201
West Delhi	202	Crvil Lines & Rohini	274
City	277	Karolbagh	337
NDMC	462	Cantonment	509

Source : Delhi 1999—A Fact Sheet, NCR Planning Board.

Moreover, there is a wide gap between the demand and supply to different types of settlements which is against the concept of social equity. About 40 per cent of Delhi's population live in JJ clusters, Slum Designated Areas and Unauthorised Colonies. About 10 per cent of the population in Delhi have no access to piped water supply and 30 per cent of the population is getting a very low level of supply (4-10 lpcd) through standposts against a planned level of 42.5 lpcd.

On the other hand is the small towns which need particular mention as they are often neglected because they fall between the definitions of coral and urban programmes. Technically, the water supply and sanitation needs of small towns are often not amenable to simple solutions such as drawing water from a spring or borehole, or digging a simple pit-latrine.Thus it is needed either to build their capacity or to find other forms of service provision, such as the local private sector.

The problem faced by the urban population is that of increasing demand-supply gap, wholly inadequate and unreliable supplies particularly in poor urban settlements, and deteriorating financial and technical performance of systems. To take only the metropolitan cities, according to a World Bank Study of 27 Asian cities with populations of over one million, Chennai and Delhi share the same rank as the worst performing cities in terms of hours of water availability per day, while Mumbai is ranked as the second worst performer and Kolkata as the fourth worst.

As to the comparisons between urban and rural areas, urban coverage appears higher for both water and sanitation. But over the next few decades almost all the world's population growth will be in urban areas in developing countries. Provision of urban water and sanitation will, therefore, become increasingly difficult and urgent. Though a great deal of money is often poured into capital schemes, insufficient attention is given on the equally important aspects of follow-up and maintenance, training and education of target groups. It is important to realise that even if a technically sound solution for water supply is provided to a community, water is only as clean as the cupped hand of the person who drinks it. This implies that there are two sides to tackle the problem of water supply and sanitation. On the one hand are the pipes, pumps, treatment stations, latrines, which may be referred to as the *hardware,* while on the other hand lies the equally important aspect of health and hygiene education, referred to as the *software.* The typical attitude of developing countries has by and large been to invest on the hardware part, failing to recognise the importance of the software side.

There has to be a radical change in the approach of the developing countries in tackling the problem. It has to be realised that those who are most affected by poor water supplies and insanitary conditions, should be given the incentive to initiate, carry out, and maintain, projects. Educating the target groups, assumes significance. The problem can be effectively tackled once people are made aware of how they themselves can take steps to improve their water supplies and sanitation, and more importantly, how an improvement to these can make a difference to their lives. It is also vital to involve local communities in finding their own solutions. The problem should neither be viewed in isolation nor will any peace-meal solution adequately address the problem. Planning for water supply and sanitation ought to be an integral part of the socio-economic development process.

Improving the management of water resources

Human consumption of freshwater threatens to push to the limits the capacity of nature to supply benefits to mankind. Therefore the issue at the forefront of water and sanitation policies must be one of management. In the first "Mayors' Asia-Pacific Environmental Summit" held in Honolulu, Hawaii (31 January—3 February, 1999) water management concerns ranked as an important consideration.

As has been mentioned above, both ground water and surface waters are being polluted by industrial wastewater discharges, sewage water from towns and cities and agricultural runoff. This is increasingly leading to degradation of river and lake ecosystems, creating or exacerbating health problems. The poor are often the first people to suffer the ill-effects of pollution because they frequently have to abstract water downstream of pollution sources. The challenge to protect water quality thus needs a major change of behaviour by industrial, agricultural and municipal polluters.

Global climate change and variability may also affect the management of water resources. The poor — whose livelihoods (often based on access to natural resources) are at risk from both the short-term natural disasters and the long-term trends associated with climate change are the most vulnerable as the majority of natural disasters involve either excess or scarcity of water. Both floods and droughts can adversely affect the survivors' livelihoods for many years. Environmental issues are often poorly integrated into national water policies and plans. Policy makers often do not understand the links between effective environmental management and poverty elimination. All these factors present a considerable challenge to manage water resources and the environment well, on behalf of poor people in particular. But despite increasing competition between expanding urban and industrial water users and those in the agricultural and coral sectors, good economic and institutional approaches are limited for dealing with matters of efficient water use and allocation.

There are opportunities for improved coordination of mission-level, water-related activities across the region, especially where the possibility of cross-border bilateral cooperation exists. The potential for expanding regional cooperation in response to the common threat of hydrological disaster also should be further supported.

Avoiding conflicts over water resources

Increasing demand for water may establish conditions for conflict over access at a range of levels: from the international river basin or aquifer level, through the sub-national level between regions of a country, down to local level conflict between communities. Trans-boundary water resources disputes have become particularly significant in recent years. Many international conflicts over water, like the one between Syria, Jordan and Israel, between India and its neighbouring countries of Pakistan, Bangladesh and Nepal, have created brouhaha. Even within India, states are constantly fighting over water rights. The altercation between Tamd Nadu and Karnataka over the Kaveri river is a prominent example.

The other contentious issue is that of large-scale dams. The Hoover dam on the Colorado has made southern California one of the richest neighbourhoods in the world while the Sardar Sarovar Dam in India has run into problems with NGOs. The government claims that the completion of this project will provide water to the 40 million inhabitants of the water deficient Kutch and Saurashtra regions of Gujarat. But the critics say that the project's benefits will be reaped by the richer areas in Gujarat and will not reach the poorer localities. Then there is the matter of the rehabilitation of the people—ultimately the ball is in the government's court. How much of its resources should be diverted to this project *vis-a-vis* other small scale projects? How should the complementarity be brought about instead of conflict? In addition the challenge lies in resolving conflicts between the people, who opt for small structures of rainwater harvesting, which are temporary and require regular maintenance in contrast to the State, which prefers large-scale structures that are permanent in nature and solve macro-level problems.

Allocating water between different users

As population grow and demand for water leads to increasing competition between different sectorial uses—for example, between major urban centres and the surrounding agricultural lands and wetland ecosystems—tensions between different uses can arise. More effective systems for deciding the value of water are necessary, for balancing market efficiency against social equity, and, thus, for allocating water between sectors.

In countries where there is a significant level of irrigation, agricultural water-consumption far outweighs domestic and industrial use. In India agriculture continues to be the single largest user of water. To be precise, the agriculture sector accounts for between 90-95 per cent of both surface and ground water in India; industry and domestic sector account for the remaining.

Agriculture has played a dominant role in strengthening India's economy. Therefore, particular attention has been given to building the irrigation facilities. The rapid expansion of irrigation and drainage infrastructure has been one of India's major achievements. From 1951 to 1997, gross irrigated areas expanded fourfold, from 23 million hectares to over 90 million hectares. There is no doubt that this development of water infrastructure has enabled the country to harness much of the country's water resources.

But even while acknowledging the vital importance of the development of water resources to achieving self-sufficiency in food grains, the overall water scenario in the country is quite grim. In other words, this achievement in the agrarian economy has been at the cost of groundwater depletion, water logging, water quality degradation and pollution and increasing salinity levels affecting large areas.

A worldwide assessment of water needs conducted by the Intemational Water Management Institute reveals that unless at least 30 per cent of the world's river flow is allocated to maintaining freshwater ecosystems, we may see the collapse of many local economies and communities.

The emerging thought that is getting ascendance with regard to water economy is that a balanced water allocation policy is needed, one that judiciously handles the distribution of water to environment, agriculture, industry, and domestic use. National level statistics for water availability show huge disparities from basin to basin and region to region. Spatially, the availability of usable water in the country varies from 18,417 cubic metres in the Brahmaputra valley of the north-east to as low as 180 cubic metres in the Sabarmati basin in the western part of the country.

Therefore establishing an institutional structure for allocating water is a fundamental role of social policy for any nation. The choice of structure is ultimately a compromise between the physical nature of the resource, human reactions to policies and competing social objectives. Individual freedom, equity, popular participation, local control and orderly conflict resolution are other important objectives which societies must juggle when choosing a structure for water allocation.

Delivering water services sustainably

Financial sustainability : On a global level there is a substantial financing shortfall across the whole water sector of both capital investment and investment for the operation and maintenance of existing infrastructure. So the challenge of financial sustainability has two aspects: how to find enough money for capital investment to reach all the currently unserved people, and how to raise enough money to cover operation and maintenance, and eventual replacement.

In recent years, the importance of the private sector to mobilize new finance has been vigorously promoted by some agencies, and criticized by others. Given the private sector's need for return on investment, well designed contracts and enforced regulation will be needed to ensure that part of improved service delivery goes to improving the level of coverage for poor people.

Institutional sustainability : The main challenge in institutional sustainability is to build good relationships between public authorities, the private sector and civil society.

Operational sustainability : Often, people are unwilling or unable to pay for irrigation or for water and sanitation because their levels of service are so poor, or because the payment systems are not appropriate or do not function. That in turn means that the income of the water agencies is too low to maintain or improve the system, leading to a downward spiral of decline.

The challenge is to reverse the downward spiral by pricing the services to recover full costs and investing the money raised in operation and maintenance to provide better service standards. This would ensure the operational sustainability of the service. This operational sustainability will be achieved if the service agencies assess the demand properly, communicate better and become accountable.

Therefore the purpose of Sustainable Water Management (SWM)

simply involves a whole new way of looking at how we use our precious water resources. The International Hydrological Programme, a UNESCO initiative, noted:

> *"It is recognized that water problems cannot be solved by quick technical solutions, solutions to water problems require the consideration of cultural, educational, communication and scientific aspects. Given the increasing political recognition of the importance of water,* it is *in the area of sustainable freshwater management that a major contribution to avoid/ solve water related problems, including future conflicts, can be found."*

Meeting the Challenges

In this section the paper outlines how we all - the people themselves, civil society, governments, the private sector can meet the challenges presented in the previous section. This section is structured according to the various actors, rather than the different subjects.

How people and communities can respond

Without the engagement of the people in the development process it is not possible to achieve any target. People play a strong role in articulating their needs in relation to their livelihood priorities. In water resources, these may relate to improving water scarcity and reducing risks and uncertainties, increasing the range of productive uses of water, or coping with natural disasters. In water supply and sanitation, the priorities may relate to protecting existing sources or finding new ones, reducing costs, improving sanitation facilities and drainage. The priorities will vary from place to place: the important point is that the people themselves should decide them, or be involved in the decision making process. Though community management, the people can work with local governments, civil society and/or the private sector to achieve the results they wish.

How civil society can respond

Civil Society Organizations (CSOs) play an important role in helping poor people to express their demands and in advocacy on their behalf. In many places, they themselves are valuable service providers, enabling poor and excluded people to determine their livelihoods, improve sustainable water resources and gain access to essential and appropriate services. CSOs also form a communication channel from the government to the people about choices of service level and the roles of the different players, including the private sector and the govemment itself and also back to the government for example about practical difficulties arising from particular policies.

How government can respond

National governments are the most important players in the water sector. Their principal task is to establish national water policies and laws. Within the framework of these national legal and policy frameworks, government agencies are responsible for allocating water between uses—such as domestic water supply, agriculture, fisheries, forestry, environmental services, industrial, transport, power and recreation.

In ranking priorities of use, and setting water prices, governments should involve stakeholders at all levels of water resources management. The government should introduce financial incentives for water users and polluters to change their behaviour: users should pay for the water they abstract and polluters' should pay according to the pollution they cause. Prudence requires the government to start recovering the cost of supplying water. Even if capital cost cannot be recovered, the operational and maintenance (O&M) cost must be recovered as far as possible. This can be achieved by empowering the communities to carry out repairs and operate accounts for local water works. In time a shift from the centralized structure to the market structure must be sought. This will ensure financial sustainability. In industry too, the present cost structure of water is rather uni-dimensional. Apart from the capital costs and those of operation and maintenance, costs of economic and environmental externalities and opportunity costs must be included as it is more expensive to clean up polluted water bodies than to minimize effect of damaging chemicals beforehand. This needs to be implemented in India through effluent tariffs and regulated discharge.

An awareness programme to inform the people about the importance of water is desperately needed. However, the government's information network needs to be developed to gauge demand if the shift from supply-oriented projects to demand-responsive projects is to be made.

The Indian Constitution places the responsibility of water management and sanitation on the respective state governments. But matters become more complicated as the use of surface water is administered by the Central Water Commission while groundwater is managed by the Central Groundwater Board. Similar bifurcation exists in different states. This leads to chaotic and inefficient use of this precious resource. Though there is a tacit assumption of cooperation, in reality there is none. What is needed is an integrated policy, which seeks sustainsibility through the prevention of unnecessary destruction of forests, regulated discharge of effluents and sensitivity towards the environment. Hence the various departments of Environment, Development, Forest and Water resources, etc., must work in tandem rather at cross-purposes.

How the private sector can respond

The private sector has been involved in various aspects of water provision and management for a long time. Though the UNDP Human Development Report 2003 states that the Public-Private-Partnership (PPP) in water and sanitation, which was at its peak in 1996, is on the decline, there are successful cases as well.

In India transmission and distribution losses is a worrying feature in the structure of water management. In urban areas the theft of electricity and water has now become old hat. To address this situation, the distribution of water should be handed over to a private entrepreneur. It is fairly reasonable to assume that in such a situation losses in distribution will be reduced and efficiency increased. The functioning of the private operator can be monitored by a regulating authority.

For the first time the National Water Policy (NWP) 2002 has clearly carved out a role for the private sector in the issue of water services, which so far has been exclusively the government's domain. The Planning Commission of India following the mandate chalked out in the World Summit for Sustainable Development held at Johannesburg in 2002 has brought out an Assessment Study on Water Supply and Sanitation. More importantly, the study by calling for an active participation of the private sector in the management of water resources reinforces one of the vital thrusts of the NWP 2002. The study views the private sector as an important stakeholder and accordingly enjoins upon the Central Government to form a hub for promoting networking among the vibrant non-government organizations and the private sector.

Conclusion

India is in the grip of a massive water crisis. This can be crippling not only for the economy but also for the people. Increasing water scarcity will spur individual and national attempts to secure water resources and result in conflicts that threaten peace and security, and consequently, sustainable development. In order to prevent this outcome and sustain water supplies, innovative non-structural approaches to water management will have to replace the large-scale supply-side approaches of the past. Non-structural management techniques integrate supply and demand-side water management through improvements in storage, use, technological, and institutional efficiency. They focus on reducing water demand, while minimizing the negative social and environmental impacts of water extraction and use. They are adaptable and rely on effective monitoring systems, as well as their collaborative and integrated success. Innovative non-structural management approaches should be the focus of water resources management for the 21st century.

We therefore need to re-think our water policy and adapt it to the constraints in supply and issues of the environment. The control of polluting agents is an important dimension in any water policy. Anything else would be a recipe for disaster.

Perhaps there is no need to be unduly pessimistic, but perhaps Coleridge's Mariner, though quoted out of context, is playing the Oracle when he says 'Water water everywhere ...Nor any drop to drink'. The question remains, will we allow it to happen?

REFERENCES

Aggarwal Anil, Narain Sunita and Sen Srabani (ed) (1999), *State of India's Environment*, The Citizens Fifth Report, Part I: National Overview, Centre for Science and Environment, New Delhi.

Bajpai, Peeyush & Bhandari, Laveesh (2001), Ensuring Access to Water in Urban Households, *Economic and Political Weekly*, September 29, pp. 3774 - 3778.

Chattopadhyaya, S. (2001) *Sustainable Water Supply and Sanitation—City Planner's Role*, Paper presented in the Biannual Conference, Nordic Association for South Asian Studies, Voss, Norway, 20-22 September.

David J. Tenenbaum (1998), Spheres of Influence: Tackling the Big Three Environmental Health Perspectives, Vol. 106, No. 5, May.

DFID (2001), *Addressing the Water Crisis*, March.

Economic Survey 2000-2001, Government of India.

Economic Survey 2002-2003, Government of India.

Fragile water resources situation -- World Bank cautions warns India, *The Business Line*. The Hindu Publications, Thursday, May 11, 2000, New Delhi.

India Infrastructure Report (2003), *The Network*, Oxford University Press, New Delhi.

Ishan Kaul (2002), Towards Holistic Management of Water Resources—The Case of India, *Margin*, Vo1.34, No.2 &3, January-June, pp. 59-85

lyer, Ramaswamy (2001), Water: Charting a Course for the Future—Il, *Economic and Political Weekly*, April 14, pp. 1235-1245.

Iyer, Ramaswamy R. (2003), *WATER—Perspectives, Issues, Concerns*, Sage Publications, New Delhi.

Kumar, Anuradha (2002), Struggle to Save Nagpur's Water Bodies, *Economic and Political Weekly*, December 14, pp. 4987-4989.

Motoyuki Suzuki (2003), Water in our future, in Hans van Ginkel & others led), *Human Development and the Environment Challenges for the United Nations in the New—Millennium*, United Nations University Press, Rawat Publications, New Delhi. pp.197-212.

Ramchandraiah, C. (2001), Drinking Water as a Fundamental Right, *Economic and Political Weekly*, February 24, pp. 619-21.

Rathore, M.S. & Reddy, V. Ratna (1996), *Urban Water Management*, Institute of Development Studies, Jaipur, Rawat Publications.

Raven, Peter H., Berg, Linda R., Johnson, George, B. (1993), *Environment*, Saunders College Publishing, Harcourt Bruce College Publishers.

Reddy, V. Ratna (2002), Water Security and Management: Lessons from South Africa, *Economic and Political Weekly*, July 13, pp. 2878-2881.

Saliem Fakir and Alexandra Baillie, *Managing Freshwater for Peace and Prosperity*, IUCN, South Africa, Downloaded from *www.elobesa.arg/earthsmmmit/fakir.htm*

Sharma Sudhiremdar (2003), Infochange India, *News and Features*, *http://www.infochangeindia.org/WaterResourcelbp.jsp*

Status Report for Delhi 21(2001), Delhi Urban Environment and Infrastructure Improvement Project (DUEIIP), Government of India, Ministry of Environment & Forests and Government of NCT of Delhi Planning Department.

UNDP Human Development Report (2003), Millennium Development Goals: A Compact among nations to end human poverty.

Versfeld Dirk (2003), Water for Livelihoods: Bringing Equity and Opportunity to the Rural Poor in South Africa, Paper presented in the International Symposium on Water, Poverty and Productive uses of Water at the Household Level, 21-23 January, Muldersdrift, South Africa, pp. 228-37.

8

Land Degradation, Management and Planning for Sustainable Development Hills

B.I. Teli

Introduction

The use of land is a down to earth index of civilization, it has been silent partner in the rise and fall of civilization. It is a non-renewable natural resource of which man depends for his various requirements, i.e. food, fibre, fuel, shelter, recreation etc. the population explosion and the ever-increasing demand of land more food and other requirement of man are pressing the use of land to its limits. The problem is very acute in old countries like India, where we have almost reached the physical frontiers with regard to the area, which could be brought under the plough without causing great imbalance in nature, accelerating erosion, floods and other calamities. The need for proper land use planning and adoption of suitable cropping pattern is most urgent today than at anytime in the history of this country. The national use of land is to ensure the maximum productivity without much harming the soil land resources, so that the present generation as well as the coming generation can live comfortably.

Even today cropping pattern followed is based on the traditional system of subsistence farming where every farmer is trying to produce everything he needs. It is in fact based on the utilization of the inherent fertility of the soil without use of much of the modern inputs. The concept of the production for market instead of home consumption has made it necessary to plan most efficient cropping pattern. It will enable the efficient utilization of soil-water-man in land management.

Study Area

Western Nayar river is the tributary of the main Nayar system which poure water in the river Ganga from the left bank. It extends between 29° 54′ 45″ N and 30°12′2″ N latitudes and 70°42′ 50″ E and 70°8′48″ E longitude.

The river originates from the Dudhatoli rage, which forms the boarder of Garhwal, Chamoli and Almora. It originates at the height of 3015m and confluence the main channel near Satpuli (Bhattkoti village) at the height of 560m. 92.5 per cent are ranges between the height of 800m and 2400m. The region is bisected by the main channel in the north and south parts. It can further be divided into 20 sub-water-sheds, which have as many as 3343 segments (2583 order) under different order. The total length of the stream is calculated to be 2274km and on a average 0.68 km per channel. The drainage density is calculated to be 2.96km/Sq/km on an average slope is 27.7 (Teli, 1991).

The watershed receives 1302mm average rainfall per annum in an as many as 76 rainy days. The co-efficient of variation is calculated to 46.8 per cent. The region has sub-tropical to cool climate. It changes with the change in altitude, direction of ridges and their aspects, intensity of forest cover etc.

The western Nayar basin has a total population of 117311 having a growth rate of 14.31 per cent sex ratio is 1223 females/1000 males, literacy rate is 38.26 per cent the workers percentage in 38.44, more than 80 per cent are engaged in primary section only, as many as 24131 households with average size of 4.86 person/households stay in 599 settlement; average size of settlement is 195 person/settlement. The density of population is high 157 persons per sq. km and S.C. population is 15.2 per cent.

Forested area covers 42.01 per cent cultivable land 37.4 per cent blank land is 20.6 per cent area. Further 7.6 per cent land is not available for cultivation. 9.3 per cent pastures and grazing land, cultivable wasteland, is 11.6 per cent. Most of the agriculture is limited to the valleys and slopes are terraced for the same. Horticulture has low extension in the region.

The watershed has as many as 1272391 fields covering 25898 hectare area calculated to be 2.03 sq.m./field. There are as many as 3885471 land owners having on an average 33 fields fragmented at least 4 to 6 sites. The numbers of cultivator is 36982 and cultivating area is 25898 ha. Hence, each cultivators has 0.70 ha. Due to the biotic pressure on the land every possible piece of the land owned by farmer has been brought under plough. The intensity of cropping is 152.4 per cent while the intensity of irrigation is 196 per cent on 3.2 per cent of N.S.A. Mandua, wheat and rice are the main crops. The productivity is low and can feed only 55 per cent of food requirement (147.104kg/per/annum). The technological inputs are very-very limited in the region (Teli,1991).

There are 1,22,696 farm level animals in the region which are calculated to be 84304 animals units. Cows, Bullocks, Buffaloes, are the main livestock which are of local quality and produced a very low person (Teli, 2002). Regarding the fuel wood it can feed only 1/3rd of the requirement and the resources are so exploited that they cannot provide the timber.

Thus, extension of cultivation on steep slopes, free grazing, natural calamities, collection of fuel and fodder for domestic needs and fire etc., help in promoting the environmental degradation. Apart from these, construction works also helps in siltation and degradation. Indiscriminate road construction has done a lot of damage to the forest and fertile land. Every km of road produces 550 cubic metre debris by land sliding and rock fall etc. killing vegetation, fertile soils, checking mountains streams as well as down streams to create floods in the plains and siltation in reservoirs. Hence the mountains are becoming poorer. The runoff measured at Jwalpa Devi is 2.85 m^3/sec.

The region has four genetic type of soil, namely red looms, brown forest soil, podsol soil and meadow soil. The nutrients in the soil are found medium, i.e. nitrogen is 2.94 per cent, phosphorus 2.71 per cent, potassium 2.75 per cent on an average (Teli, 1995).

General Land Use

Before going to detail land use survey, the researcher has tried to calculate the land use pattern on the topographical maps scaling 1 : 50000.

Table 8.1 : Land use pattern in area sq. km.

S.No.	*Land use*	*Area Sq.km*	*Per cent*
1.	Forest Area	312.6	42.01
2.	Cultivable Area	278.7	37.88
3.	Waste Land	153.7	20.67

The cultivable land is more concentrated in the valley parts of the upper tributaries. While almost all parts of lower tributaries of the basin have more cultivated area. The forest coverage is more in the upper reaches of the watershed and decreases as we move lower altitudinal zones. With the increase in cultivated area forested area decreased and with the increase in settlement density the blank and waste land decreases.

Based on the revenue data, the land use structure shows that 242.4 sq.km are is under reserved forest, which is calculated to be 32.5 per cent of the total area. While area under panchayat forest is 11.2 per cent. The land not available for cultivation is 7.6 per cent other uncultivated land accounts for 20.9 per cent and cultivated land is 27.8 per cent (Table 8.2).

Excluding the reserved forest the remaining (502.6sq.km). Statistics shows that 16.6 per cent is under panchayat forest, 11.3 per cent and is not available for cultivation, pasture land is 3.8 per cent and waste land is high as 17.2 per cent. The net shown area is 37.8 per cent which is much higher than U.P. Himalayas (13.2 per cent) the extension of the irrigation is small i.e. 4.4 per cent and limited to the velley only (Table 8.2).

Table 8.2 : Land use structure in W. Nayar basin

S.No.	*Category*	*Area*	*Percentage*	*Exluding reserved forest*	*U.P. Hills %*
1.	Area under forest				
	(a) Reserved Forest	196.10	26.32		45.10
	(b) Civil & Soyam forest	46.35	6.20		14.30
	(c) Panchayat forest	83.55	11.20	16.60	6.00
2.	Land not available for cultivation				
	(a) Barren and uncultivated land	46.20	6.20	9.20	5.40
	(b) Land put to non-agriculture	10.43	1.40	2.10	2.20
3.	Other uncultivated land				
	(a) permanent pastures and other and groves etc.	69.29	9.30	13.80	7.60
	(b) Area under misc. three crops and goves etc.	—	—	—	—
	(c) Cultivable waste land	86.25	11.60	17.20	5.90
4.	Cultivated Area				
	(a) Current fallow	01.00	00.13	00.20	0.40
	(b) Old fallow	15.60	2.14	3.10	0.60
	(c) Net sown area				
	1. Unirrigated	168.00	22.50	33.40	9.30
	2. Irrigated	22.34	3.00	4.40	3.90
	Total	745.00	100.00	100.00	100.00

Agriculture

The hill agriculture is a largely self-contained system, in which land, livestock and people are tightly integrated. It depends on land and labour. The number of working hands in a household determined the extend to which farming can be carried out all year long. All members of the family are involved in agriculture. And it is next to impossible to hire labour because of the subsistence nature of agriculture. Females are the main workforce. Besides fetching water and fodder, cook meals and carry out all agricultural activities except ploughing and harrowing operation. Male are not much hard working, though they do all types of labour.

There are two distinct cropping seasons—Kharif (June-September) and Rabi, Kharif cropping has 65.8 per cent area of the gross cropped area. Kharif crops are Rice, Mandua, Jhangora, Marsa, Kauni, Cheena, Ogal, Urad, Bhatt, Gahat, Moong, Vegetable, Beans, Chillies, Tomato etc., wheat is the main crop of Rabi Season, Barley, Masor, Musterd, Potato, Onion etc. are grown.

Mandua is the main crop of the basin followed by wheat, rice, sawah, barley and pulses. Mandua, Rice, Jhangore are mainly dry land crops, while lower parts have irrigation facilities there paddy and wheat are grown. However productivity is low. Mandua is the main crop in the upper parts while on the level and irrigated fields (Table 8.3)

Table 8.3 : Cropping pattern in western Nayar basin

Crop	*Area (in Hac)*	*Per cent*
Rice	5583.7	19.3
Maize	260.4	0.9
Mandua	9084.3	31.4
Sawah	3240.3	11.2
Wheat	7956.0	27.5
Barley	1793.7	6.2
Pulses	810.1	2.8
Others	202.2	0.7
Total	28930.9	100.00

Crop Rotation

On irrigated fields a definite sequence of wheat and rice is grown. In some cases potato is grown as a filler crop after harvesting the wheat till the transplanting of rice. In case of dry land farming tract the land is divided into two separate zones; locally known as "Sara". The sara are alternated each year for rice and mandua cultivation. Rice-wheat-mandua-fallow in a two-year rotation on one sara while other sara mandua-fallow-rice-wheat-mandua. Either the rice or mandua sara, fields remain fallow each year rotation by which soil fertility is regained and soil nutrients are used for different crop requirements.

Land Degradation

Altitudinally most of the land lies between 1200m, i.e. 62.8 per cent Nature of slope is considered as index of soil erosion. Conceptually with increase in the slope the intensity of soil erosion also increases. The data reveals that 32 per cent land area of the watershed is severely eroded. The drainage density is above the warrant level of soil erosion (0.96km/sq.km) in region. Stream order and their length also determine the land use and erosion. The existing land use, land degradation and soil erosion is evident from the Table 8.4 :

Table 8.4 : Land use and Land Degradation (Area in sq. km)

Degradation	*Forest*	*Cultivation*	*Barren land*	*Total*	*Per cent*
Slight	178.7	3.4	2.1	185.2	24.8
Moderate	93.3	261.8	21.2	376.3	50.5
Severe	40.2	12.1	128.4	180.7	24.3
Intense	0.4	0.4	2.0	2.8	0.4
Total	312.6	278.7	153.7	745	100
Per cent	42.0	37.4	20.6		

Land is the most important ecological factor. It is a result of the action and reciprocal influence of parent rocks. Climate, topography, plants, animals and age of the land. It can be represented as : $S = (g.e.b)\ \Delta t$ where s = soil, g = geology, e = evironment, b = biological influents and t = time. The problem of land degradation in the form of soil erosion, land wasting and soil duplication is age old in fact as old as the first man cut the first tree to parctice arable farming. The degradation is created by natural factor and human misbehaviour which promotes to soil erosion, deforestation, misuse of pastures and unscientific agriculture etc. the land degradation and erosion has created a number of other related problems—siltation of reservoir affected water supply electricity, transport, floods, droughts, water logging etc. The land degradation is the creeping death of earth, man, culture and his environment. When men interfere in the course of nature the rate of soil erosion is greatly accelerated. The generalized estimated soil loss under forest is 90kg per hac. per year, under hill paddy it is doubled and if the same plot is continuously cultivated for a number of years the rate of soil loss may be as high as 32 times per year.

Running water is the greatest erosive agent in the hilly region. When the heavy showers fall directly on the bare earth surface, it loosens the soil particles and take away eroding in the form of rivulets. If the drops are intercepted by the trees, plants, grass, crops etc. protect the land from erosion. The rainwater on the steep hillsides also causes mass movements of the soil due to lubricating action of water within the soil. Usually on the landscape of gentle slopes sheet erosion occurs in the region. The top fertile layer of soil is ripped off in the thin films every year by the running water over such lands. The soil nutrients are lost and yields lower down slowly. It is estimated that heavy rains double the water flow scouring capacity is increased four times, caring capacity by 32 times and size of the particle 64 times (Temporary and grist 19.58).

Biotic Pressure and Land Degradation

The biotic pressure has considerable influence on land degradation, which become problematic with human and animal interference. The indiscriminate felling of trees, land disturbed by developmental activities. Overgrazing of pastures, small uneconomic and fragmented fields, etc. promotes to denude the upper most and protective cover of surface. Not only that but the footpaths used by the man and his cattle also develop into rills and gullies, causing trail erosion. Man and cattle, particularly his pastoral flocks and hazard grazing round the year disturb the environmental balance. The village pastures are overgrazed and cattle removed every green blade of grass as fast as it emerges, will starve out. The vicious cycles continues until the land becomes a dreary waste fit to be abandoned. Annual tire followed by intensive rains are the principal causes of erosion and land degradation. Free grazing cattle scramble over steep hillsides in search of food, cutting the soil with their hooves, trampling down tree seedlings and leaving grass free tracks down which monsoon rain washes the loosen soil. The poverty of hill pasture is such that each animal unit needs at least 2-3 hectare grazing area to support the good health without causing environment deterioration, while the present pressure is 12 animal units per hectare.

The pressure of increasing population is neutralizing all effects to raise the standard of living and of nutrition, while loss of fertility is nullifying the value of all improvements made. As such the hill agriculture an arduous and unprofitable proposition. Due to lack of secondary activities most of the people are forced to leave their native place and approach to the plains to earn their bread and butter, mostly men. If the families were to accompany with their man folk, the hills would be almost empty. The region also provides maximum human resource to defence service. However, migrated men in service send their earning home by money order. Hence in this money order economy region sex ratio is very high. The availability of per capita net sown area is as low as 0.16 hec. Since there is no scope for extending the cultivable land beyond the present limit thus have to depend on present status of land. The pressure of population on NSA is as high as 6.16 persons per hectare while the carrying capacity is just half of it. The agricultural density is (non-caps) 195 persons per sq.km.

Agriculture and Land Degradation

By and large all agricultural land is terrace having slopes greater than $15^{o}20^{o}30^{o}$ and is effected by moderate to severe erosion according to the World Bank report 15° slopes get arable with sound rotational practices, if bench terracing is done, for 25° slopes terraces are recommended and for 35° slopes orchard terraces or platforms are recommended. Land beyond 30° slopes should not be under cultivation. And on this basis at least 75 per cent of land of 30° 40° slopes class needs change in land use.

In the hills farming system and cropping pattern help in water loss the system of cultivation of maize wheat helps in maximum waterloss 54.1 per cent of rainfall as against 41.2 per cent in maize-cowpea-wheat and 32.7 per cent in cowpea-wheat crops. The maximum soil loss is absorbed on the cultivated fallows while least on the grass lands (anonymous 1985) Table 8.5.

Table 8.5 : Cultivation methods and Cropping-Soil and Water Loss on 8 per cent Slope *(Rainfall 1240mm.)*

Cultivation & Cropping	*Water loss (mm)*	*Water loss as % of rainfall*	*Soil loss T/hact/year*
Up and down cultivation (maize-wheat)	670	54.10	28.5
Contour Cultivation (Maize+Cowpea+Wheat)	511	41.20	19.3
Copea-Wheat	405	32.70	28.3
Cultivation Fallow	445	35.90	44.0
Gaint Star gross	31	2.50	1.3

The agricultural fractions cannot be justified on sound and husbandry principles at least 50 per cent area of the region. Farming in this region cannot give optimum returns and presently agricultural production is not at all commensurate with the farmer's inputs, which are supplemented by government in the form of subsidies for terracing or land shaping, supply of fertilizers, etc. in spite of this subsistence farmer is inwilling to change and is looked into an economic dilemma from which he, perhaps is unable to escape.

Management and Planning

The management and planning of land in hills for sustainable development offers a new relationship between man-land and environment which view the production, consumption, function in its wider perspective for the judicious harvesting of the natural resources. Its aim is to curb the 'misuse', prevent the 'abuse', regulate the 'non use', foster the 'reuse' and 'save' and 'invest' the resources for the 'future use' thus, individual has a moral and social obligation to manage the productive capacity of land water, air and life for their integrity and stability for the aesthetic qualities of the landscape.

Erosion is the greatest destroyer of land resources in hills. Apart from soil loss, erosion creates many other problems like—siltation of reservoirs, canal, rivers, deposition of unfertile material on cultivated lands, harmful effect on water and power supply, frisking, transport and communication, loss of life, damage due to floods and drought etc. hence, the sustainable development of land is a must to maintain the ecological balance in hills. Land management programmes which engulf soil and water conservation also, cannot be effectively implemented without following the watershed approach. The conservation programmes carried out all over the nation since first five-year plan, have followed the basic principles of self-contained watershed. It is a hierological unit having unique physical characteristics, which affects the all physical landscape of the region.

There a number of measures to conserve the land from degradation. These measures are broadly grouped into four categories, namely land capacity classification, engineering measures, agronomic measures and afforestation (Teli, 1988).

Land Capability

The basic principle of land management is to use the land according to its capability and treat the land according to its need therefore the knowledge of land capability classification is a prerequisite and important for its planning, implementation and execution by conservation programmers. The land capability classification refers to systematic arrangement of different kinds of land according to those properties that determine the ability of the land to produce on virtually sustainable basis. It approaches towards the most intensive profitable and safe use, which can be made of any piece of land.

The land characteristics and its capabilities are known by their use can be made accordingly and if needed the conservational techniques could be applied it will not only help in conservation of soil and water resources but will balance between the regional and eco-cultural environment.

Engineering Measures

Engineering measures adopted in areas where the natural way of soil and water resource conservation techniques fail. It provides the better technology for the resource conservation on all types of lands. On sloping land where soil erosion becomes serious, prerequisite are the mechanical measures of soil and moisture conservation, such as bench terracing, contour/graded bundling, land shaping/grading, levelling and giving the proper slope to the terraces according to the amount of rainfall, etc. based on the situation and problem of erosion in the fields.

Eroding torrents are bit stabilized by constructing a system of cross structure placed at suitable sites in the channel. These cross-structures running the entire width of the channel between two banks are know check dams and grade stabilization sills. Those jutting out of the bank for part of the channel width are known as spurs groyness. Structures built parallel to the current are called as retaining walls, dykes, jetties, guide banks, bank revetment, leavees etc. Treatment of the main branches of a torrent with structures and progressive land treatment measures in the watershed will result in the reduction of the flood peak and its sediment land. Collapse torrents are fed by big landslides or rock falls and vertical cut in the catchments. The source of debris should, therefore, be stabilized by vegetative treatment in combination with suitable engineering measures.

Agronomic Measures

Agronomic dose not only gives higher production but also serves the purpose of erosion cannot, rainoff harvesting, storage and recycling. The problem of soil and water conservation on agriculture land is more acute, severe and important as compared to any other sector. The agronomic practices, such as type of crop and crop rotation, method frequency and interments of tillage operation, use of green and organic manure, salt, exchangeable detailions, soil moisture contents, bulk density, etc.

Agronomy do wonders in conservation of valuable and resources and their characteristics. Some of the agronomic practices are selection of suitable crop, cropping system, crop rotation, strip cropping, mulch, contour cultivation, minimum tillage with and without mechanical measures, the nature of landuse also have a direct effect on soil loss, and land degradation.There are some crops, which permit less erosion and soil loss on agricultural fields in hills, viz., (a) Legumes, such as moong, (*Phaseoulus aureus*), cowpea etc. While non-legumes, viz.

Jawar (*Corgum Vulgar*), bajra (*Penmsetum typhoideuim*), maize, etc., three crops (*Glycine max*); (b) High runoff and medium soil loss—Jhangora (*Papantum scrobiculatum*); (c) Medium runoff and medium soil loss—mandua (*Eleusina coracana*), higher (*Guizotia abyssinica*) and rice; (d) low runoff and soil loss rapier grass (*Penniseteum purpureum*), moong, groundnut, urd, soybean, stylazobium and sonnhamp; (e) low runoff and very low soil loss—rapier grass.

Mixed cropping and intercropping system has become universally accepted partice for soil conservation, rain fed agriculture, soil fertility conservation and maximizing production. The system reduces erosion by providing maximum canopy, abording beating action of raindrops and thick growth to obstruct the runoff and increase infiltration. Mix farming and erosion resisting crops help in maintaining soil fertility by penetrating and feeding roots at different depths and through symbiotic nitrogen fixation by legumes. There parctices assure/success of crops to farmers under adverse condition too. Such as in drought frost and insects etc. in the hill regions, like the present are.

Table 8.6 : Crop for land sustainability

S.N.	*Crops*	*Run-off (mm)*	*Soil loss T/Ha*	*Sustainability class*	*Remark*
1.	Maize	36.7	6.7	III	Low
2.	Jhangora	4.2	4.8	II	Moderate
3.	Mandua	23.2	3.3	I	Very High
4.	Bhatt	37.2	7.1	III	Very Low
5.	Paddy	22.9	4.3	II	High
6.	Niger	20.0	4.2	I	Very High

Mulches are also used to reduce the erosion and evaporation to improve soil infiltration and structure and to suppress weed for better crop production minimum tillage not only saw. Labour, time and money for field preparation but also aviods destruction of soil structure and binding forces if soil particles and thereby reduces vulnerability of land degradation. The weedicides result in clean cultivation by keeping area free from weeds, which is conducive to high water and sediment loss.

Apart from the above discussed agronomic measures some other measures like—manure, cow-dung, green manure, bio-fertilizers, etc. are important which not only conserve the soil and water resources but also provide a number of nutrients also. Hence, the practices of soil consort at agronomy are quite effective for preservation of splash, sheat and rill erosion on class II and III lands.

Forestal Cover

The establishment of vegetative cover is one of the effective ways of landscape management. It has already been discussed that natural fallow (i.e. with grasses and/or grasses and shrubs) gives the least and negligible runoff and soil loss with compared to other practices, it increases water table, aggregates in soil, protect against raindrops, reduces water velocity, increases infiltration capacity and fertility of soil and also provide valuable feed and forage to the livestock and human population. Trees intercept direct rainfall and reduce its velocity and the force with canopy, which raindrops strike the soil surface. The inter-caption value of Acacia in 15.5 to 26 per cent and for salshesea robusta 14 per cent but with the increase in intensity the intercaption decreases. Apart from conservation through leaf litter, dead twigs, branches etc. vegetative cover increases soil moisture, soil nutrients, organic matter and pH also.

Vegetation cover helps in stabilization of contours, fields, bunds, river and stream banks. Grass cover should be spread on the bank lands, so as to protect for bed land erosion. The ornamental plants can be used to protect the bank erosion of these waterways. While afforesting, it should be kept in mind, not to plant mono-culture but mix-culture. It not only protect the land from erosion but also provide the requirements of the localities like fuel, fodder, fibre, fruit, flower (F^5), etc., only the planting of plant is not enough but their survival should be assumed so that the maximum utilization of the resources, time, money, labour and protectional aims are fulfilled and the soil disturbed in the process of plantation is stabilized in the course of time.

People's participation in management and rehabilitation programmes be ensured at all stages—policy making, planning and implementation. Since women are practical foresters and farmers, they should be employed for conservation works.

Apart from this, the marginal, sub-marginal and cultivable lands should be brought under horticulture, cultivated fodder, etc. the development programmes should be extended through the model farms, training camps, by providing infrastructural facilities, loan, marketing facilities, seeds and seedling supply. It will not only conserve the natural resources but will also improve the land use pattern and the employment and economy of the growers. Agrostology—like crop-tree mixture and farm forestry is also one of the conservative measures of slopes while farming the land.

Constraints in Land Management and Planning

The problem of land degradation has badly affected the developmental activities in the nation. The government has taken active steps to conserve the land resource as early as 1952 by framing the National Soil and Water Resource Conservation Board which established a number of Central Soil and Water Conservation Research & Training Institutes in various parts of the nation (viz., Dehradoon for N.W., Himalayan region). Similarly all India Soil and Landuse Survey Organisation is actively engaged in such projects. In the last two plans, the concept of integrated land use planning on watershed basis was introduced. During the last decade and onward the problem has been widely talked off and government is also very serious to face the hazards of erosion, siltation of reservoirs, deforestation and ultimately environmental degradation.

The questions arise as to why soil conservation is not more widely applied? Why so much land is being wasted while millions of people are suffering from hunger? Why conservation programmes appear to be so difficult to be implemented? The simple answer in that application of soil conservation techniques and needs are much wider than just the application of such techniques and require a general policy by which soil conservation may become an integral part of wide land use and receive a wide support within a social and economic environment which is conducive to maintenance and improvement of the soil resources.

Any change imposed on land use pattern on farm will produce a number of effects on its productive capacity. These possible effects must be forecasted and if likely to cause degradation, system of management has to be designed to overcome deterioration. Increasing agricultural production should aim not only at sustaining higher levels of useful biological productivity but also at ensuring that the system is stable. This principle should be applied when uncultivated areas are turned into arable land, mono-cultures to multiple cropping, shifting cultivation to permanent agriculture and also when irrigation or mechanisation are being introduced. Therefore, conservation requirements should at the basis of land use planning decisions.

Another important constraint to the application of soil conservation is the socio-economic conditions of farming community. Each farm family earns a living from land and income is used to satisfy the basic needs of the family. Since the farmer's income is low due to low production capacity, the portion of income spent on the maintenance of the soil is insufficient to maintain the land.

A third consideration is that in most intances, community or cooperative action is indispensable. The farmers should be closely involved in the planning and execution of field programmes. New management techniques should not become a burden to the farmer, on the contrary, they should be within capacity of the farmer and he should understand its beneficial effects. It should be kept in mind that the planning is not successful without the people's participation.

Among the factors, which lead an overstressing of land resources, is right of ownership to land. Along side the assumed right of land ownership there are other historical rights, which do not fit into today's land use pattern, in the region, like throughout in India, there are rights of villagers to graze certain areas, to cut fuel wood, to take timber for agricultural implements, etc. these rights make it difficult to manage efficiently even the so called reserved forest and government policies such as land consolidation programmes, etc.

The farmer is poor and unable to take any risk due to subsistence economy. The essence of farming is to minimise risks against weather, pests and diseases. The Himalayan farmers cannot be reasonably expected to care very much about the down-stream damages, as most of these people are quite unknown about the flood situation of Gangetic plain. Therefore, soil conservation programmes will have to be tackled on a watershed basis, as a concerted action where all efforts re-enforce and complement each other.

At present, there is a wide gap between awareness of land degradation and comprehensive action to combat it—prospects for establishing comprehensive soil conservation programmes are sometimes not good because the proposals may not appeal to politicians, decision makers, planners and may not be attractive to farmers. Important reasons for the indifferences are due to lack of knowledge of the problem and its solution. Therefore, future conservation programmes can best be served by creating a better awareness of the problems and solutions and then to obtain effective results within a reasonable period.

Experience has shown that legislation alone on soil conservation is rarely successful in enforcing government's land use policies and soil conservation programmes. Therefore, there should be a greater emphasis on conservation practices and supportive research in the region.

In the Western Nayar basin, land conservation works are being implemented by State and Central schemes. Whatever little is done, is being done in a fragmentary and piecemeal manner. The programmes are being implemented by State Forest, Agriculture and to some extent irrigation departments in an uncoordinated manner.

Continuous degradation of production base and imbalance in land-water-plant, human-animal systems is leading to ecological imbalance and economic insecurity and there at the quality of our life and civilization we must therefore, link economic security with ecological security. For this we must, check erosion, manage watersheds, re-establish forest vegetation and rehabilitate denuded lands.

It leads to the fact in spite of this the poor farmers of the western Nayar basin are unable to take any risk on the name of Soil Conservation and Management. Therefore, Government and voluntary agencies should come forward and innovate the knowledge among the farmers regarding the programmes by taking them into confidence and by tailoring of the programmes to the farmer's need. This way, the people's participation will narrow down the information gaps and the farmer and villagers will not take it as a burden but they might take it as a habit of soil and water resource conservation. Thus, the seriously eroded areas as discussed earlier need prompt conservational measures and attention of both the government and the local people so that soil-water-plant conservation alone can make development sustainable.

REFERENCES

Anonymous, (1985), Conservation Techniques of Denuded Lands in Hill Region, Lecture delivered at Summer Instt. at Central Soil and Water Resources Conservation and Training Instt., Dehardun.

Teli, B.L. (1988), *Himalayan Landuse-Structure Planning and Potentialities*, National Atlas and Thematic Maping Organization, Monograph no. 8, Calcutta, pp. 55-60.

Teli, B.L. (1991), Rural Landscape Management and Planning in Western Nayar Basin unpublished D.Phil Thesis, H.N.B. Garhwal University, Srinagar, Uttaranchal.

Teli, B.L. (1995), Soil Resources : Problems and Planning in Hills. in : B.L. Kaul (ed.) *Ecodegradation of Himalayas*, Vinod Publishers & Distributors, Jammu, pp. 308-38.

Teli, B.L. (2003) Livestock Resources : Structure Planning and Potentials. Proc. of National Seminar on Central Himalaya : Environment & Development, Deptt. of Geog., H.N.B. Garhwal University, Srinagar, Uttaranchal (In Press).

Tempany and Grist, (1958), *Introduction to Tropical Agriculture*, Longmans, London, p. 70.

9

Sustainable Water Conservation and Morphology of Tanks in Mahasamund District, Chhattisgarh State

A.R. Chandraker, Jitendra Kumar and *N.K. Baghmar*

"We never know the worth of water til the tank is dry"
Thomas Fuller, Gnomologia

Introduction

The undulating nature of the district Mahasamund, makes it eminently suitable for the construction of tanks (water bodies). Tanks dominate the landscape of rural settlement in the district. Simple technology has been used to construct tanks for water conservation. On an average, in every village there are 4 to 6 tanks to conserve water for different uses. The tanks are mostly dual purpose tanks constructed mainly for 'Nistari' (domestic use) purposes though they are used for irrigation as well. Tanks water functions like thermostat for climate surrounding the settlement.

Objective

The objective of the paper is to discuss the morphology of tanks and suggest measures to conserve water. In this paper shape, size, soil, use and loss of water have been studied in detail.

Methodology

The study is based on primary data. For the analysis, 61 tanks spread over 15 randomly selected villages of Mahasmand district were surveyed. In addition 30 old persons were interviewed regarding tanks and Patwari record have also been used.

Study Area : Mahasamund District, Chhattisgarh State

Mahasamund district is situated in eastern part of Chhattisgarh State (21°32'-22°28' N and 81°58'-82°28' E) has an area of 782 sq.km (Figs. 9.1 to 9.3). There are 4411 tanks over 1092 villages. The higher physiographic positions are occupied by red coloured '*Bhata*' soil (Laterite). On the other extreme, the lowlands are occupied by dark brown or black clayey soil known as '*Kanhar*'. The district is a part of the well known monsoon region. Annual rainfall varies from 1200 to 1450 mm. Hot winds, known as '*Loo*' blow over the area in the April and May. Evaporation is maximum during May.

Discussion : Morphology of Tanks

Shape of Tanks

Out of 61 tanks 28, i.e. 45.9 per cent are rectangular in shape, 29.6 per cent and 24.5 per cent are circular and square in shape respectively (Table 9.1).

Table 9.1 : Villagewise Morphology of Tanks : Shape

S.No.	*Name of village*	*No. of Tanks*	*Rectangular*	*Square*	*Circular*
1.	Baronda	4	2	2	-
2.	Bhoring	5	2	2	1
3.	Birkoni	4	2	-	2
4.	Bamhani	3	3	-	-
5.	Belsonda	5	3	-	2
6.	Durgapali	4	2	1	1
7.	Dhanora	3	1	2	-
8.	Ghaunch	4	1	2	1
9.	Hardi	5	2	1	2
10.	Kauhakurda	3	1	-	2
11.	Machewa	5	2	2	1
12.	Parskol	5	2	2	1
13.	Sarkanda	3	-	-	3
14.	Sona sili	4	3	1	-
15.	Tumgaon	4	2	-	2
	Total	61	28	15	18

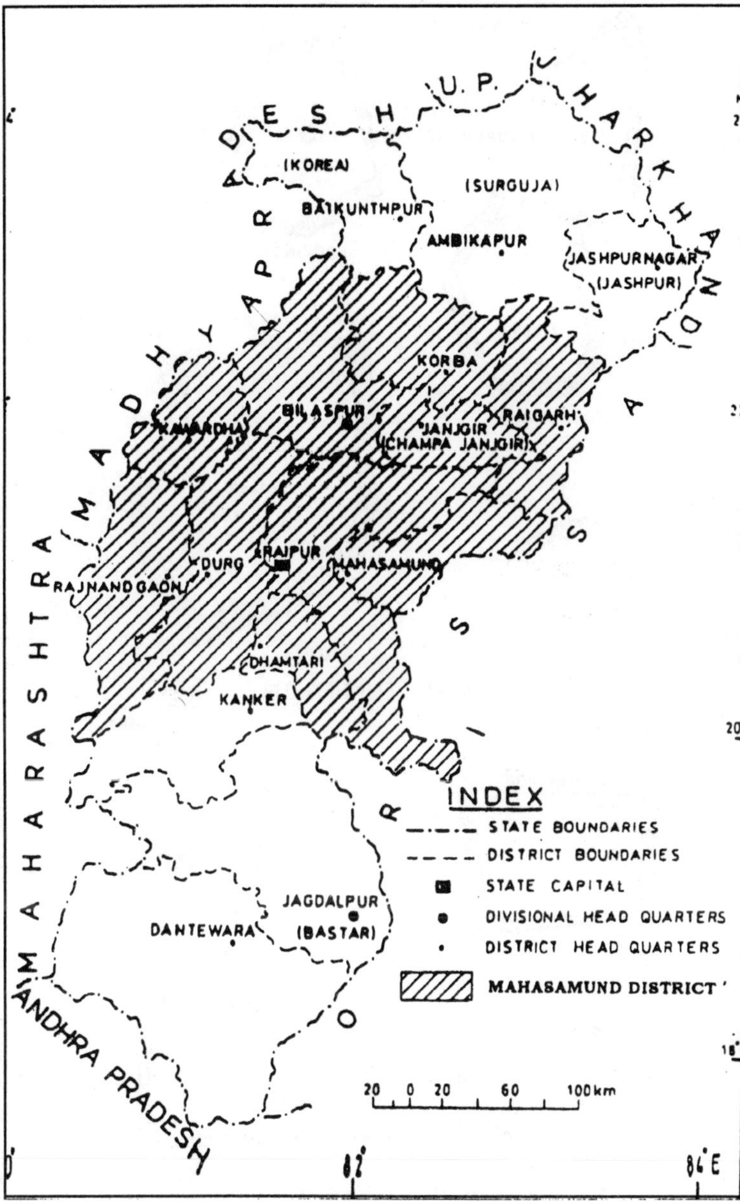

Fig. 9.1 : Chhattisgarh State.

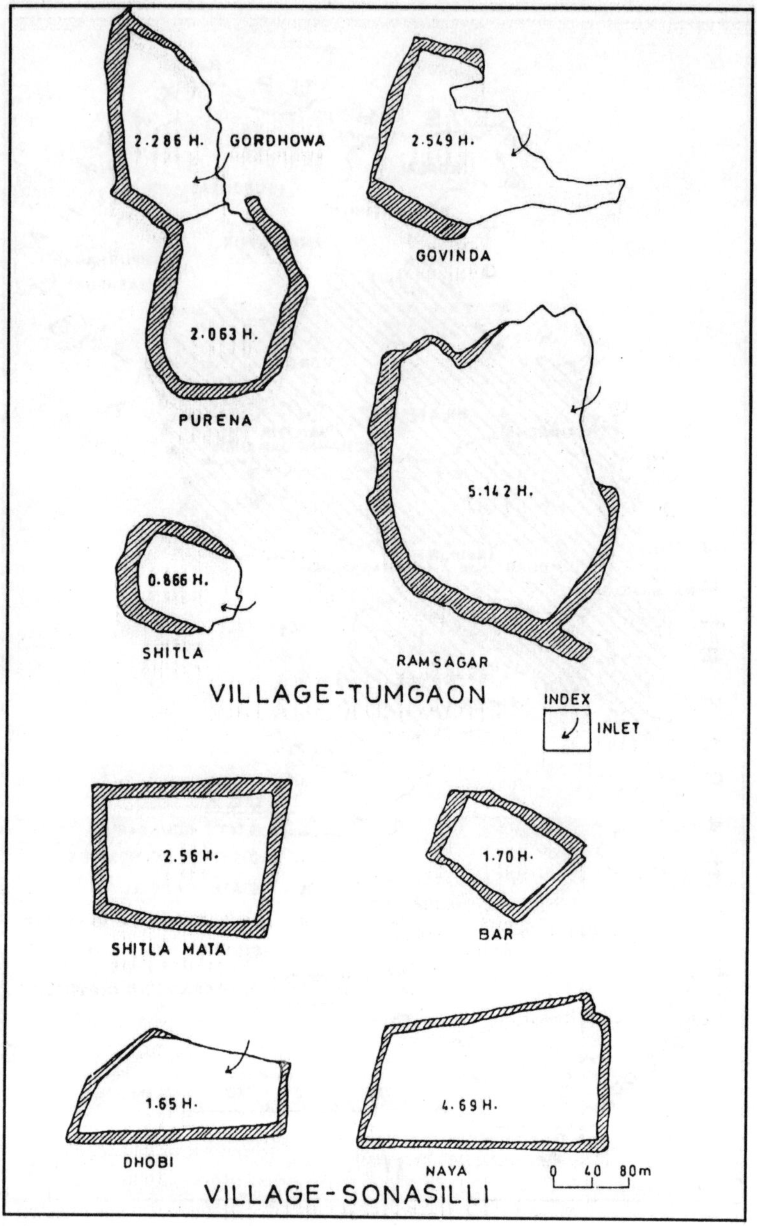

Fig. 9.2 : Morphology of Tanks Mahasamund Dictrict (C.G.) 2001.

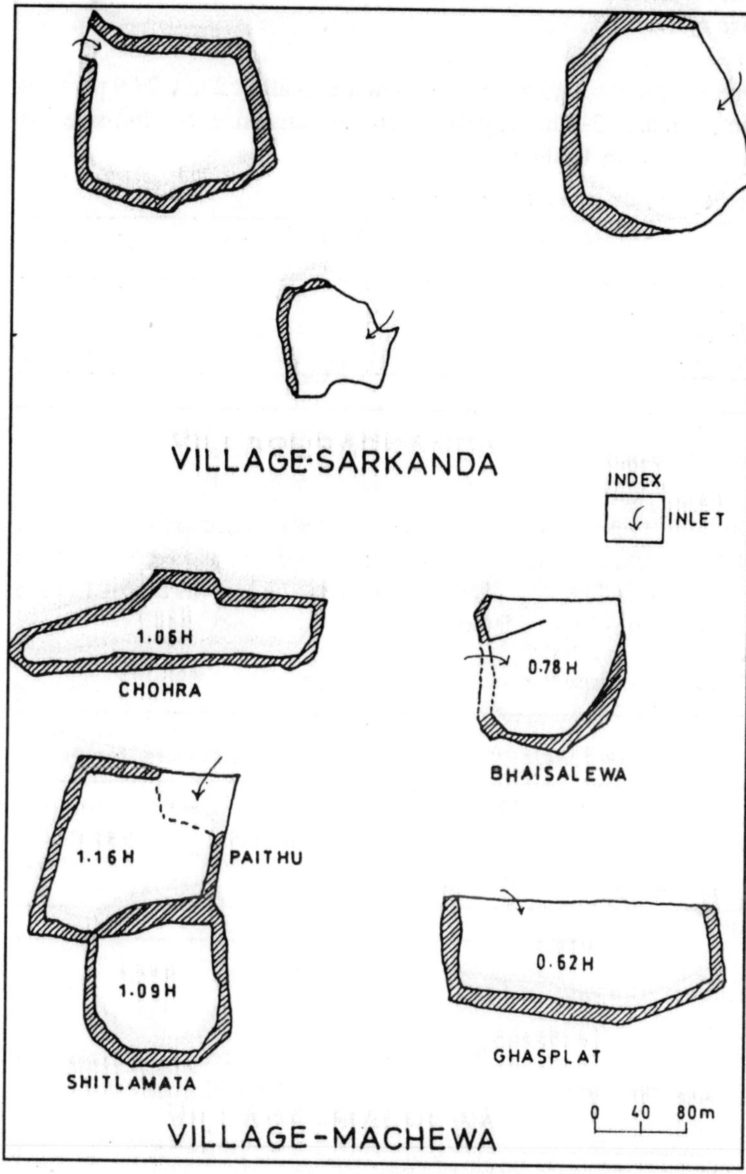

Fig. 9.3 : Morphology of Tanks, Mahasamund District (C.G.) 2001.

Size of Tank

Out of 61 tanks 34, i.e. 55.7 per cent are small (< 2 ha), 27.9 per cent medium size (2-4 ha) and 16.4 per cent large size (> 4 ha) size has been shown in Table 9.2.

Table 9.2 : Size wise : Number of Tanks

Size	*Area in ha*	*Number of Tanks*	*As per cent to Total Tanks*
Small	< 2	34	55.7
Medium	2-4	17	27.9
Large	> 4	10	16.4
All sizes		61	100.0

Soil Base of Tanks

Out of 61 tanks 12, i.e. 19.7 are constructed on Kanhar Soil (vertisol), 40.9 per cent on Matasi soil (inceptisol) and 39.4 per cent on laterite soil (entisol). Seepage of water from tanks is less in case of Kanhar soil while it is high in laterite soil.

Table 9.3 : Soilwise Base of Tanks and Seepage

Soil	*No. of Tanks*	*Per cent to total tanks*	*Seepage of water*	*Water level in summer*
Kanhar (Vertisol)	12	19.7	Low	Water available
Matasi (Inceptisol)	25	40.9	Medium	semi-dry
Laterite (Entisol)	24	39.4	High	dry
All soil	61	100.0		

Number of Bunds in Tanks

Out of 61 tanks 15, i.e. 24.6 are bounded by three sides while 75.4 per cent tanks have bund on four sides. Tanks with four sides bunds are near settlements.

Table 9.4 : Number of Bund in Tanks

No. of Tanks	*Three side Bunds*	*Per cent to total tanks*	*Four Side Bunds*	*Per cent to total tanks*
61	15	24.6	46	75.4

Depth of Tanks

Depth of tank from top of bunds varies from less than 4 m to more than 5 m. Out of 61 tanks 24, i.e. 39.3 per cent have depth from bund is less than 4 m while 21.4 per cent tanks more than 5 m (Table 9.5).

Table 9.5 : Depth of Tank from Top of Bund

Depth of Tank	*No. of Tanks*	*Per cent to Total Tanks*
< 4 metres	24	39.3
4-5 metres	24	39.3
> S metres	13	21.4
Total	61	100.0

Population and Animal Pressure on Tanks

"Water, water everywhere
Nor any drop to drink." - S.T. Coleridge

Tanks water is used for bathing and cleaning for man and animal. Few tanks are used for irrigation. Before fifty years people used the water of tanks for drinking purpose also. Now new source of drinking water is handpumps in rural area. About 90 per cent men and animal (out of 29656 man and 15414 animals of surveyed villages) depend on tanks for bathing and cleaning (Table 9.6).

Table 9.6 : Villagewise Tank : Population and Animal Pressure

S.No.	*Name of Village*	*No. of Tanks*	*Population*	*Animal*
1.	Baronda	4	1426	650
2.	Bhoring	5	3918	2450
3.	Birkoni	4	3159	2335
4.	Bamhani	3	1731	735
5.	Belsonda	5	3700	2350
6.	Durgapali	4	846	240
7.	Dhanora	3	945	500
8.	Ghaunch	4	2315	835
9.	Hardi	5	750	300
10.	Kauhakurda	3	1235	720
11.	Machewa	5	861	421
12.	Parskol	5	658	318
13.	Sarkanda	3	1350	415
14.	Sonasili	4	1815	390
15.	Tumgaon	4	4947	2455
	Total	61	29656	15414

Stage of New Water Source : Yearwise in village

Stage I

Tank : The oldest and major source for 150 years and as old as settlement.

Stage II

Well : Source of water since year 1920.

Stage III

Handpump : Source of water since year 1970.

Stage IV

Tube well : Source of water since year 1980.

In the district at present all four stages have been found in villages in different degrees, but still tank is major source of water.

The Findings of the Paper

(i) Tanks are the oldest and major source of water in village. Their size varies from 0.35 ha to 6.36 ha. Out of 61 tanks 28, i.e. 45.9 per cent are rectangular in shape, 29.6 per cent and 24.5 per cent are circular and square in shape respectively,

(ii) About 67 per cent water is wasted due to evaporation and seepage. Loss of water by seepage is maximum in laterite base tanks while it is minimum in case of Kanhar soil (vertisol). Out of 61 tanks 25, i.e. 41.0 per cent are constructed on Matasi soil (inceptisol), 34.4 per cent on Laterite soil (entisol) and 24.6 per cent on Kanhar soil.

(iii) The height of the bund of tanks varies from 1.9 m to 6.2 m On an average the breadth of bund is 2.9 m to 8.3 m. The depth of water in tanks is 1.3 m to 4.2 m. 46 tanks, i.e. 75.4 per cent have been bounded by four sides while 15 tanks, i.e. 24.6 per cent have bund on only three sides.

(iv) Every tank has its own name. The name of the tank is related with the man who constructed it.

Author's Suggestion

(i) Authors suggest to measure the amount of water in tanks during different seasons, water gauge pole to be established as soon as possible. This will help to plan the use of water for villagers.

(ii) In the district about 19 per cent land is under Barren and unculturable. Therefore, this land can be used for construction of tanks for water conservation.

(iii) In summer most of the tanks are dry, therefore, tanks should be connected with tubewell. There is such a proposal in Indira Ganga Yojana in Chhattisgarh State. In the area, rock is of limestone which wear sufficient water. This way water shortage problem can be solved in villages of Chhattisgarh State.

REFERENCES

IRRI—International Rice Research Institute, 1987 : *A Annual report for 1986*, P.O.BOX 933, Manila, Philippines.

Maheswari, B.L. and Turner, A.K., (1986) : *Some low-cast methods for seepage control in storage tank in alfisols and vertisols*, Agric Water manange 12:53-62.

Michael, A.M. and Khepar, S.D., (1992) : *Water well and pumps engineering*, Tata McGraw-Hill Publishing Co. Ltd., New Delhi.

Saksena, R.S. and Mishra, H.R. (1990) : First all India minor irrigation census—interim analytical report, *Bhu-Jal News* 5(3) : 1-6.

Sinha, A., (1988) : *Small can be beautiful : reflections on the efforts at harvesting rainwater in the tribal district of Singhbum in Bihar* (mimeo).

Singh, Jasbir, (1974) : *An agricultural-atlas of India : A geographical analysis*, Vishal Publications, Kurukshetra.

Verghese, B.G., (1990) : *Water of hope, centre of policy research*, IBH Publishing Co., New Delhi.

Vohra, B.D. (1987) : When minor becomes major some problem of ground water management, *Bhu-Jal News*, 2(1) 2-3.

Von, Oppen M., Subba Rao, K.V. (1987) : *Tank irrigation in semi-arid tropical India : economic evaluation and alternatives for improvement*, International Crops Research Institute for the Semi-Arid Tropics, Patancheru, Andhra Pradesh, India.

10

Canal Irrigation Induced Impoverishment of Soil Resources in South and Southwest Punjab

Dharam Singh and *Anupreet Singh Tiwana*

Introduction

Soils are our most valuable natural assets. Maintaining soils in a state of high productivity and hereditary esteem on sustainable basis is essential for chording the far ends of continuum of basic needs of masses. It is of utmost concern in a country like India which is if not entirely but largely dependent on agriculture. In the endless pursue of providing and producing food to the teeming population of nation, quality of the environment in which we stage an opined string of events in possiblistic and teleological manner is to be maintained. The soils, so called epidermis of the earth is to be used in a rational manner if we want to avoid the blisters of misuses and varied malfunction. Rather than leaving adverse impact on the productivity, one is to be supposed to leave imprints of better heritage for posterity.

The life supporting system and socio-economic development of Punjab is dependent directly on the proper utilization of soils (Sehgal, 1995). Punjab is facing great dilemma between the problem of increasing food production and deterioration of soil quality. Due to unprecedented swelling of human resources, there is an increasing demand for 3F—food, fodder and fibre—for the existing and the future generations. On the other hand, there is an urgent need to maintain the soil health and quality in sustainable environmental milieu. With the advent of Green Revolution of post-1960s as a comprehensive agricultural transformation involved high inputs of fertilizers, irrigation and hybrid seeds. The production did increase, but had resulted in impoverishment of soil health and that menance is still aggravated by enhancing the provocable risk to general eco-friendly conductivity.

Methodology

In under hand research work an empirical view of secondary data has been generated and is to be propounded through plausible map work. Ground truth data is collected through field traverses.

Historical Perspective of Canal Irrigation in Punjab

Punjab has centuries old prolong use of irrigation as an artificial supply of water to enrich agrarian outputs. The development of agriculture and the establishment of Indus Valley Civilization and the Aryan settlements were perhaps due to the availability of water for agriculture in this riverine state. Even during the time of the invasion of India by Alexander the Great in 326 BC, there existed inundation canals in Punjab. During the medieval period Delhi Sultans and the Mughals developed and extended irrigation facilities in Punjab to increase agricultural production (Manku, 1998). Maharaja Ranjit Singh and the Britishers ameliorated the network of canals in the 19th century. Immediately after independence, the Government of India rested her benevolent hand on the terra of Punjab to tap possible water resources.

Objective Realities and Subjective Scenario

Canal irrigation accounts for 39.5 per cent of the total irrigated area of Punjab and this is of immense importance. Most of Punjab is a flat plain with rather gentle slope. This has helped in projecting canal water to far-off places facing water dearth from canal take-off points (Mavi & Tiwana, 1993). Moreover, the absence of hard rock beds in the plains facilitates easy digging of canals. The total length of canals and distributaries in Punjab is 4600 km and approximately 70 per cent to the total length run crisscross in the south and southwest Punjab resulting in the menace of soil degradation (Map No. 10.1). The clay beds, deep in the sub-soil, acting as lower limit for the rainwater that has infiltrated through the porous alluvium resulted in the curse of salinity and excessive water logging.

To facilitate distribution and, economise water, each "*Killa*" or acre was divided into a number of compartments during the British regime. The zamindar was allowed only twenty minutes for each compartments or "*Kiaris*" for irrigation to replenish water requirements (Robbers & Faulkner, 1921). The Irrigation Commission of 1901 drew particular attention to the matter of proper utilization of canal water. The total of water rate, land revenue and "cess" per acre of cropped area varied from Rs. 8 to Rs. 10. After independence, the government has facilitated the farmers in many ways and these facilities are now posing threat to the environment. At the time the Green Revolution was adopted, sustainability was not the criteria. The only way out was to pump in the inputs so that production might rise. Owing to percolation of water through the coarse silt at the bottom and through the canal side, a considerable proportion of the water of canal is lost. This quantum of loss was estimated by Kennedy (1920) for the Upper Bari Doab Canal as 15 per cent for main canal and 6 percent for distributaries. The evils caused by percolation and flood irrigation are waterlogging conditions resulting in salanization of alluvium. The percentage of water logged and salt affected area is so far small—15,185.4 sq.km as the severely deteriorated, but the grave aspect of the querry is the fact that the of water table is going to be a persistent headache in the coming years. Crops won't rise succumb to serious damages till the water table is in the 4 or 5 feet vicinity of the surface. Even a temporary rise of the water-table to within few feet from the surface is harmful. In the districts of Muktsar, Moga, Faridkot and Bathinda, the water table levels with the land surface. The salt affected soils occur in patches rather than large contiguous tracts in south and southwest Punjab. Moreover, these patches are interspersed in alluvial plains covering parts of Mukatsar, Faridkot, Abohar, Fazilka, Bathinda, Mansa, Sangrur, Sunam and Samana Tehsils. The grid intricately woven by Bikaner canal, Abohar branch, Bathinda branch, Kotla branch, Ghaggar branch and Bhakra Canal infest with the constraints of high salt contents and water logging. Low hydraulic conductivity of subsoil cause water logging and low availability of nutrients. Presence of high contents of certain chemicals adversely affects the plant growth by causing concentration of negatively charged hydroxyl ions. The cotton production is badly affected by degraded soils through unchecked canal irrigation. Kotkapura, the cottonopolis of India has

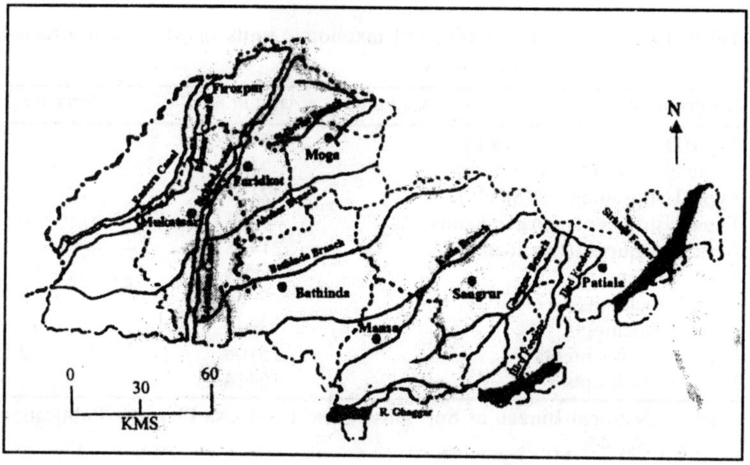

Fig. 10.1 : Network of Canals in South and Southwest Punjab.

shown cotton bankruptcy. These salinized soils have been classified as: Fine loamy, Aeric Halaquepts, Typic Ustochrepts and Natric Ustochrepts (Table 10.1).

Table 10.1 depicts that most of these soils are left barren except in sorne areas where they are cultivated as salt-tolerant crops. Presently these soils have low potential and propensities. The production potential of these soils can be enhanced by providing adequate drainage.

Remedies to the Phenomenal Jeopardy

The suggestions to face and curb the might of this gruesome problem are:

- Diversion of course of run off frorn the problem areas.
- Construction of open ditches. ,
- Field drains to drain away the excessive water.
- Planning and adoption of rational agricultural practices.
- Adopting judicious and eco-friendly water management practices.

The policies suggested are the temporary and short lived remedies. The problem of soil degradation in south and southwest Punjab is

Table 10.1 : Area of different soil taxonomic units in which salt-affected soils occur as patches

Subgroup with family (particle-size class)	*Area in ha*	*Percentage*
Typic Haplustalfs (fine-loamy)	15103	1.9
Typic Natrustalfs (fine-loamy)	9140	1.1
Typic Ustipsamments (Sandy)	42187	5.2
Typic Ustifluvents (coarse-loamy)	6623	0.8
Aeric Halaquepts (tine-loamy)	21458	2.6
Typic Ustochrepts	588095	81.8
Udic Ustochrepts	4866	0.7
Natric Ustochrepts	93946	13.1
Fluventic Ustochrepts	15108	2.1
Vertic Ustochrepts	16414	2.3

Source : National Bureau of Soil Survey and Land Use Planning Publication No. 45.

complex in nature. The plans with blinkers are not judicious for sustainable development of the state. Unfortunately farmers are being encouraged to economic viabilities for rice cultivation in waterlogged areas. No doubt, Punjab will get rid of waterlogging problem but after few years the water requirement for new varieties of rice will increase and this will halter skeleton the ground water resources. Moreover, the yearly adoption of rice cultivation will result in increase in iron deficiencies in soils.

Conclusion

An academic endeavour to protrude a concerned approach for the menace created by canal irrigation in south and southwest Punjab has been made. A restropective glue of the recent past of canal irrigation induced impoverishment has been adhered to irrational use and misuse of soil resources *vis-à-vis* mounting pressure of persistent ever-increasing requirements. Rhetorical suggestions in a winnowed voice have been made to lessen the impact in totality.

Acknowledgements

The authors are thankful to Mr. R.P.S. Chowdary in the preparation of the article. They express their deep gratitude to Prof. H.S. Mangat,

Department of Geography, Punjabi University, Patiala, and Prof. V.K. Asthana, Department of Geography, Kurukshetra University, Kurukshetra, who encouraged and inspired the present study.

REFERENCES

Bhangava, G.P. and Bhattacharjee, J.C. (1982) Morphology genesis and classification of salt affected soils. *Review of soil research in India.* Publication 12th International Cong. Soil Science, New Delhi.

Kennedy, J. (1920) *Canal Irrigation in Punjab*, Civil and Military Gazette Press, Lahore.

Krishna, M.S. (1968) *Geology of India and Burma.* Higginbothams, Madras.

Manku, D.S. (1998) *Geography of Punjab*, Kalyani Publishers, Ludhiana.

Mavi, H.S. & Tiwana, D.S. *Geography of Punjab*, National Book Trust, India.

Nayyar, V.K. (1990). Micronutrients in soils and crops of Punjab, P.A.U., *Tech. Bull.*

NBSSLUP (1981) *Soil Survey and Land use Plan of Patiala District*, Punjab. Report No. 441.

Roberts. W. & Faulkner, O.T. (1921) *Punjab Agriculture*, Civil and Military Gazette Press, Lahore.

Sehgal, J. (1995) *National Bureau of Soil Survey & Land Use Planning,* Publication No. 45, Nagpur.

Sharma, P.K. and Dev, G. (1985) Pedogenesis of soils in different physiograhic units in a riverine plain in Punjab. *J. Indian Soc. Soil Sci.*

The Second Citizens' Report (1984-85), Centre for Science and Environment, 1996, New Delhi.

Venkatratnam, L. and Rao, K.R. (1977). *Application of remote sensing techniques for soil studies.* NRSA documents No. 22.

11

Land Degradation Problems
Their Impacts and Remedial Measures for Sustainable Development in Indian Desert Region

G.S. Chauhan

Introduction

In the last few decades, degradation has become the most contentious ecological issue of almost all over the world especially of arid and semi-arid lands. The land degradation phenomenon has recognized as a global problem and it is on the top priority in the United Nations Environment Programme (UNEP) agenda because it has drawn the attention of almost all environmentalists, soil and agricultural scientists. Actually, desertification and land degradation are the key issues in the developing countries where declining land productivity is threatening the food production and environmental sustainability. So, there is a quite big challenge before world community to combat the demon of hunger and it can be only solved by managing and conserving our land resources especially degraded lands which are covering considerably a vast area of the globe. According to the agricultural scientists nearly 15 per cent of land has undergone degradation due to several human-induced activities. The issue of land degradation is also important as far as India is concerned. There is a severe problem of land degradation in South Asia in general and Western Rajasthan in particular.

In case of India, out of total geographical area (329 million ha) about 141 mha is subject to water and wind erosion. Apart from this, about 34 mha of the country is affected by special land degradation problems like water logging, alkaline and acidic soil, salinity, ravines and gullies, shifting cultivation, riverine and torrents etc. According to the Draft Report on 'Status of land degradation in India' by Ministry of Agriculture, the area affected by water and wind erosion has been assessed as 67.6 mha in 1994. However, presently the area under the same problem has further extended. Water logging and salinity are the major causes of land degradation through special problems constituting over 40 per cent of the land area affected by such problems.

In comparison to the other parts of the country, the land degradation problem is very critical and severe in Rajasthan particularly in the Indian desert region. Because all the arid and semi-arid regions of Western Rajasthan have agrobased economies and rapidly increasing human and cattle population have exerted heavy pressure on the limited and rapidly diminishing land resources. Hence, the problem of land degradation is rampant and very serious in both the arid and semi-arid areas of Western Rajasthan.

The other areas of dry lands where the problem of degraded land is rather severe and critical include the sand dominated and canal irrigated areas which together account considerable a huge area of Rajasthan. Thus, a vast area of Western Rajasthan is affected by the menace of soil and land degradation. In fact the Rajasthan state is one of the worst affected states of India, with regard to land degradation and soil erosion problem. According to an estimate about 16.7 mha area of the state is severely suffering from the problem of soil erosion only. Hence, keeping in mind the extent and magnitude of the problem that afflicts the desert region's large part, an urgent need has been felt to treat all such type of degraded and wastelands which arc lying without use since long time. By treating and managing all such degraded lands, productivity of various kind of crops can be increased up to the greater extent to satisfy the needs of contemporary society of 21st century.

The term 'land degradation' defines reduction in the productivity, capability and capacity of land due to certain processes mainly by human and biological activities. But the land degradation which has been defined by the UNEP is as 'land degradation' in arid and semi-arid and dry sub-humid areas resulting from adverse efforts of 'human impacts', and constutes a major environmental problem.

Study Region

The Indian desert region is one of the four distinct physiographic divisions of India, the three others being the Himalayas, the Indian plains and Peninsular India. The Aravalli ranges divide the desert region in two parts i.e. Western Rajasthan and Eastern Rajasthan. The eastern Rajasthan is comparatively more damp and fertile region from the Western Rajasthan. In fact, these ranges acted as natural abatement to resist the western desertic conditions to spread onward. These ranges work as an agent to protecting the south-eastern Rajasthan from attaining typical aridity. The Indian desert region which represents a major agro-climatic zone of the country consists of twelve districts. Although, there were eleven districts in the study region initially but in 1995, one more district has been created from the Sri Ganganagar district, which is right now known as Hanumangarh district. The region covers near about 2,08,751 sq. km area of Rajasthan state. It possesses around 61 per cent area of the state. The region is the world's most crowded desert, the population density is 30 to 77 persons per sq. km depending upon places, occasionally going down as low as 13 persons per sq. km. A very large part of population leads to nomadic or semi-nomadic life in all these parched lands.

The major characteristics of this hostile and dry region are severe wind erosion, high temperature, scanty and erratic rainfall, frequent droughts, high rate of evapo-transpiration, devoid of perennial river and sparse natural vegetation. The varieties of sand dunes are extended over a large area. The sand dunes are very common and peculiar feature of these dry zones. The sand dunes cover approximately 60 per cent of the area of these barren lands. Whereas droughts occur quite frequently accompanied by very low relative humidity and high wind velocity as strong as 136 km per hour. Dust storm locally termed as '*andhi*', are quite habitual in these hostile arid and semi-arid areas. The soils of the arid zone are mostly sandy to sandy loam consisting excessive permeability and these are generally suffering from lack of moisture, nutrient and micro-organism which limit the crop production and plant growth.

Land Degradation Problems and their Causal Factors

The issue of land degradation is very significant as far as the arid and semi-arid areas are concerned. Hence, there is a necessity to explore the status of our understanding on the various causes of land degradation hazard. There are number of causal factors for land degradation problems. Any single factor may not be directly liable for such a critical problem. In fact, land degradation is a result of many factors or a combination of factors which damage the soil, water and vegetational resources and restrict their uses or productive capacity. The most pertinent and dominant factors of land degradation in the study region are severe wind erosion, water logging, salinity, alkalinity, seepage, mining and quarrying activities, accelerating human and livestock population, defective and faulty land use followed by mismanagement of irrigation water. Apart from these, there are several other factors of equally importance for the menace of land degradation (Fig. 11.1) which have been directly and indirectly responsible for the land degradation hazard in the arid and semi-arid regions of western Rajasthan.

While the increasing demographic pressure has put the negative impact on land use pattern with rising demand of land resources for food, fuel, fibre and infrastructure development. This situation has too led to degradation to land resources. Frequent recurring droughts is on another important causal factor for the same problem. Because dry land regions are highly prone to the climate variability and in the desert eco-systems drought is a regular part of the natural cycles affecting productivity leading to land degradation and desertification. It is also relevant to mention here in the same context that the problem of land degradation and desertification is more serious in the desert margins where the resilience to adverse weather condition is very less in comparison to the semi-arid and dry sub-humid regions.

Despite lack of accurate figures, the extent of problem is clearly immense, and human population inhabiting in the arid regions which are most susceptible to degradation, are highly vulnerable to impoverishment as a consequence of the deterioration in their environment and ultimately responsible for the imbalance of ecological system of dry lands. The land and soil resources of Indian dry lands are facing numerous problems. But the most serious problems of these lands are mainly severe wind erosion, salinity, alkalinity, water logging, water erosion and shifting sand dunes, which have been discussed in detail along with their spatial extent in the following

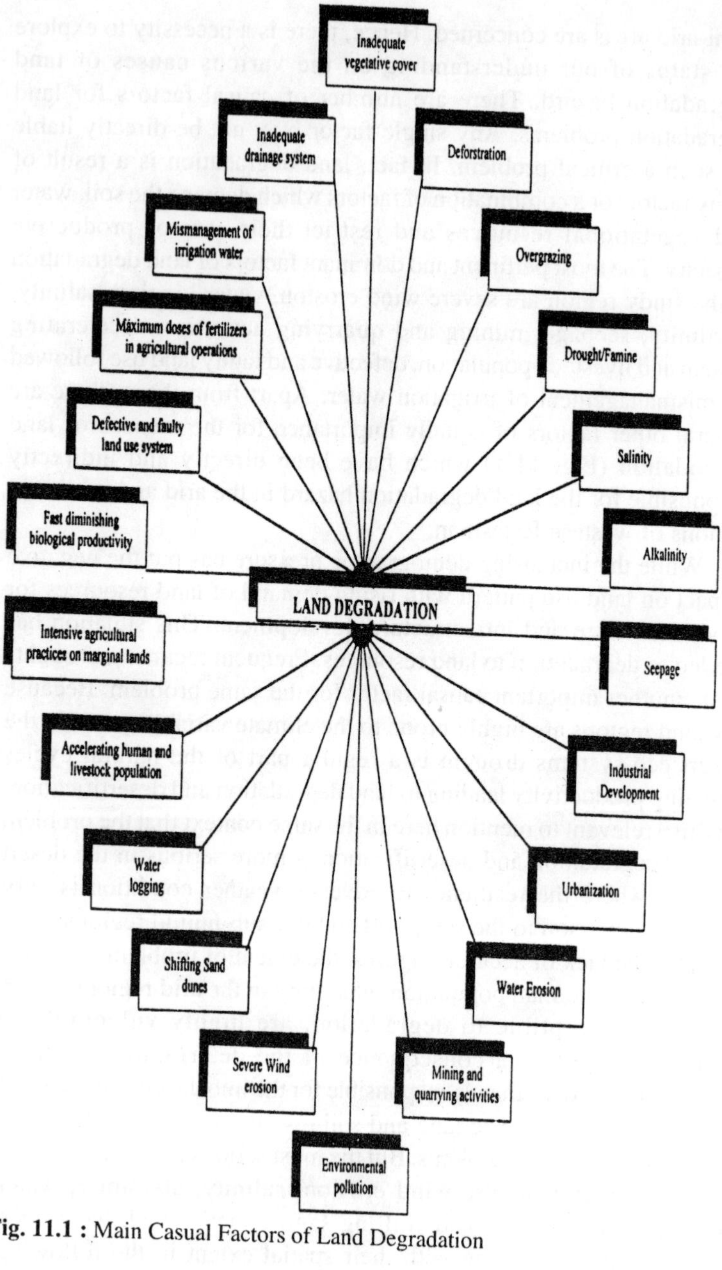

Fig. 11.1 : Main Casual Factors of Land Degradation

discussion. Besides these, various other problematic lands, their extent and magnitude have to be highlighted in the ongoing discussion.

Spatial Extent of Degraded and other Problematic Lands

With the vast geographical area, large population, complex and harsh climatic conditions, the Indian desert region is exposed to various types of land degradation problems traditionally. In fact, the process of land degradation related with natural climatic conditions in the area is quite prevalent and it is continued since long. But this process has got more spatial momentum from the second half of the 20th century due to vital and immense human activities and climatic changes. As it has already been mentioned that the problem appears solely but it can be seen from various dimensions and viewpoints. The land degradation hazards like salinity, water logging and severe wind erosion etc. are not a rare phenomena in the dry lands arena particularly in the arid and semi-arid areas. The Indian dry land is probably amongst the country's most hazard-prone areas with regard to land degradation.

The entire region is continuously suffering from the menace of various kinds of land degradation. The arid and semi-arid areas are presently suffering from tremendous and significant types of land degradation hazards. But out of them, following are some of the major type of land degradation problems that are universally rampant in the study region are as under.

Severe Wind Erosion

The wind erosion is a severe and a very complex problem of dry regions because of scanty vegetation, low and erratic rainfall and high wind velocity. The problem is most extensive in almost all arid districts of western part of the study region and currently several parts of the dry lands are badly affected by this problem. The tremendous human activities are straightly accountable for the same menace. Moreover, the most commonly responsible factor for inducing wind erosion are those remove massive activities operated by human and livestock which often vegetation cover and those that destabilize natural soil surfaces. According to an estimate about 129.26 lakh ha land of the country is affected by this problem. Out of it about 106.23 lakh ha (83.18%) land of the dry zone is affected by this hazard only. The estimated figures clearly depict that the total estimated wasteland (11.34% of the arid region) occupy maximum area under this problematic category.

Owing to the unpredictability of the rains, farmers are constrained to plough their fields during the dry summer months, thus, promoting wind erosion which further promote the process of land degradation on large scale. The wind blows away dust and silt on an average five to eight metres. The major dust storms sweep the desert domain each year during May and June. It is estimated that near about 5334 million tonnes of soil is eroded annually in the country by the various processes of land degradation.

Waterlogging

The area affected due to water logging in the country is about 8.51 m ha. In canal irrigated area, land degradation is a twin problem of water logging and salinity that have taken place extensively at number of places. It has been observed either in those areas where the intensive canal water is used for agricultural purposes or in those areas which are low lying and where rain water remains standing for a longer period. Apart from these two pertinent factors, there are other factors as well which trigger off the problem of water logging. The water logging hazard is very critical and serious in Pali district located along the foothills of Aravalli ranges. In the down south about 91.9 per cent area of the district is water logged followed by Ganganagar, i.e. about 4.28 per cent (Table 11.1). Apart from these two districts, Bikaner district is also suffering from the same hazard and around 3 per cent area of the district is currently under this problem. While in these latter two districts this problem has been observed quite alarming especially in those parts where Indira Gandhi Nahar Pariyojana water is introduced. In these areas due to the availability of irrigation water farmers are using excessive pesticides and insecticides in order to get better output. Thus, causing land degradation and environmental hazards. Along both sides of the canal at so many places problems of water logging can be seen very clearly and at numerous locations it

Table 11.1 : Indian Desert Desert Region : Spatial Extent of Barren Land and other Problematice Lands *(Area in Hect.)*

District	*Barren & Uncultivated land*	*Salt Affected*	*Water logged area*	*Marshy Swampy area*	*Gullied Ravinous area*	*Sandy Area*	*Land with or without Scrubs*
Barmer	49,375	31,000	-	-	-	6,55,313	94,562
	46,48	25.27	-	-	-	14.40	7.37
Bikaner	-	1,367	2,148	-	7,226	11,22,890	1,43,213
	-	1.11	2.96	-	12.4	24.67	11.20
Churu	351	2,006	-	-	-	28,945	4,147
	0.33	1.64	-	-	-	0.64	0.32
Ganganagar	-	-	3,098	-	-	2,90,993	46,334
	-	-	4.28	-	-	6.39	3.61
Jaisalmer	7,620	20,880	620	-	-	21,43,500	7,28,810
	7.17	17.02	0.86	-	-	47.10	56.84
Jalore	15,357	27,265	-	-	8,437	35,429	9,218
	14.46	22.23	-	-	14.52	0.78	0.72
Jhunjhunu	3,218	-	-	-	5,344	5,494	5,461
	3.03	-	-	-	9.20	0.12	0.43
Jodhpur	9,219	17,353	-	-	7,383	2,36,600	1,68,867
	8.68	14.17	-	-	12.7	5.2	13.20
Nagaur	970	18,155	-	-	5,602	21.078	67,157
	0.91	14.5	-	-	9.64	0.46	5.24
Pali	15,140	2,190	66,560	3,125	16,560	7,315	9,375
	14.25	1.78	91.9	100.00	28.5	0.16	0.73
Sikar	4,963	2.405	-	-	7,563	3.420	5.121
	4.67	1.96	-	-	13.00	0.07	0.4
Study Region	1,06,213	1,22,651	72,426	3,125	58,115	45,50,977	12,82,265
	23.70	68.62	100.00	100.00	14.39	98.94	49.16
Rajasthan State	4,48,120	1,78,716	72,426	3,125	4,03,784	45,99,531	26,08,297
	100.00	100.00	100.00	100.00	100.00	100.00	100.00

Source : Agriculture *In Rajasthan, Some Facts (1997).* Directorate of Agriculture, Rajastha, Jaipur.

has reached the critical point. About 72,426 ha (1997) area of the Indian desert region and about 7 mha area of the country has been estimated water logged while the area extensively covered by the Alkaline soils is very vast and it is nearly 3.58 m. ha followed by acidic soil, i.e. about 4.50 m. ha including saline soil is about 5.50 mha in the country.

Salinity is the result of the concentrated salt in the soil. Because of high rate of evapo-transpiration the upper crust of the land appears in the form of white colour resulting no use of such lands for agricultural operations. Out of the total salt affect area of Rajasthan state (1,22,651 ha) about 68.62 per cent is concentrated in the Indian desert region (Table 11.1). Apart from Ganganagar and Jhunjhunu districts almost lands of all districts of the Indian desert region are suffering from the menace of salinity. The problem has become more complex at number of places. The vast lands of Barmer, Jalore, Nagaur, Jaisalmer and Jodhpur are most salt affected districts of the dry region which account more than 90 per cent of total salt affected area of the study region. The similar problem is more acute in those areas where saline water is used for irrigational purposes in absence of good quality water. The estimated figures show that out of the total area affected by saline and alkaline lands, 7.28 lakh ha (10.16%) by area of the desert region falls in this category.

Water Erosion

Water erosion plays a significant role in the better-endowed rain fed areas. Water erosion is a considered a pertinent process of soil erosion in those tracts of dry regions which experiences occasional high rainfall in the canal irrigated area. Water erosion means not only increased loss of top soil but also for water and nutrient storage capacity. In dry lands, limited natural vegetation covers relatively thin soil and high rainfall intensities all promote runoff during precipitation events and therefore erosion. Apart from this, the intensive human actions too lead to accelerated water erosion in dry lands. Nearly 7.11 per cent area of the arid zone is affected by the water erosion out of the total degraded land in the country. Near about 1 per cent of semi-arid zone and about 7 per cent of the arid zone are subject to water erosion degradation.

Shifting Sand Dunes

Shifting sand dunes is perhaps one of the oldest feature of dry lands. It is a severe and predominant problem of the dry lands which further damages the properties and nutrient values of existing land resources consequently such physical hazards further excites the process of land degradation. Shifting sand dunes is a quite common feature of the arid and semi-arid areas and it has been observed mainly in those tracts where sandy soils are concentrated on large scale. The figures reveal that the problem of shifting sand dunes is very rampant particularly in the three districts, i.e. Jaisalmer, Bikaner and Barmer. However, it also taking place in certain other parts of the desert region but in rest of the districts its extent and magnitude is relatively less.

The expansion of sandy area is relatively very high in above mentioned districts. According to the figures of 1997, out of the total sandy area of desert region, about 47.10 per cent area is concentrated in Jaisalmer district, 24 per cent in Bikaner followed by 14.40 per cent in Barmer district (Table 11.1).That is why the problem of shifting sand dunes in these districts is relatively very high. Interestingly the same problem has also been analysed in those areas where the lands are found devoid of scrubs and vegetation cover. Under this category as well, all the above stated districts have been placed. A large percentage area of these same districts have been noticed without scrubs or with very poor vegetation cover. The similar type of phenomenon can also be seen in those parts of the dry lands where barren and uncultivated lands have been spreading. Because of the high temperature, low pressure is created resulting wind starts to blow. Due to blowing of wind at a rapid speed and devoid of natural vegetation the sand dunes often shift from one place to another. About 18 per cent (7.00 mha) area of the desert is totally affected by the shifting of sand dunes. The problem is more serious in the extreme western parts of desert where the vegetational cover is negligible and sand dunes concentration is very high.

Apart from these problems, other key problems of the region are poverty, degraded pastures, paucity of water, scarcity of fodder resources, frequent recurring droughts followed by intensive agricultural practices on marginal lands which ultimately trigger off the process of massive land degradation.

Impact of Land Degradation

Land degradation is very timely and important to all the countries across the world rich or poor owing to its adverse impacts on land productivity, food, livelihood security estimate change and eventually the quality of life.

The related statistical figures depict the localized and national impact of degradation which is now becoming quite clear. Loss of top fertile soil, pollution, shortage of water and biomass have categorically declined the agricultural productivity in many parts of these barren lands. The problem of land degradation poses a very serious threat to environment and its quality. Loss of fertility, severe deflection of soil nutrient, expansion of wasteland due to gullies and ravines. Such hazards further led to the environmental degradation and ecological crisis. The rise and growth in environmental degradation status of the concerned sterile regions further onward excites the phenomena of ecological imbalance.

In the current situation the human induced soil erosion has affected the one-fourth of the inhabited areas in the world. With regard to the impact of land degradation on agricultural productivity is very severe and critical. In most of the dry areas especially in the canal irrigated areas where the productivity of various crops was noticed relatively very high, that has tremendously declined either due to the existing problems of water logging and seepage or due to salinity or alkalinity. In such areas mainly in Ganganagar, Hanumangarh, Bikaner and some parts of Jaisalmer district the fertile soils which are located along the canal's sides or in the vicinity of IGNP, there the agricultural lands have lost their fertility and productivity. Similarly the shifting sand dunes have also the massive visible impact on the agricultural fields, grass lands, tanks, wells and reservoirs. At several places where the problem of shifting sand dunes is rampant, in such areas aeolian deposits often takes place on good and fertile agricultural lands and other water bodies. Moreover, land degradation damages and hit the ecological system mainly the flora and fauna wealth of the concerned areas. Due to this problem the physio-chemical properties changes, soil acidity rises, and drainage network is badly hampered and disturbed.

The indirect impact of land degradation such as droughts also hits the poor hardest. Because of recurring droughts people mainly the poor sections of society are forced to migrate from these dry areas to the other areas for their livelihood and sustenance of life. In such situation quite often social and water crisis coexists. Poverty in these cases is a direct outcome of economic destruction.

Besides the environmental degradation, health hazard also takes place in various parts of these parched lands. Due to the menace of water logging in many parts of the canal command areas, the natural vegetation has destructed on large scale and in many places it is severely depleting day by day. In totality, all such problems of land degradation pose a very severe and great impact not only on the agricultural productivity, soil nutrients value, vegetative cover, hydro-geological cycle, settlements and water bodies but almost on all anthropogenic activities that are operated by men. Precisely, it may be stated that these all forms of land degradation are indirectly pushing all the valuable land resources into a very deteriorating condition from where perhaps it would be more difficult and impossible to rejuvenate and renovate all such degraded lands. Hence, keeping in mind their massive impact on various socio-economic statuses and agricultural prosperity, all such degraded and problematic lands should be cured and treated very scientifically and technologically for the socio-economic betterment and food sufficiency for future generations.

Following are some of the key remedies which will not only help in achieving the sustainable development but also help to check the problem of land degradation in the study region:

- Massive plantation of indigenous plant species should be done along the road sides, railway lines, streams and canal sides.
- Afforestation programme should be launched to plant the trees on wastelands including hill slopes.
- Shelter belt plantation, wind brakes and fence line hummocks should be encouraged to resist the process of severe wind erosion.
- Biological and mechanical methods should be adopted to prevent soil erosion, i.e. strip cropping, crop rotation, application of green manure, vegetation cover and secondly, contour terracing to control gully erosion.
- Training should be given to the farming community to adopt the methods like mixture of tree, shrub, creeper and grass seeds that will help them to stabilize shifting sand dunes and to prevent Aeolian deposits in the cultivated fields, ponds, wells and other water bodies.
- Irrigation system should be improved by introducing (a) series of conveyance system (b) lining of distribution system and (c) improving efficiency by introducing a levelling sprinkler and drip irrigation.
- The plantation has the ability to lowering water table, so, trees

like *Eucalyptus, Caesarian* and *Lacunae* should be planted especially in the water logged areas.

- Efforts should be done to strengthen the embankments and vulnerable points of streams, canal and river system. To raise green and forest plantation should be encouraged especially on the foothill areas.
- Barriers should be fixed across the wind direction in order to check the menace of severe soil erosion.
- Conjunctive use of surface and ground water may prevent the ongoing problem of water logging.
- In the salinity dominated areas massive tree plantation programmes should be introduced especially in those areas which are of saucer shaped.
- In the seepage affected areas, wells should be constructed along the canal sides at a distance of at least 50 m, and after storing seepage water, it should be again dropped inside the canals through pump sets in order to drain out seepage water.
- More and more rural development programmes with certain aims and objectives should be set in motion in order to manage and conserve the valuable land resources.
- Balanced use of fertilizers and chemicals should be done in order to get the better output of all sorts of crops.
- Rural masses should be given incentives and they should be encouraged to adopt agro-forestry and promote farm and social forestry.
- Farmers community should be advised and suggested to use intensive green manure especially cow dung and animal waste to maintain the soil fertility and its biological productivity.
- More stress should be given on local tree species, which may check soil losses, and efforts should be made to develop grassland and vegetative cover to slow down the process of land degradation.
- Well accepted strategies including some of these rational should be adopted to conserve and management of rain water like water harvesting, check dams and tanks by breaking up hard strata which will further help in ground water recharge.
- In saline dominated areas for agricultural operations, canal water should be provided through water pipeline.

- Khadin cultivation be promoted in the recurring drought areas where canal water is not supplied to produce the rain fed crops and other purposes.

Concluding Remarks

Extreme land degradation has been analysed and observed in many parts of the study region. Undoubtedly, it has assumed serious proportion in various parts of the dry region affecting food production and productivity. Due to the problem, the existing land resources which are very rich potentially for biotic production are suffering from one type or the other type of land degradation.

Land degradation has emerged as a critical and major issue not only in the dry lands but all over the country. As compared to the other parts of India, the problem is more acute in the Indian desert region. According to an estimate about 174 mha or approximately 53 per cent of total geographical area of the country is affected by soil erosion and other types of land degradation. Out of the total problematic area around 82 per cent area suffers from water and wind erosion and rest suffers from problems like water logging and alkalinity. Because of the degraded lands production of most of the crops have been reduced to a greater extent. The bunch of all these quoted hazards have also been engrossing our agricultural scenario but further threatening not only environmental degradation but our food security and sustainable development. The land has limited carrying capacity of population beyond which there will be degradation and loss in productivity due to excessive use. Our land resources are facing unprecedented threat of land degradation and destruction.

So, action is needed at all levels, i.e. national, regional and local levels to prevent further degradation. Action plans should be made to regenerate the degraded lands. On the same line we need positive steps to extend the forest cover and to rejuvenate degraded lands. By managing and conserving all our land resources particularly curing our degraded lands we cannot only combat the demon of hunger but we can also solve the problem of food grains.

Therefore, we must have to look for proper rehabilitation measures to arrest the process of land degradation. Keeping in view, the gravity of the problem especially its spatial extent and magnitude, various programmes and policies have been implemented. The programmes which have been launched for the sustainable development of the desert region are afforestation, development schemes for pasture lands, shelter belt plantation programme, effective integrated schemes for proper land use, stabilization of sand dunes and selection of the suitable species for plantation.

REFERENCES

Abrol, I.P. & Bhumla, D.R. (1971) FAO World Resource Rep. 41 : 42.

Burch, G. *et al.* (1987) Biological and Physical Phenomena in Land Degradation in R. Dumsday (ed.), *Land Degradation Problems and Policies*. Cambridge University Press, London: 27-48.

Chopra, Kanchan. 1990. *Land Degradation: Dimensions and casualties in Agricultural Development in Punjab*, Issues in Resource Use and Sustainability, Vikas Pub., New Delhi : 19-35.

Dhruvanarayana, V.V. & Rambabu (1983), Estimation of Soil Erosion in India. *J. Irrig. Drain. Engg.* ASCE, 109 (4) : 519-34.

Gupta, J.P. and Gupta, G.N. (1981) A note on Wind Erosion from a Cultivated field in Western Rajasthan, *J. Ind. Soc. Soil. Science.* 29 (2), 278, 279.

Lal, R., Hall, G.F. and Miller, F.P., (1989), *Soil Degradation: I. Basic Processes Land Degradation and Rehabilitation* 1:51-69.

Mensching, H.G. *et. al.* (ed), (1984) *Resource Management in Dry Lands*, Rajesh Pub., New Delhi.

Sehgal, J. & Abrol, I.P. (1992) Land Degradation Status : India, In *Desertification Control Bulletin* (UNEP) Pub. : 24-31.

Shankar, Vinod (1977) Measuring Desertification Through Plant Indicators In : *Desertification and its control,* ICAR, New Delhi, pp. 192-193.

Shankarnarayan, K.A. (1977) *Impact of Overgrazing on the Grasslands.* Annals of Arid Zone 16 (3), 349-259.

Shankarnarayan, K.A., (ed) (1988) Wastelands Development and Their Utilization, Scientific Pub., Jodhpur.

Shankarnarayan, K.A. (1998) Ecological Degradation of the Thar Desert And Eco-Regeneration in Ishwar Prakash (ed). *Desert-Ecology, Scientific,* Pub. Jodhpur : 1-14.

Shankar, Vinod and Gupa, J. N. (1992) Restoration of Degraded Rangelands In J.S. Singh (ed.) *Restoration of Degraded Lands: Concepts & Strategies,* Rastogi Pub., Meerut, pp. 115-56.

Shankar, Vinod and Singh, J.P. (1999) Land Degradation and Rehabilitation Strategies Involving Agroforestry In : K.R. Solanki *et al.* (eds.), *Sustainable Rehabilitation of Degraded Lands Through Agroforestry*, NRCAF, Jhansi, pp. 12-17.

Singh, J.S. and Jha, A.K. (1992), Restoration of Degraded Land: An overview In J. S. Singh (ed.) *Restoration of Degraded Lands: Concepts and Strategies.* Rastogi Pub., Meerut, pp. 1-10.

Singh, P. (1994) Land Degradation - A Global Menace And Its Improvement Through Agroforestry In P. Singh, P.S. Pathak and M.M. Roy (eds.) *Agroforestry Systems for Sustainable Land Use*. Oxford & IBH Pub. Col., New Delhi, pp. 4-20.

Singh, R.B. and Chauhan, G.S. (1997) Environmental Regeneration of Wastelands Through Agro-forestry in Rajasthan, *Geographer*, Vol. XLIV, No. 1, pp. 33-48.

Singh, R.B. and Chauhan, G.S. (2001) *Drought Disaster in Indian Desert Region: Impact Assessment and Mitigation Strategy*, Paper presented in National Seminar on Drought Disaster Management on 21-22 November, Institute for Resource Management and Economic Development, Delhi.

12

Water Harvesting Practices at Village Pipaldhara, District Nagpur (Maharashtra)

An Impact Assessment

S. Girgaonkar and *N. Charthad*

Introduction

The Village Pipaldhara is situated in 20° 58′ N and 78° 54′ E in the Hingna tehsil of Nagpur district of Maharashtra (Fig. 12.1). It is a small village and occupies a total geographical area of 716.03 hectares. The land utilization pattern of the village reveals that 70.50 per cent area is under cultivation while 24.48 per cent is reported as culturable waste. The area comes under moderate drought prone zone (Girgaonkar, 2002, p. 7). Population of the village is 658 persons (2001).

Objectives

The objective of this investigation is to assess the impact of water harvesting practices on ground water level and agriculture.

Data Source and Methodology

The present micro level study is based on the primary data collected during the 1st week of November 2003. The study is confined to three farm sites at the village Pipaldhara. Direct observations, intensive field investigations, questionnaire, interviews of the farmers, local persons and officials of the Department of Agriculture engaged in the water harvesting activities have generated primary data. This has been supplemented by data obtained from various secondary sources. Meteorological data (1975-2000) have been collected from the Regional Meteorological Centre for Sonegaon Airport, which is the nearest observatory station from the village Pipaldhara. The collected data from all sources have been compiled, tabulated and analysed.

The Problem

In the village ground water is the only source of water for various uses. The village with moderate (1119 mm) rainfall, combined with very high rainfall variability during rainy months (over 35 per cent), rainfed nature of agriculture, inadequate irrigation (15%) have made cultivation difficult. Failure of crops, low per hectare yield, drying up of orange trees have become a regular feature. Lowering of water level due to disbalance between use and natural replenishment of ground water has further compounded the problem. Inadequate water availability for agriculture itself become a stumbling block.

Water Harvesting

Under National Watershed Development Project, therefore, various low cost methods of water harvesting have been adopted to recharge the ground water and to create irrigation potential (Table 51.1).

Table 12.1 : Water Harvesting Structures at Pipaldhara

Name of Structure	*Farm Pond*	*Dug Pond*	*Shunken Pond*	*Cement Nala Bund*	*Earthen Nala Bund*	*Vanrai Bund*
Total Number	8	27	5	12	14	7

Source : Office of the Agricultural Officer, Department of Agriculture, Hingna.

Physical Potentials of the Region

The area of the village is occupied by the Deccan Trap Basalt. It is found in a layered sequences of different lava flows. Eighteen such flows have been identified in the village (District Resource Map, GSI). Spot investigations have revealed that the area is underlain by highly weathered basalt (2 to 3 m thick) followed by highly fractured basalt rock at the bottom. Ground water occurs in this highly weathered mantle and cracked zone and forms a shallow acquifer. In the village soil varies between light grey brown clayey shallow soils to medium black and deep black cotton, clayey nature soil (Table 12.2).

Table 12.2 : Soil Classification of Village Pipaldham

(Area in hectares)

Shallow Soils (0 - 40 cms)	*Medium Soils (40 - 250 cms)*	*Deep Soils (250 cms +)*
59.52	380.11	40.00

Source : Project Report of Pipaldhara Watershed under National Watershed Development Project. Area includes part of the village lying in the Watershed.

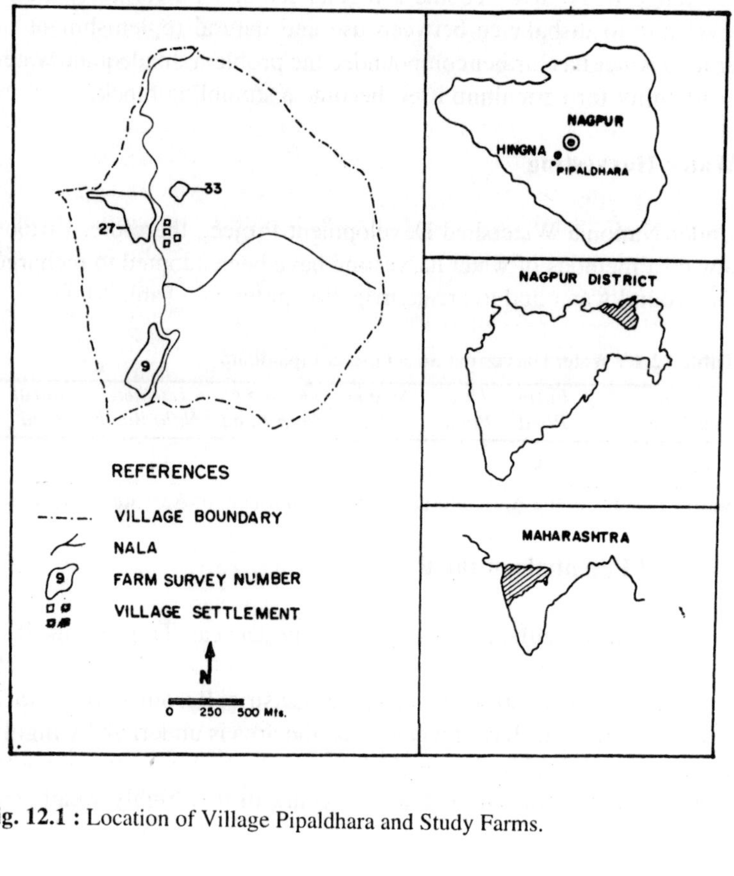

Fig. 12.1 : Location of Village Pipaldhara and Study Farms.

The field observations revealed that deep black cotton soil is only confined to nala banks. Major part of the area of the village comes under medium black cotton soil. Predominance of clay fraction in all types of soils indicates slow natural infiltration.

The climate of the village is characterized by dry sub-humid type with hot summers and mild winters. The annual average rainfall is 1112 mm distributed over 60 to 70 rainy days. From middle June to end of August is the period of high rainfall. In the month of September, rains are occasional. Annual rainfall is quite dependable with 21 per cent rainfall variability. However rainfall variability of rainy season months exceeds 35 per cent. It signifies long dry or wet spells during rainy season. High temperatures of 45° or more are witnessed during May, while low temperatures of 8° to 10° C are reported in the months of December to January. The village does experience a condition of drought. There is possibility of drought in every 5.5 years (Jain, S.K. and Jain, P.K., 2003, P.1D).

The village shows critical condition for ground water development and potentials. The post-monsoon season ground water level and trend for the last 5 years (Table 12.3) indicates there is much scope for artificial ground water recharge. Since five years, continuous downfall of water level suggests the natural recharge is not sufficient to saturate the aquifer optimally.

Table 12.3 : Post Monsoon Water Level (b.g.l.) at village Pipaldhara (October 1998 to October 2003)

Year	*1998*	*1999*	*2000*	*2001*	*2002*	*2003*
Water level (m)	1.50	1.60	2.10	2.55	2.68	2.15

Source : Ground Water Survey and Development Agency, Nagpur, b.g.l. - Below Ground Level, m - Metre.

Impact Assessment of Water Harvesting Structures at Study Farms

Site 1 — Farm Survey No. 9

The farm is located south of village Pipaldhara (Fig. 51.1) and ideally situated by the right bank of local nala. The size of the farm is 6.60 hectares. Lowering of water table and inadequate supply for irrigation was the basic problem. Therefore nearly 48 in of straight flow of nala, adjacent to the farm has been utilized to recharge the ground water. Nearly 40 m long Shunken Pond with 1 : 6 slope at upgradient side and 1 : 12 slope at down gradient side acrd maximum depth of 3 m has been designed and dug in the nala at the cost of Rs. 22000. The work was completed in the month of March 2000. As a result in the following rainy season rainwater flow was intercepted and resident time of nala water has increased. Besides that recharge period which is limited to monsoon has been extended further. Non-stop percolation of nala water has increased the water level of the adjacent wells by 3 to 4 m. During the time of visit it was observed that water level of the Shunken Pond and the well was same (2.2 m b.g.l.). It shows continuous inflow of water and acceptance of artificial recharge.

Slight slope and big size of the farm has also provided an option of farm ponds. Two farm ponds have been constructed at the cost of Rs. 24000 each, with 20 m × 20 m and 3 m. deep in March 2001, at the other end of the farm. Investigations revealed that these two farm ponds were fully filled-up three times in the rainy season. Thus nearly 72 lakhs litres of water has been received by the desaturated aquifer for recharge. At the time of visit both the ponds were half filled (6 lakhs litres of water each). Thus availability of water in large quantity in the farm itself has boosted the moral of the farmer in this moderate drought prone zone.

Before the commencement of the scheme jowar and pulses with 400 and 900 kg per hectare yield respectively, were the dominating kharif season crops. Rabi cropping was absent. Discussions with the respondent owner farmer revealed that there were incidences of crops getting spoiled due to inadequate availability of water in the recent past. Before 5 years his crops were even dried up. He was not therefore able to get what yields he could have obtained by properly irrigating fields.

With the availability of water since two years jowar has been replaced by cotton. Now cotton and pulses have become the dominating kharif crops. The fanner has also oriented sizeable area of his farm towards horticulture and vegetables (Table 12.4). Lime, sweat lime, mango, oranges have been planted last year which need perennial watering. Gram and vegetables dominate the rabi season. Availability of irrigation in time has resulted an increase in yield rates by 150 to 200 kg per hectare (Table 12.4). The respondent farmer has accepted that the rise in the yield of crops has been recorded because of ensured water supply at the right time.

This total transformation of cropping pattern is the result of persuation and establishment of water harvesting practices. Through these low cost efforts sustainability to ground water and agriculture has been provided.

Site 2—Farm Survey No. 33

This roadside farm occupies an area of 1.25 hectares (Fig. 12.1). Desaturated shallow aquifer of the existing dug well was the basic problem. Sufficiently thick, highly weathered basalt indicated sufficient underground storage space available for recharge.

Initially farmer was not ready to provide land for the construction of Dug Pond. But later he got motivated by the stunning results of other neighbouring farms. Already existed unused dug well ideally located at the down gradient side in the lowest spot provided an option of catching surface rainwater runoff by gravity flow for recharging well, through a small size storage pond. The size of pond was kept 5 m×5 m and 1.5 m deep. Rainwater runoff from the field made to drain into the pond. In order to make pollution free water before recharging the well a small filteration pot measuring 0.60×0.60 m and 0.60 m deep was constructed at the well side corner of the pond. Simple graded filter system with boulder at the top (0.20 m) followed by medium size metal pieces (0.20 m) in the middle and mixture of gravel and sand (0.20 m) at the bottom was used. Further a small P.V.C. pipe was placed to establish direct linkage of unpolluted water and to transport filtered water to the well. The work was completed in June 2002.

Discussion with the respondent farmer has revealed that the recharge well got filled up 9 times during this rainy season. Within a year of commencement water level of the well has come up to zero to 2 m from ground level during monsoon and post-monsoon period. During the time of visit the pond was 1/4 filled (12500 litres) with water and the depth to water level in the well was hardly 1.8 m bgl. Rise in water level clearly indicates acceptance of recharge.

Now the total demand of irrigation water of this small size farm is met from this saturated dug well. Discussion with the owner farmer

Table 12.4 : Salient Features of selected study farms

Sr. No.	*Particulars*	*Site-1*	*Site-2*	*Site-3*
1.	Survey Number	9	33	27
2.	Farm area (Hectare)	6.60	1.25	7.70
3.	Problem	Lowering of ground water level	Totally desaturated aquifer	Lowering of ground water level and drying of orange trees.
4.	Method of artificial recharge	* Shunken Pond-I ** Farm Pond-2	Direct recharge of well	Vanrai Bund
5.	Year of commencement	*March - 2000 ** March - 2001	March 2002	September 2002
6.	Rise in ground water level	3 to 4 m	4 to 5 m	2 to 3 m
7.	Past cropping Pattern			
a.	Kharif	Jowar, Pulses, vegetables	Cotton, Pulses	Cotton, Pulses, soybeans.
b.	Rabi	Nil	-	-
c.	Horticulture	Nil	-	700 orange trees.
8.	Present cropping pattern			
a.	Kharif	Cotton, Pulses (Tur) vegetables	No change	Cotton, Pulses (Tur), Soybeans. vegetables
b.	Rabi	Gram, Vegetables	-	Gram, wheat, vegetables
c.	Horticulture (No. of trees)	Sweet lime - 600 Lime - 300 Mango - 90 Orange - 600	-	700 orange trees.
		Past		*Present*
9.	Per hectare yield of crops considering all surveyed farms (Approximate)	Pulses - 950 Soybean - 750 Jowar - 450 Vegetables - 350		1200 1100 - 450

Source : Fieldwork - November 2003.

has revealed that a whole agricultural system of the farm used to suffer during drought years and at the time of long dry spells. Availability of water in right time did not spoil his crops this year. This year he is expecting a bumper crop of cotton and pulses with 20 to 25 per cent rise in yield per hectare. So far he has not made any changes in the existing cropping pattern.

Site 3 —Farm Survey No. 27

The farm (7.70 hectares) is located to the north-east of village Pipaldhara (Fig.12.1). A small nala flows to the east of the farm. Lowering of water level and drying up of orange trees were the major problems. Ground level deepens fast in the subsequent months, due to heavy withdrawal for irrigation. Since three years before commencement of the scheme owner farmer has experienced poor ground water condition and therefore inadequate irrigation.

Vanrai bund, a kind of a check dam has been constructed by public participation on the adjoining nala which has dependable flow up to the month of December. Cost of construction was merely Rs. 1600. As a result nala flow has been blocked and water has been stored for recharging adjacent well. A well gets continuous water for recharge from the bund. Thus immediate impact is felt on ground water level. Site investigations revealed that water is continuously pumped for 10 to 12 hrs for irrigation, but within the next 2 to 3 hrs water level again restabilizes.

Harvested water therefore has brought relief to the farmer and continuous irrigation is possible now during non rainy months. Thus through this simple and low cost water harvesting structure sustainable irrigation potential has been created. Enquiries revealed that emergencies created during long dry spells and routine water requirement is now easily fulfilled. Since two years agricultural economy of the farm is less affected. Before the introduction of the scheme jowar and pulses dominated the kharif cropping pattern and nearly 700 orange trees. Economy of the farm had suffered a severe set back due to drying up of orange trees. Winter cropping was absent. This year wheat and gram have been introduced along with vegetables in the rabi cropping. Standing kharif crops were good quality pulses and cotton.

The immediate impact is felt that now water is available in abundance in the farmers field in this moderate drought prone zone. With the introduction of rabi cropping area under irrigation is on rise and assured water supply to orange orchards. Increase in irrigated area and reasonably good yields of oranges as compared to recent past lead to an overall rise in the income of the farmer. Thus the farmer has been benefited by this simple low cost scheme.

Suggestions

Repeated desiltation is suggested for continuing the benefit out of these low cost structures. Here it is the responsibility of the farmer to clear siltation at regular intervals and should undertake minor repairs as and when necessary, otherwise it may hamper created irrigation potential. For continuing the benefit it is felt that there is an urgent need to organize farmers problem solving camps particularly related to siltation of ponds and other structures, otherwise in due course of time it may prove fruitless.

Conclusion

Thus it has been discussed that how introduction of water harvesting practices can improve ground water condition and crop situation in this rainfed moderate drought prone zone. These three farms have registered a significant change and improvement in ground water level and crop yields due to recently developed water harvesting practices. Significant irrigation potential has also been created. An effort has been made to commercialize agriculture through low cost water harvesting methods.

It is felt that adoption of water harvesting practices has made justice to the farmers and as a whole to the society and the environment. It has improved the environmental condition by arresting falling of ground water level. By using these practices efficiency and sustainability of agriculture and ground water level can be easily achieved under rainfed conditions.

Thus it is concluded that appropriate low cost and socially accepted water harvesting practices seems most logical approach of rain water management under rainfed conditions to improve ground water situation and optimizing agricultural production.

It is felt that water harvesting practices in the village Pipaldhara has set the standards of methods of rainwater management under tainted conditions. The village has become a success story in the region. There is an urgent need to replicate the system. Water is the precious gift of nature and it therefore should be used judiciously and with proper planning.

Techniques of Water Harvesting Structures Used in the Article

Shunken Pond

A saucer like huge pond is dug in the bed of a small nala where the course of a nala is straight.

Vanrai Dam

Sand and soil filled cement plastic bags are arranged one upon another and is used as a barrier across small nalas. Thus a low cost check dam is created.

REFERENCES

Girgaonkar, S.A. (2002) : *Drought Assessment for Vidarbha and Rural Development*, Unpublished.

Jain, S.K. and Jain, P.K. (2003) : *Rainwater Harvesting*, Central Ground Water Authority, Central Ground Water Board, Central Region, Nagpur.

13

Policy to Achieve Sustainable Land Use

Deep Shikha

Introduction

With the growing pressure of population communities and governments have become more concerned about the sustainability of development. The word 'sustainable' implies the notion of equilibrium. An equilibrium between economic growth and environmental preservation. We are already grappling with the problems and consequences of unsustainable development, over exploitation and depletion of natural resources and land degradation. Bunch of evils like decrease in productivity, shift in insect-pest and weeds spectrum, increasing deficiency of secondary and micro-nutrients in the soil profile, problems of water logging and salinity under irrigated areas, water population and continuous depletion of water table has been engrossing our agricultural system threatening our food security.

Such problems enhance the need of more careful planning so that no scope could be left to repent. To achieve the targets of sustainable development of micro-level planning is required so that a unit of a region can use its resources according to its own requirements by maintaining ecological balance.

Identification of Problems

As it is apparent in spite of all the efforts made by the government desired results are not achieved. Our farmers and officers face variety of problems which require immediate attention. These problems are identified after a general discussion with farmers, agricultural officers, BDO's and SDO's. The problems are :

- Mal-distribution of inputs such as seeds, fertilizers and pesticides. These facilities, in fact, are limited to the richer and awared farmers. Smaller farmers do not get or do not try to get these facilities. Reasons behind this are innumerable, such as : lack of money, lack of courage or approach, illiteracy, village politics, cast politics etc.
- Our farmers are even today devoid of technical knowhow regarding farming. Most of the farmers are still practising the unsystematic ways to get their yield although this facility is made available by the government at a stone throw distance.
- There is no doubt that the government has made many efforts to provide irrigation facility to the farmers but problems regarding irrigation are still there with a big question mark. At some places farmers are spoiling their yields by over irrigation and some farmers do not get water to irrigate their fields. The problems of water logging and seepage have become acute rather increasing day by day threatening the agricultural growth.
- Electricity supply is another giant problem being faced by the farmers. One can say that this problem is increased by the farmers themselves by the misuse of electricity. Yet most of the farmers don't get this facility when they require.
- Many advance level researches are being done by our experts. Unfortunately only a small number of farmers are able to apply such skills. Majority of them are working in their fields according to age old knowledge and skills because of so many reasons.
- The size of holdings in our villages is so small that machines cannot be used in these fields. Rich farmers having big fields are able to apply modern tools and techniques.
- Most of our farmers do not have enough money to buy or arrange quality seeds and other modern tools and techniques. Results of the loan facility made available to the farmers are not encouraging. Defaulty allocation of loans, misuse or wrong utilization of money cheating etc. are some common practices

which defeat the purpose. As a result poor farmers are still poor and they are not able to improve their condition.

- Our farmers have to face another serious problem, i.e. marketing of products. They do not get the proper cost of their products. This affects their economic condition and ultimately their food security. Farmer's dependence on government for the storage of their products give middle man a chance to exploit them by not giving the proper cost. A vicious circle of government officials and traders is existing that does not allow the farmers to get the proper cost and majority of farmers have to sell their products at the rates decided by the traders.
- Sometimes wrong decisions of government of importing food grain or sugar at the time when the yield in our farms is ready, also make the farmers suffer as government godowns are not available for most of the farmers, and in the absence of their own storage facility they have to sell their products at throw away prices.

Fore-said problems make one realize that the government plans are not benefiting the people for whom these are made. This is taking place possibly because either there are some loopholes in the plans or the failure of the implementation machinery. As a matter of fact government has made many efforts to improve the condition of farmers especially the poor farmers. A general survey reveals that their growth is quite slow although a little change is visible. We cannot wait and watch such a slow growth if we want to match our growth with the growing pressure of population.

Micro-Level Land Use Plan

Government has its own limitations to implement the plans as it cannot employ the number of people according to the size of population. Secondly, in spite of various efforts to encourage peoples' participation the success is not yet achieved in this regard. Keeping in view the fore-said problems and limitations a plan to accelerate growth is presented here in the form of a MODEL which suggests the joint efforts of people concerned, experts and government machinery.

With an effort to synthesize government machinery and the people

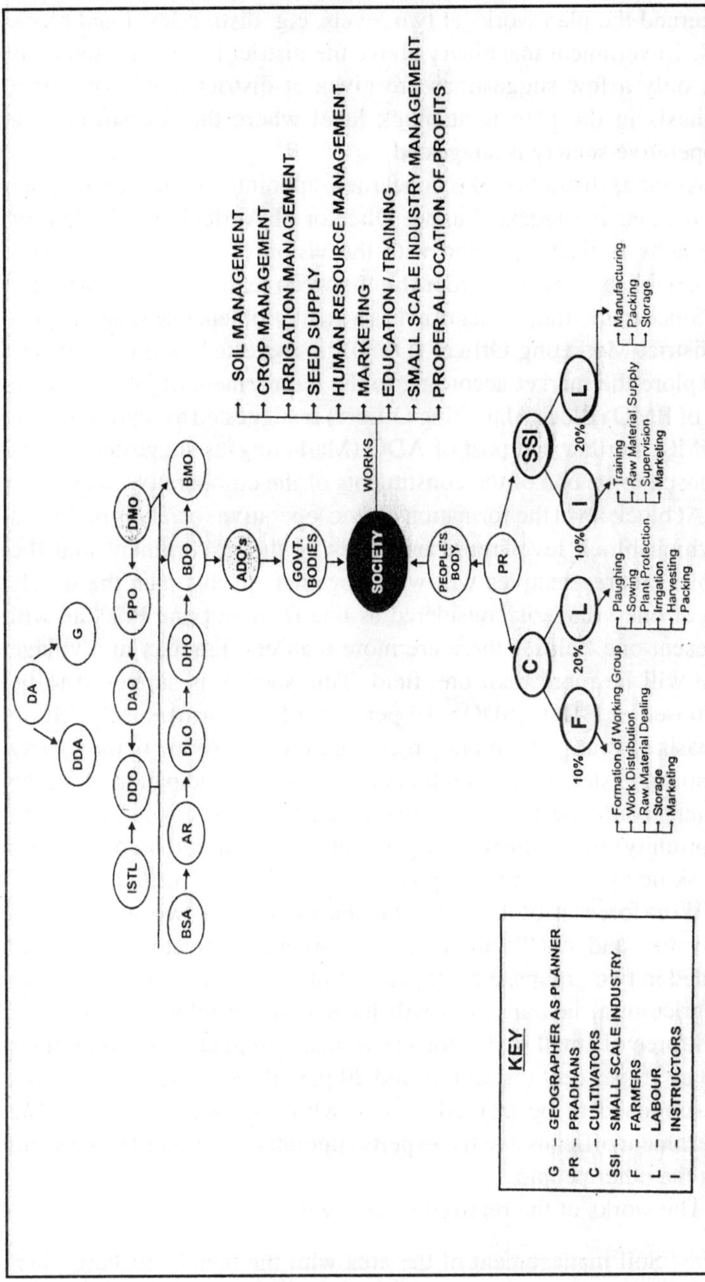

Fig. 13.1 : Policy to Achieve Sustainable Land Use

concerned the plan works at two levels, e.g. district level and block level. Government machinery above the district level will remain as it is, only a few suggestions are given at district level. The main emphasis in the plan is at block level where the formation of a co-operative society is suggested.

As far as district level is concerned, appointment of a geographer as a planner is suggested under Director of Agriculture. Student of geography is well equipped with the vision of regional disparities who establish a causal relationship between man and environment.

Since marketing is identified as a giant problem one separate post of District Marketing Officer (DMO) is suggested who is supposed to explore the market according to the requirement of his area. One post of BMO (Block Marketing Officer) is suggested as representative of DMO. Further one post of ADO (Marketing) is suggested who is supposed to be one of the constituents of the co-operative society.

At block level the formation of a co-operative society is proposed in which block level representatives of the government and the peoples' representatives will work together. In this plan the usable land of one village is considered as one field and one Pradhan will represent one field. If there are more than one Pradhan in a village there will be more than one field. This society is supposed to be composed of BDOs, SDOs, 10 per cent of the farmers, selected on the basis of stratified sampling method, (Recornposition of the society is also suggested after a certain period so that participation of all the farmers could be ensured. This will check the concentration of opportunity) in few hands, 20 per cent of agricultural labour, 10 per cent skilled people and 20 per cent labour for small scale industry.

Workforce at block level will be divided in two groups, e.g. cultivators and small scale industry workers. Cultivators are further divided in two groups, e.g. 10 per cent of the farmers and 20 per cent of agricultural labour. Both will have separate allotment of work. Workforce of small scale industry is again divided into two sections such as 10 per cent instructors and 20 per cent labourers. Instructors are supposed to be trained people who will get training by the government officials and the experts appointed. These instructors will train the other people.

The works of the referred society will be as follows :

- Soil management of the area with the help of incharge Soil

Testing Lab is to be done by this Society. Necessary instruments will be made available and farmers' training to deal with minor soil treatment techniques will be organized at block level.

- Crop management under the supervision of experts will be done by the Society so that proper crop in proper time could be sown to avoid the problems related to the crop combination, soil deterioration and market demands.
- Proper guidance of experts and execution of their directions by the government officials is must for the irrigation management. Steps to improve or check the misuse of this facility will be one of the major tasks of the society.
- Society will also look into the matter regarding the supply of quality seeds at block level which will be on the recommendations of DDA.
- Development can never be sustainable without proper management of human resource. Hence generating employment opportunities according to the ability of the people and availability of local resources will also be one of the main-tasks of the society.
- Promotion for small scale industry is another important work of the society. Setting up of the factory based on the raw material available, quality control, etc are to be dealt under this head.
- Compulsory education and training to develop the skills is also to be looked after by the society.
- Proper allocation of profits will also be one of the main concern of the society.
- Development of own storage facility of block level so that the products could be sold in right time, at right place with good deal of profit will also be the main task of the society in support of the growth of the area and of the people as well.
- In this way government and peoples' body will work together to identity the problem and to find out and apply the solutions at village level before the problem is in critical state.

Benefits of the Plan

With the help of the model discussed in the paper targets of sustainable development through sustainable land use will be achieved and area would be benefited in following ways :—

- Soil management will remain in the hands of experts (ISTL) and society as well, since the profit would be shared by the people, they will be interested to check further deterioration and to improve the quality of soil. Decisions regarding selection of crop, will be in the hands of experts. This will help to produce according to the requirement or demand and the farmers will be able to avoid the wastage of their energy and money.
- Supply of tools and instruments by the society will check the wastage of money by buying extra tools and instruments in an area.
- Possibility of over-watering and under-watering would be reduced as irrigation would be done under the supervision of experts (i.e. the trained people) and the crop would be watered in time in right quantity.
- The plan makes it easy for the government to supply good seeds to the fields according to the requirement of the area.
- Steps taken in the plan to generate employment opportunities will solve the problem of unemployment which will ultimately assure the food security to the people of the area.
- Set-up of small scale industries based on local resources and agricultural wastes as suggested in the plan will definitely promote the over-all development of the area.
- As marketing was seen one of the biggest challenges in our general field survey, an additional set up of government and peoples body regarding marketing has been suggested. Concerned officials will explore the local, regional, national and international market for the industries developed in the area and peoples body will look after the manufacturing according to the requirements of the market with the help of guidelines given by the government body. This will definitely improve the economic condition of the people and promote the sustainable growth.
- Since proper allocation of profits is based on the skills/ability of the people, awareness regarding tire quality of work would

be developed. People will learn not to adjust with the quality if they want to get a good deal of profits. Incentives to the government officials would make them to work sincerely.

Conclusion

In this way the plan presented here to achieve the targets of sustainable land use will definitely give way to the sustainable growth. If a village develops properly, an area, a region and ultimately the nation will develop. To attain the growth with ecological balance we have no other option left.

REFERENCES

Food and Agricultural Organization of United Nations (1992), *Production year book* 1991, Vol. 45, United Nations, Rome.

Bhalla, G.S. & Alagh, Y.K. (1979). *Performance of Indian Agriculture: A Districtwise Study.* Sterling Publisher, New Delhi.

Dantwala, M.L. (1978) Future of institutional reform and technological change in Indian Agricultural development. *Economic and Political Weekly*, Special Number, August 1978.

DES (GOI) *Agricultural Statistics at a Glance 1998, 1999 and 2000*, New Delhi.

Singh, G. B. (1998) *Natural resources management for sustainable agriculture in 21st century.*

14

Knowledge of Degraded Agricultural Land and Watershed Management

Syeeda Khatoon

DEGRADED AGRICULTURAL LAND

What is land degradation?

Land degradation (or "desertification") is used in the broad sense to mean impoverishment of the land by human activities and by natural causes (such as climate). The human activities may be referred to as casual agents. Land is considered "degraded" when its productivity is diminished[1] —Boradly speaking soil degradation may be of two types.

1. Physical degradation
2. Chemical degradation

These two types, of degradation mostly occur simultaneously. Chemical degradation is generally confined to humid areas having good amount of rainfall. This occurs in two ways :

- Chemical matter are dissolved and transported with surface running water.
- The leaching of chemical matters and nutrients down the soil profile thus creating the upper soil infertile and changing the existing soil structure.

Physical degradation is common in almost every part of the world whether humid or arid/semi-arid soil is eroded as well displaced away by the action of natural agents like water, wind, glacier etc. This mainly occurs as surface removal of soils. Sometimes negligible amount of finer soil or colloids also move down the soil profile and are deposited within the profile. It is cemented with an admixture of different chemical matters by physical degradation process.[2]

Desertification is not a new problem. It is a worldwide phenomenon affecting all the continents of the earth and gradually decreasing the natural resources of land. Desertification is the spatial extension of desert like conditions as a result of man's impact on the eco-system of semi-arid regions and involves a complex physical and geographical process having negative consequences on land use and ultimately leading to the strong disturbances of the natural ecological equilibrium.[3]

Desertification has been a problem in some areas since about 7000 BC. In fact, it is believed that land degradation contributed to, if not caused, the decline of some of the world's great civilizations, including those in Mesopotamia, Egypt, Greece, North Africa, and China. Desertfication is occurring everywhere in the world and even the most highly industrialized nations are not exempted from it. The United States, for example has had to face the degradation of irrigated lands and increasing problems with its ground water reserve. But the United States had considerable success in improving and restoring its forest and rangelands. The USSR has had similar success in Asia and has also effected improvement on its irrigated lands. In Mediterranean Europe, forest and woodlands have been brought into better condition. In Australia the advance of desertification has generally been halted. It is in the developing part of the world that desertification is accelerating most rapidly.[4]

Causes of Land Degradation

Land degradation caused by agriculture takes many forms and has many causes. Some of the most important types of land degradation include:

(1) Degradation related to overgrazing by livestock.
(2) Degradation related to soil erosion, here related to inappropriate cultivation practices.

(3) Degradation attributable to soil salinization (a build up of salts in soil that results from irrigation in certain situations).
(4) Degradation attributable to waterlogging (another problem related to irrigation).

Extent of agricultural land affected by degradation (globally)

The World Resources Institute (and the United Nations, in a 1990 survey) estimate that the productivity of 17 per cent of the earth's vegetated land surface has been decreased human activity. (This is equal to the land area of China + India, or about 1.2 billion ha). This figure includes land other than agricultural land. With regard to agricultural land in particular, it is estimated that about 10 million ha of crop land is lost each year (globally) to various forms of degradation combined with conversion to non-agricultural lands.

A 1990 survey by the Unified Nations resulted in the following estimates :

1. About one-fourth of the world's total crop land is affected by degradation severe enough to restrict its productivity.
2. About 15.6 per cent of this degraded agricultural land is "strongly" degraded—land whose "original biotic functions" (e.g., nutrient cycling) are largely destroyed.
3. Another 51.7 per cent of the world's degraded agricultural land is moderately degraded that is, exhibiting greatly reduced agricultural productivity.

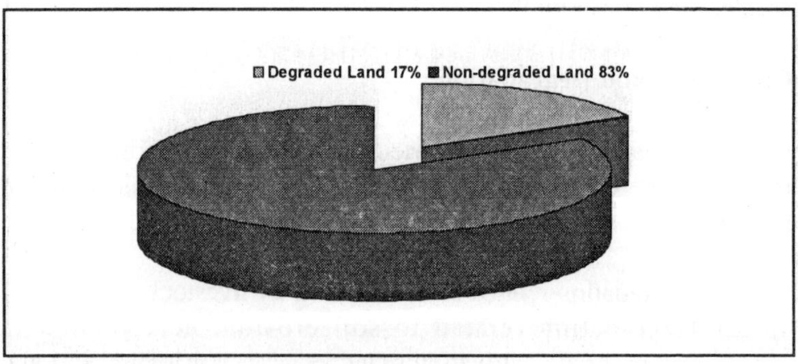

Fig.14.1 : Global degradation of Agriculture Land

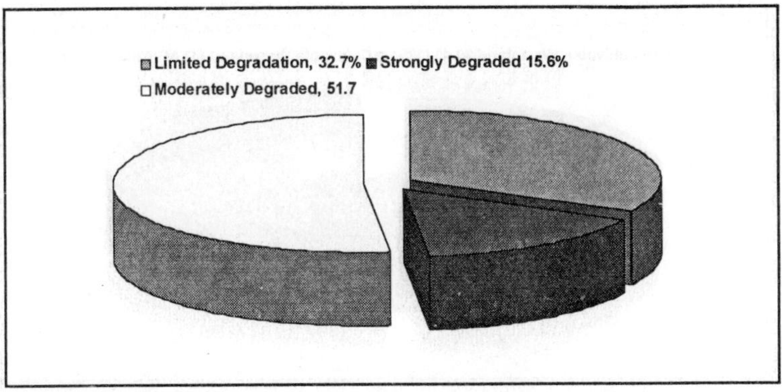

Fig.14.2 : Degraded Land

Applying estimates of productivity loss due to degradation to the categories given in the UN study, production averaged 17 per cent less on degraded lands than it would have been without the degradation. (This may have contributed to the slow down in grain yield growth over the past decade.)

Degradation is unevenly distributed around the world, varying by nation and by climatic region. For example, according to UNEP (the UN Environmental Programme) 60 per cent of the agricultural land in non-humid areas of the world is already affected by desertification (i.e. experiencing some loss of productivity).

Hence nearly 40 per cent of the world's agricultural land is seriously degraded, which could undermine the long-term productive capacity of those soils, according to scientists at the International Food Policy. Research Institute (IFPRI), who carried out the most comprehensive mapping to date of global agriculture.[5]

Extent of Agriculture Land Affected by Degradation in India

According to the inter-ministerial working group constituted by Prof M.G.K. Menon, the wastelands mean degraded land which can be brought under vegetative cover with reasonable effort and which is currently under utilized and land which is deteriorating for lack of proper water and soil management or on account of natural causes. Wasteland can result from inherent/imposed disabilities such as, by location, environment, chemical and physical properties of the soil or financial or management constraints (NWDB, 1987).

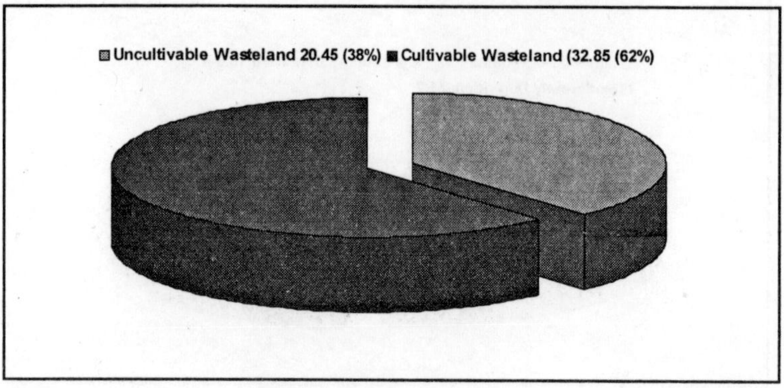

Fig.14.3 : Wastelands in India (in Milion Hectares)

It has been reported by NRSA (1985) that in India out of total 53.30 million ha of wastelands 32.85 million ha or 61.5 per cent is under cultivable wastelands and 20.45 million ha or 38.5 per cent under uncultivable wastelands. The greatest extent of wasteland in the country is in Rajasthan, 12.94 million ha, followed by Uttar Pradesh, 431 million ha, and Gujarat, 33 million ha.

Based on the visual interpretation of landsat data on 1:1,000,000 scale, the salient characteristies of the following major types of wastelands in India have been highlighted by Gautam and Narayan (1988).

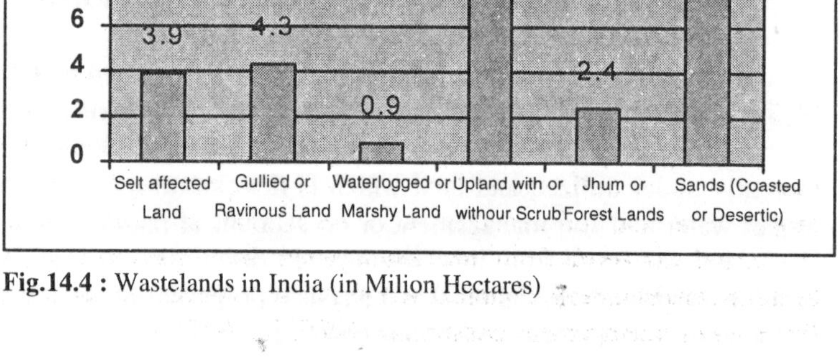

Fig.14.4 : Wastelands in India (in Milion Hectares)

Salt Affected Land

It has been estimated that 3.9 million ha of land is salt affected. The highest concentration of this type of wasteland is in Gujarat followed by Uttar Pradesh and Punjab. The Category of wasteland in the states like Himachal Pradesh, Jammu and Kashmir, Assam, Tripura, Manipur, Mizoram, Arunachal Pradesh, Meghalaya and Nagaland is either totally absent or is concentrated in small pockets due to hilly terrain. These wastelands appear in white to light blue tone and fine to mottled texture on landsat false colour composites mosaic of Rajasthan and adjoining regions.

Gullied or Ravinous Lauds

This category of wasteland comprising of an area of 4.3 million ha is largely associated with the rivers like the Yamuna, the Chambal, the Tapi, the Krishna, the Tungabhadra, etc. The gullied ravinous areas could be easily separated and mapped from the interfluve areas on-landsat false colour composite image. Gullied lands appear in light yellow to bluish green tone whereas the interfluve areas due to the stand of crops in wet season exhibit red tone on above landsat image.

Waterlogged or Marshy Land

This category of wasteland constitutes 0.9 million ha of which 75 per cent is in West Bengal, Uttar Pradesh and Bihar. Impeded drainage conditions, high water table seepage from-canal and reservoirs and presence of impervious layer at shallow depth are responsible for the development of this type of wasteland. Waterlogged or marshy areas in different shape and size are distinctly visible on landsat false colour composites mosaic of Rajasthan and adjoining regions.

Upland with or without Scrub

This is the most dominant category of wasteland, which occurs throughout the country and constitute 11.8 million ha of land. It largely occurs in Maharashtra followed by Andhra Pradesh, Karnataka, Madhya Pradesh and Rajasthan. These wastelands appear in light yellow to dark yellow tone with red patches on landsat. False colour, composites mosaic of Rajasthan and adjoining regions. In summer season it becomes difficult to separate the fallow lands from uplands due to similar reflectance. In wet season when the fallow lands are put under cultivation; the uplands with or without scrub could be easily separated and mapped front fallow lands.

Jhum or Forest Lands

This category of wastelands by and large occurs in the North eastern region comprising of Assam, Arunachal Pradesh, Mizoram, Manipur, Meghalaya, Nagaland and Tripura. The total area under this category is 2.4 million ha and Manipur has the largest area followed by Madhya Pradesh, Assam, Mizoram, etc. It is easy to distinguish forest blanks amidst forest on land sat false colour composites musaic due to their white tone, shape and size whereas the forest areas appear in red tone.

Sands (Coastal or Desertic)

This is second largest category of wasteland which comprises about 10.5 million ha of land. The coastal sands lie along the eastern and western coast of Peninsular India whereas the desertic sands occur in North-western parts of the country. The maximum area under desertic sands is concentrated in Western Rajasthan followed by Punjab and Haryana. Both coastal and desertic sands due to their location, shape, size and patterns are distinctly visible on landsat false colour composite mosaic of Rajasthan and adjoining regions in white to grey tone with red patches.[6]

Impact of Land Degradation

The evidence compiled by IFPRI suggests that soil degradation has already had significant impacts on the productivity of about 17 per cent of the globe's agricultural land. Combining the new map of agricultural land with existing expert assessments of soil degradation suggests drat almost 75 per cent of crop land in Central America is seriously degraded, 20 per cent in Africa (mostly pasture), and 11 per cent in Asia.

"The economic and social effects of agricultural land degradation have been much significant in developing countries than in industrialized countries," says Dr. Serageldin, chairman of the Consultative Group on International Agricultural Research (CGIAR). "These are precisely the regions where the greatest growth in food production will be needed, but where all indications are that achieving such growth will be the most difficult."

The analysis of the world's agro ecosystems comes from satellites, maps, and tabular data sets. IFPRI's scientists have undertaken this project in partnership with the World Resources Institute (WRI) as part of a larger international initiative, the Millennium Ecosystem Assessment, a comprehensive multi-year, scientific assessment.

According to Dr. Per Pinstrup-Andersen, IFPRI Director General, threats to the world's food production capacity are compounded by three disturbing trends :

(a) 1.5 billion additional people will be on the planet by 2020, almost all in poorer developing countries;
(b) The natural fertility of agricultural soils is generally declining; and
(c) It is increasingly difficult to find productive new land to expand the agricultural base.

Competition for water will further magnify constraints to food production. While inputs and new technologies may succeed in offsetting these declining conditions for the foreseeable future, the challenge of meeting human needs may grow ever more difficult over longer periods of time.[7]

Combating Desertification

Action against Desertification cannot however yield results unless the government of respective country perceives it as an integral part of their established plan for social and economic development. The key to combative measures is to be found in proper land use remote sensing coupled with ground surveys can be extremely useful for monitoring desertification and related natural resources.[8]

The measures like appropriate levelling, terracing and

embankment of fields, safe disposal of runoff, gully, plugging, afforestation and growing of crops, which have soil binding capacity, regulation of grazing etc. are appropriate for sustained development.[9]

The measures may reduce erosion to an acceptable level. They may be selected for their ability to deal with specific process of erosion which have been shown to be important in a particular area. They nary be appropriate to the local farming system in terms of their level of technology and compatibility with existing farming practices. They may be economically justifiable either by cost benefit analyses in the short term or by retaining productive soils in the long term.[10]

Hence sustainable land use means the best use of the land secured with the formulation and implementation of comprehensive action plans based on mapped percentage, land used inventory and large scale land use maps. These plans taken into account sustainability, participation of rural masses and the interest of the Nation which depend on the local situations and the problems.[11]

Most of India is blessed with abundant monsoon rains, its water supplies are strained by unchecked population growth and rapid development. The over-pumping of underground aquifers and the relentless cutting of trees, which trap and store water, have upset the hydrological balance turning the subcontinent into a refuge of floods and droughts whatever water is available and is sometimes contaminated by sewage and industrial waste exposing those who drink it to diseases.[12]

To educate farmers on efficient water use and management, a pilot project demonstration is being laid out is some villages. The construction of field channels levelling of land and adoption of cropping patterns are to be demonstrated. Then the task of the IADP project staff engaged on water management programmer is four fold.

1. First to select a village and to persuade the villages to participate in this programme.
2. Second to undertake a thorough survey of this village for laying out the outlets, drop structures and field channels.
3. Third to install the new system.
4. Fourth to undertake responsibility for repairing and maintaining the entire system.

Once a village has been selected for the programmes by enlisting the co-operation of the farmers a 500-1000 acre area can be taken up for laying out the system for demonstration. The IADP water management staff provides all technical help such as the survey for the laying out field channels on in scientific basin and designing the outlets and drop structure.[13]

WATERSHED MANAGEMENT

What is Watershed Maim ement?

A watershed is a drainage area on earth's surface from which runoff, resulting from precipitation flows past a single point into a lesser stream, river, a lake or the ocean. In other words, it is a land surface (body of soil) bounded by a divide which contributes runoff to a common point. A positive water accretion to its upper boundary is in the form of precipitation and a negative accretion is in the form of evaporation. There can be drainage laterally or vertically, when water runs out of it, and leakage, when there is a perched water horizon. Generally Watershed and Drainage basin are used synonymously.

Watershed Management involves management of the land surface and vegetation so as to conserve and utilize the water that falls on the watershed, and to conserve the soil for immediate and long term benefits to the farmer, his community and society.[14]

Watershed Management implies the wise use of all soil and water resource so as to provide a clean, uniform water supply for beneficial use and to control damaging overflow. The term is very nearly synonymous with soil and water conservation with the difference that the emphasis is given on flood protection and sediment control also besides maximizing crop production.[15]

Social and hydrologic factors are perhaps the most important aspects of Watershed Management.

The Concept of Management by Watersheds

Management by watersheds will require land and resource management programmes to be changed. First, watersheds must he recognized as basic ecosystems and managed as ecosystems to secure favourable conditions of water flow. All programmes must be conducted to achieve this goal.

Second, a truly interdisciplinary approach is crucial to the programmes success. Watershed scientists must lead the way, but all others must be part of the watershed management team.

The watershed is a basic ecosystem. It synthesizes inputs of water and energy with its physical, chemical, and hint is systems (geology, soils, terrain, and vegetation) to produce an array of landforms and stream channels and a range of streamflow and sediment yield characteristics. The balanced range of conditions sustained by the hydrologic function of a healthy watershed includes :

- Integrity of streamflow such that the magnitude and frequency of floods and base flows is not upset. Healthy watersheds behave like a "sponge" and exhibit high rates of infiltration that results in minimum surface runoff. Most precipitation soaks into the soil, which tends to retard flooding, recharge ground water, and regulate streamflow
- Integrity of (lie fhivial system expressed as stable stream networks and channels and a balance between runoff and sediment yield. In healthy watersheds; the stream network is not expanding through gully erosion, streams are not aggrading not degrading, channel form is maintained over time, and stream banks are well vegetated.
- Integrity of water quality and aquatic habitat. Healthy waters exhibit healthy streams that support productive, diverse, and stable populations of aquatic life and display a balanced range of habitat features such as depth of pools, composition of substrate, and squence of pools and riffles.[16]

Objectives of Watershed Management

Main objectives of the Watershed Development Programmer are to conserve and improve, the natural resources, economic upliftment of the marginal conservators and landless labourers by increasing the employment potential through long-term sustainable agricultural farming.[17]

The overall objectives of the watershed development programmes may be outlined as below :

1. Recognition of watersheds as a proper unit for wise utilization and development of all lands. The land should be treated in accordance with its peculiar need and by method that will control soil erosion, conserve water, encourage wildlife improve farm income and prevent flood damage to agricultural lands.
2. Retardation and prevention of floods through small multipurpose reservoirs and other water impounding structures at the head water of streams and in problem areas.
3. Provision of an abundant water supply for domestic, industrial and agricultural needs.
4. Abatement of organic, inorganic and soil pollution.
5. Broad expansion of recreational facilities, i.e. picmics and camping sites with more lakes and stream suitable for boating, fishing or swimming.
6. Utilisation of natural local resources for improving agriculture and allied occupation or industries (small and cottage industries) so as to improve socio-economic conditions of the local residents.[18]

Planning Watershed Management

Following discussion should be useful in preparing a sound watershed Management Plan.

A. *Inter-disciplinary team work in watershed, management planning :* Watershed planning is a coordinated analysis by a team of technicians representing various disciplines. The principle disciplines are hydrology, geology, engineering, soil science, agronomy and economics. Preparation of a work plan requires the input of considerable effort by various technicians, scientists and economists.
B. *Size of tie manageable watershed :* The evident question of how big the watershed could be for planning can be solved by past experience and national guidelince. The size can be adjusted depending upon the problem involved. In gully control and drainage problem areas the watershed units may be smaller than in flood problem areas. Where floods are

foremost problems, several smaller units may be combined until upstream problem areas are included within the watershed areas.

C. *Demarcation of priority watersheds on the basis of vulnerability of erosion :* A big watershed may he divided into several smaller viable sized sub-watersheds. The watersheds may be given priority ratings on the basis of erosion indices and delivery ratios for locating highly vulnerable mini-watersheds or sub-watersheds. The high priority rated subwatersheds require prime attention.

D. *Preliminary data for management planning :* Principle factors, that affect the operation of individual watersheds, and which must be studied before management plan could he developed are — shape of the watershed. Topography and slope of the land soils and their characteristics, amount of precipitation and storm patterns, land use and cropping pattern on watershed, and size of the watershed. Necessary data for interpretation includes :

(a) Meteorological information
(h) Hydrological Information
(c) Soil and Land use data
(d) Economic and social data
(e) Summary of management plan : The plan should contain:

1. Description of the watershed
   (i) Physical data
   (ii) Economic data
2. Watershcd problems
   (i) Flood water damage ,
   (ii) Sediment damage
   (iii) Erosion damage
   (iv) Problems related to soil and water management
   (v) Problems related to crop management
3. Works of improvement to be installed
   (i) land treatment measure

(ii) Structural measures
(iii) Crop management measures
4. Comparison of benefits and costs

Socio-Economic Aspects of Watershed Management

A watershed is the natural unit for economic management and, therefore, in its management beside the government view point, the view point of individuals and communities, who live in the watershed should also be considered. The people have their needs, custom values, and therefore the measures taken for treatment of the watershed should have compatibility with the needs of the people. Proper management of watershed therefore requires not only fair knowledge and application of scientific techniques, but also substantial investment of labour and capital whose returns may be high only in the long run and thus incompatible with the perspective of the farmer. How to motivate the farmers to make the necessary investment and sacrifice is a socio-economic problem.

The most important economic aspect of watershed management is to determine whether the investment decisions (on such programmes are rational. There are alternative techniques of soil conservation, water harvesting, storage and conveyance of water, alternative cropping patterns and technology package for making optimum use of available water, etc. It is the job of the economist to evaluate these alternatives and suggest the efficient one. Several techniques like benefit cost analysis, interval rate of return, etc. are available for evaluating such projects.

Therefore to study the economic feasibility of watershed management and safeguard the individual benefits of the beneficiaries as well as the social benefits, the following have to be taken into consideration :

(a) Social objectives and interests of general public as custodian of natural resources
(b) Development of backward areas
(c) Growth of ancillary industries
(d) Channelisation of income of agricultural sector to productive use

(e) Production of wages consumer goods
(f) Encouragement of cooperative objectives
(g) Generalization of employment.[19]

Watershed Management in India

Soil and water conservation programme is being implemented in critically degraded watershed ranging from 2000-4000 hectares in areas. The areas being delineated by All India Soil and Land Use Survey, New Delhi. This in 1988-89, an area of 23.04 lakh hectares has been treated under this programme at the cost of Rs.279.47 crore. This scheme, however, has continued up to the 7th Plan period. The conservation work and the area thus served are surveyed and evaluated by the Soil and Water Conservation Division of the Ministry of Agriculture, Administrative Staff College of India, Hyderabad and Agriculture Finance Corporation, Mumbai, etc.

Another centrally sponsored scheme of Integrated Watershed Management in catchment of flood-prone rivers of the country was launched during the 6th Plan period. In this plan period eight rivers of the Gangetic Basin covering seven states and one union territory was treated. By this about 4.21 lakh hectares of land was treated till 1988-89 covering an expenditure of Rs. 78.18 crore.[20]

Through a range of schemes and programmes, the government is investing over US $ 500 million per year into the rehabilitation of micro watersheds. Within semi-arid areas, one may often find co-existing (but rarely interacting) programmes under the auspices of several different agencies, including the ministry of Rural Areas and Employment and the Ministry of Environment and Forests, as well as various bilateral and multi-lateral donors.

Under the first of these, the National Watershed Development Programme for Rainfed Areas was formulated in 1990. This programme has a budget of Rs. 1338 crore under the Eighth Five Year Plan and by 1994 has covered 2554 micro-watersheds.[21]

REFERENCES

1. "IFFRI", "WRI" & "CGIAR," *Degraded Agriculture Lands Threatens World's* Food Production Capacity (www. wri.org; www. oriar.orgifpri).
2. De, N.K. Ghosh P. *India, 4 Study in Soil Geogrpahy*, Sribhumi Publishing Company, Calcutta, 1993, pp.137-38.

3. Jain, J.K. *Combating Desertification in Developing Countries* : Scientific Publishers, Jodhpur, 1986, Preface (X).
4. Jain, J.K. *Combating Desertification in Developing Countries : Ibid.* Preface (X).
5. Dawnloaded from internet site; www.orst.edu.
6. Patel, A.N., Singh, Surendra, *Remote Sensing Principle and Application* : Scientific Publishers, Jodhpur, 1992, pp.137-39.
7. Down loaded from internet site www.ogiar.org.ifpri.
8. Jain, J.K. "Combating Desertification in Developing Countries" *op.cit.*, Preface p. (XIII).
9. Singh, Ravindra Pratap (Ed.) : *'Environment and Charge'* : Singh Y Imo : '*Soil Land use and Conservation*; Rajesh Publicatons, New Delhi, 2002; p. 112.
10. Morgan, R.P.C.; *Soil Erosion and Conservation,* Edited by Davidson, Longman Group. U.K. Limited, 1986, p. 253.
11. Singh, Jasbir (Professor Emeritus) *Agriculture development in South Asia : A Comparative Study in the Green Revolution experience*, National Book Organization 1997, p. 578.
12. Singh, Gaganchand : 'Water Harvesting for Community Development'; *op.cit.*, p. 97.
13. Kumari, P.K. *Economics of Water Management*, Heritage Publishers, New Delhi, 1997, p. 12.
14. Tripathi, R.P., Singh, H.P., *Soil Erosion and Conservation*, Wiley Eastern Limited, 1993, p. 245.
15. *Ibid.*, p. 247.
16. Maxwell, James : *Ecosystem Management by Watersheds*, Regional Hydrologist, USDA-Forest Service, Rocky Mountain Region, Lakeview, Co. (www. watershed.org).
17. Singh Ravindra Pratap; 'Environment and Change'; Kushwaha, R.A.S.; 'River Basin Planning for Sustainable Development'; *op.cit.*, p. 119.
18. Tripathi, R.P., Singh, H.P., *Soil Erosion and Conservation*; *op.cit.*, p. 247.
19. *Ibid,* p. 263.
20. De, N.K. and Ghosh, P., *India : A study in Soil Geography, op. cit.*, p. 182.
21. Farrington, John; Turton, Cathryn; James, A.J. (Edited by); 'Participatory Watershed Development (Challenges for the twenty-first century)'; Oxford University Press, New Delhi, 1994; p. 7.

15

Land Degradation in South Asia
Imperative for Sustainable Agricultural Practices

Sanjay Kumar Singh, D.N. Bajpai and *Nalin Kumar Singh*

For entire humanity, very few natural resources are as valuable as land, without which even the existence of human being on this planet cannot be imagined. All the essential commodities needed for the survival of life is obtained from land itself. But this valuable resource is under severe strain. Land is a renewable resource so long as it is used according to its potential. The long-term sustainabiliry of land productivity and its potential depends mainly on the methods, types and intensity of land use. Unplanned and unscientific exploitation of land leads to its degradation and non-renewability.

South Asian region is confronted with severe problem of land degradation. A well carved region, namely South Asia comprising India, Pakistan, Nepal, Bangladesh, Bhutan, Srilanka and Maldives has distinct physical and cultural homogeneity. Most countries of the region share similar ecosystems—for example, the Himalayan ranges, Thar desert, Indo-Gangetic Plain, long coast lines or the coral island ecosystem are common to more than two countries of the region. The farmers of the region dance to the tune of monsoon down pour and the Himalayan rivers provide nectar to a large population of the region. The geographical unity, the spatial proximity and the cultural legacy have all carved South Asia into a separate and distinct region.

All the countries of the region are developing economies and have almost similar type of problems. These countries constitute one of the most densely populated regions of the world and are amongst the poorest with regards to developmental indicators. Though structural changes have been pursued by these countries during the post-independence era, agriculture still dominates over the nascent industrial sector of the region.

The countries of South Asia have agro-based economies. The escalating human and cattle population has posed a threat to the limited and rapidly diminishing land resources of the region. South Asia, with around 23 per cent of the world population and only 3.4 per cent of the world's total geographical area is one of the most densely populated regions of the world (UNDP, 1998). This high demographic pressure of the region has posed a major developmental threat and has manifested its impact on the land use pattern which has undergone continuous changes due to dynamic and complex interplay of socio-economic, political and technological forces. With a high total population of 1.29 billion (UNDP, 1998) and high growth rate of population, there is an increasing demand of land resources for food, fodder, fuel, fibre and infrastructure development.

LAND USE PATTERN IN SOUTH ASIA

The land use pattern of South Asia clearly indicates the impact of increasing population. Major part of the land is devoted to cropland and still there is pressure to expand agriculture to feed the burgeoning population. In South Asia croplands constitute 50 per cent of the total land area which is five times the world average. In countries like Bangladesh and India this figure is 69 and 57 per cent respectively (FAO, 1994).

In case of permanent pasture land the situation is equally alarming which is merely 5 per cent of total land area, i.e. one-fifth of the world average. Forest and woodland constitute 20 per cent of the total land area of the region. In countries like Bangladesh, Maldives and Pakistan the situation is even worse where this percentage is merely 15, 3 and 5 respectively. Bhutan is the only country in the region which has more than 50 per cent of its are covered by forest and woodland (Table 15.1).

Land Degradation

Heavy population pressure on land and its unscientific exploitation has led to degradation of land resource. The major factors and forms

Table 15.1 : Land Use in the South Asia, 1992

Country	Total Land Area (000ha)	Arable and Permanent crop land		Permanent Pasture		Forest and wood land		Other Land	
		Area (000 ha)	% age of total land	Area (000 ha)	%age of total land	Area (000 ha)	%age of total land	Area (000ha)	%age of total land
Bangladesh	13017	9044	69	600	5	1890	15	1483	11
Bhutan	4700	134	3	273	6	2560	54	1733	37
India	297319	169650	57	11400	4	68500	23	47769	16
Maldives	30	3	10	1	3	1	3	25	83
Nepal	13680	2354	17	2000	15	5350	39	3976	29
Pakistan	77088	21110	27	5000	6	4050	5	46928	61
Sri Lanka	6463	1905	29	439	7	2100	32	2019	31
South Asia	412297	204200	50	19713	5	84461	20	103933	25
World (rounded)	13100000	1440000	11	3420000	26	3880000	30	4370000	33

Source : FAO, 1994.

being expansion of agricultural land and its intensive use, reckless deforestation, overgrazing, unsustainable agricultural practices, toxic effects of pestidices and chemical fertilizers, ill-planned irrigation practices, water logging, salinity, soil erosion, desertification, natural hazards etc. In South Asia the present situation in the result of wrong developmental strategies pursued during the post-independence period by various governments. According to Dent, India has the most degraded land as percentage of total land area (49.8 per cent) in South Asia, followed by Pakistan (17.3 per cent), Sri Lanka (10.8 per cent) and Bangladesh (7.0 per cent) (Dent, 1999).

The percentage of degraded agricultural land are as high as 75 in Bangladesh, 61 in Pakistan, 44 in Sri Lanka, 26 in Nepal and 25 in India (Table 15.2).

Table 15.2 : Total Area by Degree of Land Degradation in South Asia

Country	*Light*	*Moderate*	*Severe*	*Total*	*Total (As Percentage of Agricultural Land)*
		(1000 Hectares)			
Bangladesh	6187	1080	0	7267	75
Bhutan .	36	0	4	40	10
India	2935	20128	21941	45005	25
Nepal	429	759	0	1188	26
Pakistan	7530	8243	0	15773	61
Sri Lanka	35	158	838	1030	44

Source : FAO, UNCP, UNEP (1994).

Causes of Land Degradation

The causes of land degradation may be both natural and manmade. Some of the man-made causes are as follows :

Deforestation and Land Conversion

The forest land is being convened into croplands. In South Asia the annual rate of deforestation during 1981-90 is 1.1 per cent whereas for countries like Bangladesh, Pakistan, Nepal and Sri Lanka it is as high as 3.3 per cent, 2.9 per cent, 1.0 per cent and 1.3 per cent respectively. For India and Bhutan this figure is 0.6 per cent (WRI 1996).

Overgrazing

South Asia has large cattle and sheep population. Bangladesh has the highest cattle/buffalo density of 153 per square kilometre and that of sheep/goat is 99 per square km. In the region, India is next in this regard with figures of 80 and 44 respectively (SAARC, 1992). But the pastureland constitute only I1 per cent of the total land area in South Asia. Except in Nepal, where it is 15 per cent, in other countries it is merely 5 to 6 per cent of the total land area leading to a heavy pressure on crop and forest land for cattle rearing.

Unsustainable Agricultural Activities

Fertiliser consumption has increased sharply is South Asian countries in order to increase food production. In India between the years 1960-80, the consumption of three main types of fertilisers has increased by 5 times from approximately 22 lakh tons to 110 lakh tonnes (Govt. of India, 1990, Annual Report). And over 80,000 tonnes of plant protection chemicals were used in the mid-1980s in India against 2000 tonnes in mid-1950s.

The use of HYV seeds and chemical fertilizers are continuously increasing in all the countries. For example, chemical fertilizer use has increased by four times from 327 thousand tonnes in 1970 to 1260 thousand tonnes in 1984-85 in Bangladesh (Asaduzzamam). Similarly, the average use of fertilizers in Pakistan was 65 kg per hectare in 1985-86 as against 36 kg in 1977-78 (Khanna, 1991).

Cultivation on steep slope is another factor of land degradation, which can be controlled by terracing and contour farming.

In traditional farming system, shifting cultivation is a major cause of land degradation. In India alone, shifting cultivation is practiced on 4.96 million hectare and in Bangladesh about 800,000 people depend on shifting cultivation in the northern and eastern hills where land degradation rates are quite high (SAARC,1992). In eastern and southern parts of Bhutan also shifting cultivation is prevalent.

Mining and Quarrying

A great amount of precious vegetation and soil resources are removed during the mining operation. In India mining is major source of land degradation particularly in eastern and central parts where most of the mineral deposits are concentrated. Land subsisdence and debris production also cause land degradation. Mining on hill slopes disrupts the natural drainage. The gem-mining activities carried out in river beds, banks and upland areas of Sri Lanka lead directly to soil erosion.

Types of Soil Degradation

There is definite proportion of various salts, minerals, organic matters, gases and water in the soil. A change in the proportion of above materials due to various factors is known as soil pollution and land degradation. Soil degradation mainly occurs due to water and wind erosion, desertification and physical and chemical degradation of soil.

Water Erosion

Water is a major source of soil erosion and it is found in various forms such as sheet erosion, rill erosion, gully erosion, riperion erosion, land slide erosion and splash erosion (Table 15.3). All countries of South Asia suffer from water erosion. In the counties like Sri Lanka, Nepal and Pakistan 46 percent, 34 per cent and 20 per cent of the agricultural lands are affected by water erosion respectively. Whereas in other South Asian countries the degree of water erosion is light, in India 91 per cent of erosion is of moderate and severe type; and in Sri Lanka severe erosion constitutes 78 per cent of the total water eroded land (United Nations n.d.).

Table 15.3 : Area Affected by Water Erosion in South Asia

Country	*Area affected (1000 hectares)*	*As per cent of Agricultural land*
Bangladesh	1504	15
Bhutan	40	10
India	32773	18
Nepal	1592	34
Pakistan	7204	28
Sri Lanka	1074	46

Source : FAO, 'Selected Indicators of Food and Agriculture Development in Asia-Pacific Region1982-1992,' Bangkok, 1994, p. 24.
UNEP,'Our Planet', Vol. 6, No. 5.

In India 'very high to severe' erosion occurs in north-eastern Himalayan region, Siwalik hills, Western Ghats and the vertisol soils of the peninsular region. Whereas in Sri Lanka major contributing factors to soil erosion are : deforestation on steep slopes of the watersheds, shifting cultivation, removal of grasses and other protective cover and gem mining in upland areas (SAARC, 1992).

Wind Erosion

Wind erosion mainly occurs in dry area and in areas devoid of vegetation. Main factors affecting wind erosion are : velocity of wind, duration of wind and ground protection. In South Asia, wind erosion is prevalent in India and Pakistan only. In India, wind erosion is most common in the arid regions covering 31.7 million hectares of land of which 61 per cent is in the Thar area of western Rajasthan. Here sandy soils have low organic content and have low water retaining capacity and hence are vulnerable to strong winds. Wind erosion is also common in parts of Haryana, Gujarat and Punjab. Ploughing by tractors has also aggravated the situation (Table 15.4). The loss of soil vary under different site conditions- from 407 tonnes/hectare in fallow land, 932 tonnes per hectare in ploughed land using country plough and 2837 tonnes/hectare in ploughed land using tractors (Dhir, 1990).

Table 15.4 : Areas affected by Wind Erosion in South Asia

Country	*Area affected (1000 hectares)*	*As per cent of Agricultural land*
Bangladesh	0	0
Bhutan	0	0
India	10796	6
Nepal	0	0
Pakistan	10740	42
Sri Lanka	0	0

Source : FAO, 'Selected Indicators of Food and Agriculture Development in Asia-Pacific Region 1982-1992,' Bangkok, 1994 p. 24.
UNEP, 'Our Planet', Vol. 6, No. 5.

Wind erosion is a serious problem in the Cholistan, Thal, Thar and Kharan deserts of Pakistan. 42 per cent of the agricultural land in Pakistan are wind-eroded. In the Cholistan desert, 81 per cent of the area is severely affected by wind erosion (United Nations n.d.).

As per latest estimates, in India out of 173.6 million ha of degraded and eroded land, 83 per cent is subject to water and wind erosion while about 17 per cent is subject to various other types of land degradation (SAARC, 1992).

Chemical Degradation

Chemical degradation of soil is caused by improper use of chemical pesticides and fertilizers. In addition to it, misuse of irrigated water, lack of drainage and salt water intrusion lead to a rise in ground water tables, accumulation of salts and solidification of soils. A study shows that 938 million hectare of land are covered by salt affected soil in India which includes 5.5 million hectare of saline and 3.88 million hectare of alkaline soils (Govt. of India, 1990, Agriiculture in Brief). Bangladesh comprises approximately 0.9 mha highly eroded soil, 0.95 mha salt affected soils, 1.78 mha highly acidic soils and 0.23 mha acid sulphate soils (SAARC, 1992). In Pakistan the area affected by soil salinity is between 10 to 12 mha, while in Bangladesh it is 1.3 mha and in Sri Lanka 0.7 mha.

Physical Deterioration of Soil

Faulty canal irrigation is the main cause of water logging in India. In addition to this, surface flooding due to natural topography and drainage congestion cause widespread submergence and water logging. Nearly 17 states of India are affected by water logging and a total area of 8.58 million hectares is waterlogged (Govt. of India, 1990, Agriculture). Many areas of north western region of India are experiencing a rapid rise of water table due to over-irrigation leading to salt deposition.

A survey conducted by WAPDA shows that is Pakistan out of 16.3 mha of irrigated area surveyed by it, in 55 per cent of the area water table was at less than 3 metre depth (United Nations n.d.). In Bangladesh water logging is caused by the perennial floods.

Desertification

In South Asia, mainly India and Pakistan are affected by the problem of desertification. In India desert constitute 12.13 per cent of the total area where as 19.17 per cent of the total area is covered by semi-desert (Negi, 1999). Thar desert lying in the western part of India also covers some portions of Pakistan. In recent years, the desert is spreading gradually in other areas. The spread of desert has resulted in soil erosion, loss of forest cover, covering of fertile soil by sand, sedimentation of rivers and lakes, drying up of drinking water sources, declining food grain production, loss of pastureland, increasing area of wasteland etc.

Out of total geographical area of 329 million hactares in India, wastlands claim 175 million hactares (53.19%), of which 150 million hactares is affected by water and wind erosion, 4 million hectares by shifting cultivation, 6 million hectares by water logging, 8 million hectares by salinity and alkalinity and 7 million hectares by river action and other factors (National, 1986).

SUGGESTIVE MEASURES FOR SUSTAINABLE AGRICULTURAL PRACTICES

For human being land is the most valuable and basic but non-renewable resource. In the countries of South Asia, where agriculture is the main source of livelihood, land resource management is of utmost importance. The most important aspect of land resource management is to check the misuse of land resources and put it to optimum and sustainable utilisation. Some of its aspects are :

1. *System of Stabilising Population and Land Resources* : due to ever increasing population, per capita agricultural land availability is decreasing day by day. Stabilisation of land resource is possible only by an effective Family Planning System and awakening among masses.
2. *Land Reform* : since there is little scope for horizontal expansion in agriculture, it can only be done by implementing land reform measures. It will help in equitable distribution of land resources which will put land resources to optimum use and also in eradication of poverty.
3. *Reclaimation and Development of Degraded/Waste land* : In South Asia vast area of land is classified as degraded/waste

land. For example in India alone the net cropped Area (166 million ha) is less than the total degraded land (175 million ha). So there is need to develop these so called wasteland by using proper technologies. For this detailed chemical and physical analysis of the soil is done for soil zoning/ classification. After that land levelling, bunding, land shaping and drainage control is facilitated for uniform spread of water for assured irrigation. In addition to it, inorganic treatment and organic amendments are also applied. For example for reclaiming *usar* land of Alkali soil, inorganic treatment is done by using chemicals like calcium salts, calcium chloride, gypsum or by using acids (like sulphuric acid) or iron sulphate, aluminium sulphate, lime sulphate etc. Organic amendments to these soil are done by decomposition of cattle manure and plant residues in the soil. By this soil liberates CO_2 and organic acid which improves soil permeability and thus increases water table aggrigates. High/medium salt tolerant crops and vegetable should also be planted on the land.

4. *Land Use Planning* : Land use planning is the first step of agricultural planning. It starts with land evaluation. Land evaluation was first done in USA to show the risk of soil erosion and for sustainable land uses. With the help of land evaluation we are able to know which land is suitable for agricultural purposes and which are for non-agricultural purposes (cattle grazing, forest etc.). Optimum land use decreases the risk of further degradation. After that land is classified according to suitability. Now-a-days with the help of Remote sensing data, soil mapping, Forestry and vegetation mapping, Hydrological mapping, Land use mapping etc. are done with precise accuracy. It helps in decision making process.
5. There is need to conserve the precious soil cover. For soil conservation different areas are classified on the basis of intensity of soil erosion and then priorities for soil conservation and land reclamation are determined. Soil conservation methods include crop-management as well as mechanical methods. Proper crop-management ensures to cover the soil with crop for the maximum period to halt slash, wind and

water erosion, to increase the rate of infiltration of rain water and to check surface runoff. Inter-cropping, mixed cropping, trash farming, crop rotation and crop combination and social forestry are some of the crop management methods. Mechanical methods include contour farming or terracing in hilly regions, formation of earthen check dams, filling of rock and boulders and plantation of shrubs, grasses and trees etc. to check rill and gully erosion in ravines and bad lands.

6. *Afforestation* : In South Asia forest cover is only 20 per cent which should be increased to at least one-third of the total geographical area. Afforestation should be encouraged on degraded land in command area, Drought Prone Area, Desert area and by road side plantation etc. Sand dune stabilisation, pasture development and social forestry should be encouraged with the help of local community. This is necessary for increasing soil fertility, halting soil erosion, for ecological balance and also for meeting fuelwood requirements of rural population.
7. In dry areas there should be proper watershed management. For this micro watersheds are recognised and managed to check the rainwater within watershed area to recharge the ground water aquifers and to maintain soil moisture.
8. All parts of the South Asian region are not equally endowed with water resources. To ensure availability of water in all areas, the proposal to form National Water Grid is an important step. By this all the major rivers of the country would be inter-connected and water would be diverted from water surplus to water defecit areas. Such co-operation is also needed among all the countries of the region.
9. In order to check the spread of desert and for stabilization of sand dunes, intensive afforestation should be done and Green Safety Belts should be formed in and around desert areas.
10. There is need for an integrated and holistic approach for land conservation involving inter-agency coordination at national and regional levels.
11. The action oriented environmental strategy necessities a boarderless system to optimize man-nature-development equation, where environment and development need to be treated as partner and not as opponents.

12. There is also need for environmental awakening and people's participation to check land degradation in South Asia. The microlevel grassroots efforts must be combined with macro level initiatives at national and regional levels.

REFERENCES

Asaduzaaman, M. Rural Development in Bangladesh. *South Asian Journal*, Vol.1, No. 4, p. 393.

Dent, F.J. (1999) Land degradation in Asia and the Pacific. In : *Environment and Agriculture*. FAO, Bangokok.

Dhir, R.P. 1990. Land Degradation in Arid Zone. In : I.P. Abrol *et al.* (eds.) *Technologies for Wasteland Development*. ICAR, New Delhi. pp. 93-107.

FAO, (1994) *Selected Indicators of Food and Agricultural Development in Asia Pacific Region 1982-92*, Bangkok.

Government of India (1990) *Annual Report of Department of Agriculture 1989-90.* Ministry of Agriculture, New Delhi. p. 150

Government of India (1990) *Agriculture in Brief*, 23rd edition. Ministry of Agriculture, New Delhi.

Khanna, B.S. (1991) *Rural Development is South Asia : Pakistan*. Deep and Deep Publication, New Delhi. p. 82.

National Waste Land Development Board (1986) Discussion papers, Minutes and Decisions, p.5.

Negi, P.S. (1999) *Eco-development and Environmental Geography.* Rastogi Pubilication, Meerut, p. 246.

SAARC : Regional study on the causes and consequences of natural disaster and the protection and preservation of the Environment, 1992. Kathmandu.

UNDP (1998) *Human Development Report, 1998.* Oxford University Press, New York.

United Naions, n.d. *State of Environment in Asia and the Pacific : 1995*. Economic and Social Commission for Asia and the Pacific & Asian Development Bank, New York, p.5.

WRI (World Resource Institute) (1996) *World Resources 1996-97.* Oxford University Press, New York.

16

New Trends Towards Sustainable Agriculture Land use : A Case Study

S. Hasan Ansari

It has been noted that new trends towards sustainable development has been started in sixties of the past twentieth century in our country. Before revolution the issue of land, productivity, food security, climate change, environmental security and eventually the quality of life was very important to give attention. The economic conditions and low income per head was on very high. The bigger industries were running in loss and the unemployment was on very alarming state. All these problems were forced to think about new trends towards sustainable agricultural land use in which new techniques for seeding, sowing, ploughing and harvesting have to be adopted. New agricultural policies, institutional development, community development works, innovations and new agricultural developments pave the way of transformation of traditional ploughing, harvesting, thrashing, storing and the use of chemical fertiliser and insecticides and pesticides. Another factors and causes of land degradation are growing population, shrinking farm size, environmental policy, legal and institutional framework for environmental protections. Indian farmers were producing one tonne per hectare, and the challenge to get another tonne of food grain from the same farm, while land was shrinking, water and soil quality degraded. It is need of the hour to use currently available knowledge and technology to alleviate the problem. Hence to get maximum benefit of new trends towards sustainable agricultural land use policies, assured irrigation facilities, timely availability of chemical fertilizers, use of HYV seeds of wheat, rice and insecticides, pesticides at the time of maturity are necessary. An attempt has been made to see the role of innovation of new technology for sustainable agricultural land use in a district of Rajasthan as a case study.

Aim and Objective

The main aim of the paper is to highlight an impact of new agriculture technology on the land use through its changing nature of agricultural implements, fertilizers, irrigation facilities, HYV seeds, insecticides pesticides in the area under study.

The objective is to present a real picture of the land use, productivity, size of land and the use of fertilizers and how are the farmers still engaged in agricultural activities to get more and more yield from the same farm in a situation like power supply and irrigation facilities are not ensured in the district of Dhaulpur of Rajsthan State.

Approaches : The present study has been considered in three steps, i.e. first step is related with the introductory remarks of the study area. Second step highlights the changing nature of agricultural implements, power supply for agricultural activities, the biotic and mechanical energy during 1986-87 and 1995-96. The third step envisages how the size of land holdings is associated with the use of new implements, HYV seeds and chemical fertilizers for sustainable agricultural land use by producing high yield from the same field.

Study Area

Before observation of impact of new technology on agricultural land use, it is necessary to introduce the study area. The district of Dhaulpur of Rajasthan state has been selected for a case study. It is located from $26^{o}22'$ to $26^{o}\ 57'$ latitudes and $77^{o}\ 14'$ E to $78^{o}\ 16'$ E longitudes which is surrounded by district of Bharatpur in North-west, Karauli in the west, Uttar Pradesh in the North-east and Chambal river on the south. It accommodates 749479 persons inhabited in 3 towns, 4 Tehsils, 4 Panchayat Samitis and 551 rural settlements spread over an area of 3008.98 km^2. The ratio of rural and urban settlements is recorded as 82.81 per cent and 17.19 per cent respectively. Out of 29.58 per cent workable forces, 71.25 per cent and 6.53 per cent are agriculturist, and the agricultural labourer; while other workable people are engaged in trade and commerce and other services sharing 4.26 per cent, 8.12 per cent respectively. The cultivators of the area under study are used 50942 plough (*Hal*) but due to introduction of new technology in the agriculture activities the number of diesel engine has been increased from 306 to 17805 during 1986-87 to 1995-96 and the number of tractors has also been increased from 665 to 2971 even after the number of plots has again been divided and the average size of farms decreased during 1986-87 and 1995-96. All these statements have been analysed and the results has been given in sequence.

Observation and Analysis of Data

Table 16.1 records agricultural implements of Dhaulpur district of Rajsthan State during 1986-87 and 1995-96. It reveals that there was 68108 and 47716 number of implements during 1986-87 and 1995-96 respectively which indicates that the number of agricultural implements has been decreased by 20392 in a decade. If we go in a detail observation, it may be found that the maximum number of diesel engine increased by 17499 for lifting of ground water for irrigation of the cultivated land. The number of plough (*Hal*) and carts has also been decreased by 29432 and 2917 respectively while the number of tractor and electric pump sets has been increased by 2306 and 538 respectively in a gap of 10 years, i.e. 1986-87 and 1995-96. It may be observed that tractors, electric pump sets, diesel engines are taking place of the traditional means of plough (*Hal*) and agricultural implements in the area under study.

Table 16.1 : Agricultural implements in Dhaulpur District

Name of Implements	*Years*		*Increase in !number*	*Increase/ decrease in percentage*
	1986-87	*1995-96*		
Plough	50942	21510	-29432	-57.78
Carts	77	4160	-2917	-41.22
Diesel Engine	306	17805	+17499	+5,718.62
Electric Pumpset	683	1221	+538	+78.77
Tractor	665	2971	+2306	+346.77
Total	68108	47716	-20392	

Source : Census of Animals, Dhaulpur-1999.

Table 16.2 : Assessment of Energy (in horse power) per unit per hectare in Dhaulpur District

Name of source	*Energy (in horse Power)*
Human	0.20
Bullock	0.50
Bufallow	0.90
Camel	2.00
Diesel and Electric Pump	5.00
Crusher	7.50
Tractor	50.0

Source : Annals of the Rajasthan Geographical Association, Vol. III, 1983.

Use of Energy

The changing nature of use of energy/power for agricultural activities has been calculated by transferring human, animal and agricultural implements in the form of power energy into two categories, i.e. (i) Biotic Energy and (ii) Mechanical energy. Table 16.3 shows the assessment and use the of energy in agricultural activities in the district of Dhaulpur during 1986-87 and 1995-96. It uses of mechanical energy i.e. to say modern agricultural implements.

Table 16.3 : Use of Energy in Agricultural Activity in Dhaulpur District

Classification of Energy	*Used Energy*				*Change in 10 years*
	1986-87		*1995-96*		
	In Horse power	*In percentage*	*In Horse Power*	*In percentage*	
Human	26587.80	15.55	32039.00		-14.81
Animal	42944.60	25.12	22358.50	7.49	-19.63
Machine	101457.50	59.33	244047.50	81.77	+22.44
Total	170989.90	100.00	298445.00	100 00	

Source : Based on Human and Animal Census Report, Rajasthan, 1999.

Table 16.4 : Use of Biotic Energy in Gross cropped Area (000ha) in Dhaulpur District

Tehsil	*Years*		*Changes/Transformation*
	1986-87	*11995-96*	
Baseri	527.93	499.68	-28.25
Bari	461.61	393.36	-68.25
Dhaulpur	521.38	240.52	-280.86
Rajakhera	402.03	534.46	+132.43
District	489.32	379.86	-109.46

Source : Based on Human and Animal Census Report, Rajasthan, 1999.

Mechanical Energy : It has been observed from Table 16.5 that the cultivators have been using mechanical energy in the cultivated land of 268.75 (000 ha) during 1986-87 and 1697.96 (000 ha) during 1995-96, which shows drastic change in the use of mechanical energy, i.e. 1429.21 (000 ha) in a decade. On comparing figures from one tehsil to another, it may be noted that the highest use of mechanical energy found in Rajakhera tehsil, i.e. 2325.64 (000 ha) and the lowest in Dhaulpur tehsil, i.e. 986.78 (000 ha). There is a significant point to note that Rajakhera tehsil used more energy of both nature, i.e. biotic and mechanical energy depending on the small size of land holdings but in other cases of 3 districts, the mechanical energy has been used for more production of the crops. However, it may be observed that the farmers of the district of Dhaulpur adopted mechanical energy for cultivation which led to adopt the HYV seeds, chemical fertilisers in the area under study.

Table 16.5 : Use of Mechanical Energy in Gross cropped Area (000 ha) in Dhaulpur district (Rajasthan)

Tehsil	*Year*		*Transformation*
	1986-87	*1995-96*	
Baseri	131.15	1786.40	+1655.25
Bari	199.28	1514.68	+1315.40
Dhaulpur	374.22	1361.00	+986.78
Raja Khera	277.79	2603.43	+2325.64
District	268.75	1697.96	+1429.21

Source : Base on Census report of animals, Rajasthan, 1999.

Cultivated Land under HYV Crops

Transformation of Traditional to HYV seeds may become possible after the introduction of green revolution, i.e. early ripe seeds of short stature, more fruits of seeds, assured water supply, chemical fertilizers, insecticides and pesticides. Table 16.6 reveals that Sarson has taken place of main crop followed by wheat and Bajra which shows due to lack of irrigation facilities and power supply, the cultivators did not adopt HYV seeds of wheat. Hence those crop required-very less amount of moisture and chemical fertilizers, viz., Sarson (79.14 per cent increased in area) and Bajra (1.00 per cent increased in area) adopted after a gap of 10 years.

Table 16.6 : Area of HYV crops in Dhaulpur (per cent of crops over total agricultural area)

Crops	*Year*		*Decadal change*
	1986-87	1995-96	
Wheat	93.39	100.00	+ 6.61
Bajra	99.00	100.00	+1.00
Sarson	2.17	81.31	+79.14

Source : District Agriculture Officer, Dhaulpur (Rajasthan), 1999.

High Yield Variety Seeds

The detail analysis of HYV seeds may be explained by going through Table 16.7 which shows that 520900 kg of HYV seeds were used in 1986-87 and 689500 kg during 1996-97 in the district. On an average 2.92 kg per ha and 3.88 kg per ha of HYV seeds have been used during 1986-87 and 1995-96 respectively. Which indicates that a nominal increase of HYV seeds, i.e. 0.96 kg in a gap of 10 year, but it has different results at tehsil level. The highest amount of use of HYV seeds has been recorded in Dhaulpur Tehsil, i.e. 4.50 kg per ha and the lowest in Baseri tehsil, i.e. 0.23 kg per ha after a period of 10 years. It indicates that except Dhaulpur tehsil, other three tehsils have been still adopting traditional seeds due to irregular supply of electric power and non-availability of assured irrigation at the time of requirement.

Table 16.7 : Use of HYV of seeds in Dhaulpur District (Rajasthan) during 1986-87 and 1995-96

Name of Tehsil	*Use of All HYV Seeds*				*Change In kg*
	1986-87		*1995-96*		
	(in kg)	*(in kg)*	*(in kg)*	*(in kg)*	
Baderi	131246	3.37	158263	4.20	+083
Bari	98142	2.54	104502	2.77	+0.23
Dhaulpur	216612	3.02	321613	4.50	+1.48
Rajakhera	74900	2.57	105122	3.41	+0.84
District	520900	2.92	689500	3.88	+0.96

Source : Office of the District Agricultures, Dhaulpur, Rajasthan.

Use of Chemical Fertilizer

Above discussion clears that HYV seeds did not spread much in the area under study. Table 16.8 also support the statement that the used chemical fertilizers has been increased from 4433 kg to 7909 kg during 1986-87 and 1995-96. It shows a very less amount of chemical fertilizer (3476kg) was used in cultivated land after a gap of 10 years. Hence there is no doubt to say that the cultivators of the study area neither try to know about new agriculture technology nor want to adopt it due to un-assured supply of electric power and irrigation facilities. The size of land holdings may also be another important problem to use chemical fertilizers and new agricultural implements.

Table 16.8 : Use of chemical fertilizer in Dhaulpur District (Rajasthan) during 1986-87 & 1995-96

Name of Tehsil	*Use of chemical fertilizer*				*Increse per ha (In kg)*
	1986-87		*1995-96*		
	(in kg)	*kg/ha*	*(in kg)*	*kg/ha)*	
Baseri	953	24.53	1524	40.45	+15.92
Bari	946	24.50	1449	38.39	+13.89
Dhaulpur	1952	27.36	3754	52.49	+25.13
Rajakhera	582	19.93	1180	38.38	+18.45
District	4433	24.87	7909	44.50	+19.63

Source : Office of the District Agriculture, Dhaulpur, Rajasthan, 1999.

Land Holdings

Table 16.9 indicates 6 categories of land holdings of the area under study viz., (i) margined (< 1 ha), (ii) very small (1-2 ha), (iii) small (2-4 ha), (iv) medium (4-10 ha), (v) large (10-12 ha) and (vi) very large (>20 ha). It shows that the highest number of plots was found of the marginal farmers, i.e. 47156 and the lowest of the very large size farmers, i.e. 71 whose share comes only 13.12 per cent and 1.28 per cent of the total number of plots during 1986-87. But after a 10 year gap the number of plots of the same category has been again increased by 1622 and 43 indicates that the size of plots decreased from 0.51 ha to 0.48 ha and from 33.08 ha to 28.05 ha of the marginal and very large size farmers as Table 16.10 is also proved it. Similar results may also be seen in very small and small category of farmers while the case of medium and large size cultivator, the number of plots decreases from 29.81 per cent to 28.13 per cent and from 6.71 per cent to 5.82 respectively as Table 16.l0 reveals the same results. It means marginal and small land holding farmers sold their land to either medium or large size of farmers due to non-economy to them. In case of very large size farmers, they did not get suitable profit of their product and even after introduction of new technology of ploughing and irrigation they did not want to adopt HYV seeds due to unassured supply of water for irrigation in time. Hence it is again proved that green revolution has not been much effective in the area under study.

Table 16.9 : Size. of Land holding in Dhaulpur district during 1986-87 and 1995-96

S. No.	*Classfic-ation of land holding (ha)*	*1986-87*			*1995-96*			*Cha-nge*
		No.	*Area in ha*	*%age*	*No.*	*Area in ha*	*%age*	
1.	Marginal (<1)	47156	24154.83	13.12	48778	23322.62	14.02	+0.90
2.	Very small (1-2)	25629	36596.70	19.87	23850	33708.54	20.26	+0.39
3.	Small (2-4)	19149	53790.42	29.21	17758	49661.65	29.85	+0.64
4.	Medium (4-10)	9485	54890.21	29.81	8246	46793.06	28.13	-105
5.	Large (10-20)	946	12364.84	6.71	761	9676 94	5.82	-0.89
6.	Very large (<20)	71	2348.63	1.28	114	3198.21	192	+0.64
	Total	102436	184154.63	100.00	99507	166360.66	100.00	

Source : Office of the Agriculture Officer, Dhaulpur, Rajasthan, 1999.

Table 16.10 : Averge size of land holding in Dhaulpur district during 1986-87 and 1995-96

S.No.	*Size of land holding (in ha)*	*Averge size (1986-87)*	*Average of landholding size 1995-96*	*Change ill per cent*
1.	Marginal (<1)	0.51	0.48	-0.03
2.	Very small (1-2)	1.43	1.41	-0.02
3.	Small (2-4)	2.81	2.80	-0.01
4.	Medium (4-10)	5.79	5.67	-0.12
5.	Large (10-20)	13.07	12.72	-0.35
6.	Very large (>20)	33.08	28.05	-5.03

Source : Office of the Agriculture Officer, Dhaulpur, Rajasthan, 1999.

Concluding Remarks

It may be concluded with remarks that an impact of new technology towards sustainable agricultural land use is seen in the category of very large size of farmers who are adopting HYV seeds, chemical fertilizers, tractors and electric pump sets for assured water supply in time; but in the case of other category of farmers did not want to adopt new technology due to shrinking of lard size and redivision of plots in the divided family. It has also been seen that the marginal/ very small farmers sold their land due to non-economical. While in the case of large and medium size farmers who are adopting new techniques in agricultural activities gradually becoming prosperous by producing more and more crops from the same farms and also maintained fertility of their farms for sustainable agricultural land use in 21st century.

17

Resource Utilization Pattern and Development in Hills

A case for the Pindar Basin of Garhwal Himalaya India

Vishwambhar Prasad Sati

Introduction

The management of natural resources in the mountain environments has emerged as one of the most significant challenges to human understanding and organizational ability in the current century. Although the mountains and uplands constitute about 20 per cent of the surface of the earth, it is difficult to find an area not affected by their environmental characteristics. The upper parts of the mountains are always covered by snow and they are the sources of major rivers in the world. The vital cultural and ecological importances of these mountain waters further enhance their significance in satisfying the needs of all living creatures. Of somewhat less significance ecologically, but nonetheless of great economic importance, are the uses of the mountains for forestry, agrohorticulture, mineral extraction, livestock rearing, tourism, and recreation.

Pindar Basin that constitutes the major part of the Alaknanda Basin in the Garhwal Himalaya, is the most ecologically fragile zone. The series of landslides, which washed away the thousand tones of soil per year, reduce the rate of soil fertility. Thus the whole basin is landslide prone area. Every year, one or more landslides occur. The basin can be divided into three physiographic zones according to their elevation. The highly mountainous area, which is always covered by snow, is the major sources of rivers, like Pindar, Kail and its tributatries. Alpine meadows and small patches of field terraces characterize the second zone. Where, mainly herbs in the high altitude are grown, and agriculture in the middle zone is practiced quite extensively. This zone is called 'Danda'.[1] Along these two practices, horticulture is slowly but steady being practiced throughout the basin. Even some quality fruits are grown. In the low-lying area, which is the third physiographic zone, traditional subsidence cereal farming is being practiced extensively. But the production of the crops is very low. This zone is called 'Gangar'.[2] The topography of the basin prevents the developmental process and optimum utilization of resources. Besides, the main valley root, from Karanprayag to Gwaldom and Mundoli, the entire basin is lagged behind, in terms of, transportation facilities. In spite of these huge impediments, there are many fertile patches in terms of agricultural practices, horticultural practices, horticulture, herbs culture, potato production, and various other food grains; pulses, oilseeds etc. are intensively and extensively grown. Only, the need is of developing a trend for sustainable development of these huge resources. The present study discusses on the resources available in the region and their analyses to enhance the economic status of region.

Study Area

The Pindar Basin comprising of 18266.0 km^2 extends from 30° N to18^o N latitude and 79° 13′ to 80° E longitude. It consists of the eastern part of Garhwal Himalaya with height ranging from 800m to 6800m. River Pindar originates from the 'Pindari Glacier' in district Almora (32 km.) and flowing an approximate 100 km. with its numerous tributaries, confluences into the Alaknanda river at Karanprayag in Chamoli district (68 km). The watersheds of Ramganga in the south, the Saryu in the east, Mandakini in the north and Alaknanda in the north-west, giving it a distinct socio-geographical identity, separate the study region. This basin consists of the four communities develop-mental blocks of the Chamoli district, viz., Karanprayag, Narainbagar, Tharali, and Deval in ascending order from lower to higher altitude and a part of Kapkot block in Almora district.

Methodology

The present study is based on primary and secondary data collected through questionnaire and schedules and from various sources respectively. District statistical handbook of District Chamoli is also used. Maps have also been drawn for interpretation of data.

Pattern of Resource Utilization

The Pindar basin is characterized by heterogeneity in terms of landscape, resources and their utilization paterns. Right from the valley region to the highly elevated duelling zones, the climatic conditions regulate the all types of utilization pattern. While, the valley region is considered for developing service centres and the wide patches of the cultivated fields are characterized by growing cereal crops; the middle and upper belts are known for growing varieties of crops, vegetables, fruits, herbs, rearing livestock, goats, sheep; manufacturing wooden, Ringal, and Bamboo goods. The production of potato, apple and citrus fruits, milk and milk made goods is gaining impressive mark since the decade of nineties. In this regard, some patches of Ghaish-Blan, Kurer-Kwarad, Sol-Dungri, Shubhtal-Jhaltall have the progressive position. The cash crops are taking the place of cereal farming. Consequently, the economic level of the populace is undoubtedly being developed. The resource utilization pattern is varied from place to place, based on infrastructural facilities, particularly transportation and extension services. The undeveloped agricultural practices are also due to the remote-ness of the villages from the main stream. Settlements are sparsely distributed. But the highest population is found in the middle altitude. In the high and low altitude, the population is very low. Because, the high altitudinal zone has very low temperature, while, in the low-lying areas the land for settlement and agricultural practices is less. The study is being done on the following manner.

Water Resource

The Pindar river is the main drainage system and the biggest river of the region. There are many other perennial sources, which are the tributaries of the Pindar. The main tributaries are Kail Ganga, Pranmati Gad, Meing Gadhera, Kewar Gadhera, Chopta Gad, and Ata Gad. Besides these there are so many small perennial streams, and natural springs. But due to uneven and rugged topography, water crises are prevailed in most of the villages—summers in particular and winter in general. Since, the villages are located on the hill patches and water cannot be utilized for any purposes. The Pindar River has many ideal locations for constructing dams. The two best locations are near the Bagoli and Simli. These two sides have already been proposed for dam construction and many times work is started during the winter and summer but during the rainy season, the finished works are swept away by the river. Micro hydle schemes can be introduced, on the other hand, to mitigate the problem of drinking water and electric supply, even for irrigation purposes. The old practice of water mills which locally known as 'Gharat system' can be reintroduced as a form of floor mill and electricity generation. Still these are in practice in some of the areas. Almost, even the dominance of electric mills the water mills are also in functioning in almost all the small tributaries. In some areas, the water mills come to practice only during the rainy season because during the other seasons the volume of water remains less to meet the need for running water mills. The author to counting of these water mills has conducted a survey and it was found that more than 200 water mills are still in practice in the entire basin. Since the practice of water mills is being traditionally done so this system is needed to improve, which could get the new strength.

There are quite diversities in the course of river from outlet to inlet. Since, the river flows with its youthful stage. The streams make water fall, V shaped valleys, gorges etc. The wide and stretched valleys of these streams constitute the fertility of land.

Forest Resource

The Pindar basin in very rich in terms of forest and its diversity. Right from the valley region to the highly elevated Alpine meadows, locally known as Kharak or Bugyal, the diversities in plants are found extensively. In the middle altitude, Pine (Chir) are found while in the upper reaches, temperate confierous forest mainly Banj (oak), Buransh, Tilonj, Devadar, Surain etc. are found extensively. Except these forest types, many other fodder plants like, bhimal, khadik, are also grown along with the agricultural fields. Oak is the predominant flora of the moist temperate forest growing on the elevated region of the basin. Kharsu (*Quercussemicarpifolia*). Tilonj (*Q. Dilitata*), Rianj (*Q. lanuginose*) and Banj oak (*Q. leucotricophora*) are the four main species, which have a traditional edge over the other species in this region. Out of these four species Banj oak (*Q. leu*), which being a very rich fodder and ideal fuel has been an inseparable part of folk life in hills, is the one, which starts growing from 1676 mt with Rhododendron trees. It is found up to 2286 metres mixed with the other species of the same genera. The oak forest of the Pindar basin appears to be thriving due to the imposing presence of pine trees. A note in the *Hindustan Times* by HT correspondence, 'will oak disappear from Nainital' is a real story in the hills of Uttranchal.[3] According to the altitude and forest types the basin can be divided into the following zones :

1. Low-lying river valley (below 1000-metre elevation from masl), mainly along the Valley of Pindar River, right from Karanprayag to Tharali, where the height does not reach above 1000 metre. Presently, a forestation is carried on by the department of forest in this region to avoid the landslide. These plants are of Eucalyptus, poplus, bamboo etc.
2. Middle altitude (between 1000 to 1600 metre elevation from masl). Pine is found extensively. There are many patches, where dance pine forests are found. They are starting from Bagoli to the valley of Pindar along with both sides its tributatires. Such as, the slope of Kaver Gadhera, Mieng Gadhera, Pranmati Gad, etc. The present trend of pine trees distribution is that they are invading the coniferous regime. Pine trees invaded most of the oak dominated areas and environmentalists advocating that this process is being existed due to temperature incensement in the atmosphere.
3. Temperate belt (between 1600 to 2000m elevation from mean sea level). The area constitutes Love-Kush top, watershed of Pindar and Ramganga, Kanpur Garhi, Kankhrakhet, Shubhtal and Chhaltal, Kurur Kwarad, Sol-Dungri etc. The region is very rich in terms of soft forest wood, which is very useful

for many purposes. Main trees found in this area are oak, buransh, tilonj, surai, devdar, etc. This region is inaccessible from the reach. Thus, the woods are unutilized.

4. Highly elevated region (between 2000m to snow line). This region is accounted for herbs. Extensive fields of herbs are found. These are locally called Kharak or Bugyal. The main bugyals are : Khankhrakhet, Bedni bugyal, Roopkund, Shail Samunder etc. Bedni bugyal and Roopkund are world famous spots and they can be developed as the tourist places. Different kinds of herbs are found in these bugyals.

Forest based small industries can be developed in these areas. Particularly, where these forest are found. They are not used properly in terms of small-scale industries. While every year thousands of tones of timber is used in form of fuel-wood as well as building materials. They are also used for fodder need.

Agricultural Resources

Subsistence agriculture is the main stay of the people inhabited in the Pindar Basin. Most of the workable force is engaged in the agriculture and its allied practices. But this practice does not fulfil the food need, therefore, to cope with food scarcity, the people of the region have out-migrated to the other parts of India. The production and productivity of the crops are very low. Cost benefit analysis reveals that the output by the agricultural production is low, even, below the level of input. Cereal crops are dominant in the cropping pattern. The production of fruits is getting impressing level, but the land devoted to fruit plants is very less. The climatic conditions for production of fruits are congenial. There are two fruit belts demarcated. These are Kalimati-Janlechatti and Gwaldom-Lolti. Many other small fruit belts are also located there, which can be developed for intensive fruit production.

The above table reveals the most of the crops are grown in the entire basin, mainly cereals, oilseeds, and pulses. Presently, the production of potato is being increased. There are ideal conditions for production of cash crops like potato, pulses, oilseeds and different types of vegetables; therefore, these cash crops can be grown

Table 17.1 : Area under Different Crops in the Pindar basin

	Block (area in ha)							
	Karanprayag	*%*	*Narain Bagar*	*%*	*Tharali*	*%*	*Deval*	*%*
Paddy	2238	47.7	1632	43.72	1594	61.2	570	34.3
Wheat	2778	51.7	1733	46.4	780	29.9	777	46.8
Barley	166	3.0	190	5.1	61	2.3	62	3.7
Masur	7	0.1	10	0.3	6	0.2	5	0.3
Pea	-	-	4	0.1	4	0.2	-	-
Arhar	9	0.2	15	0.4	6	0.2	6	0.3
Pulses	24	0.4	73	1.9	58	2.2	17	1.0
Oilseeds	14	0.3	31	0.8	30	1.2	12	0.7
Potato	130	2.4	45	1.2	67	2.7	211	12.7

Source : District Statistical Diary of Chamoli District.

extensively. Though paddy and wheat are leading crops of the area (more than 80% of cropped area) their yield is very low owing to some localized factors (traditional views and old methods of ploughing and harvesting). The harvesting period of the paddy is variable according to elevation but the ploughing period is the same everywhere. In the Gangarh, the paddy harvesting begins one month earlier than the Danda. The plouging period is March-April both in the Gangarh and the Danda and the harvesting period is the early September in the Gangaıh and October in the Danda.

Landuse Pattern in the Pindar Basin

Land is one of the most components of ecosystem and it constitutes main part of economy of the Himalayan environment. Land use pattern is varied according to relief and slope factors. The following table reveals the land use pattern in the Pindar basin.

Table 17.2 : Land Use Pattern in the Pindar Basin

(Percentage of Geographical Area)

Block	*Forest land*	*Pasture land*	*Agricultural land*	*Horticultural land*	*Cultivable land*
Karanprayag	52.2	3.3	28.4	8.5	0.7
Gairsain	74.0	2.4	12.6	6.9	2.9
Narain Bagar	67.2	8.9	18.5	7.1	0.3
Tharali	25.2	0.5	4.8	0.2	0.4
Deval	80.8	1.6	5.7	6.5	1.4

Sources : District Statistical Handbook of Chamoli.

The above table reveals taht forest covers the highest percentage of the total geographical land followed by agricultural land. The percentage varies from one block to another. There are three crop seasons Ravi, Kharib, and Jayad, Under the Ravi crops, wheat, oilseeds, barley, masur etc. are grown, while, under the Kharib crops, rice, koni, jhangora etc. are grown. Pulses and vegetables are grown with the Jayad season.

Cropping Pattern

"The kind and sequence of crops grown over a period of time on a given areas of soil can be described as the cropping system. It may be a pattern of regular rotation of different crops, one growing only one crop after one year on the same area" (Johnson and Peter, 1984). The overall cropping pattern of the Pindar basin is typical and underdeveloped agricultural economy. Here, nearly 90 per cent of the total cropped area is devoted to subsistence food crops mainly grown from domestic consumption and local market. Commercial or cash crops occupy a very negligible portion of the cropped area. Due to spatial differences in the agronomic, economic, and cultural conditions, variations in the cropping pattern in the region can be seen, but on the whole these variations are of very little significance. Rice and wheat dominate the agricultural realm from one corner of the region to the other, obviating the culture of the other crops (Table 17.2). The influence of the monsoon on the cropping pattern is very dominant, with the result that the total cropped area about 70.75 per cent is under Kharif or rainy season crops. In the region what ever may be the type of soil or the amount of rainfall the dominance of food grains in the cropping pattern is everywhere obvious. A case study of cluster of five villages has been done for interpretation of cropping pattern. These clusters of villages are located almost in the centre of the basin. Agriculture is the main occupation of the villagers. Of the total population, 49 per cent people are engaged in the primary works, mostly in subsistence agriculture. The method of agriculture is traditional. Agricultural implements are mainly made of wood and oxen plough the fields. The seeds are sown using spray method. The main crops are paddy, wheat, manduwa, pulses and millets, and in the vegetable, potato, pumpkin, cucumber, brinzal, etc. The production of potato has been recently increasing. The people are self-reliant for their food grain needs. Agriculture is totally dependent upon rainfall, because of inadequate facilities of irrigation. Under the cropping pattern, the total agricultural land is divided into two parts : Tally and Mally Sar, which are called 'Sar System'. Since March-April till September-October, there are grown the Kharif crops in the Tally Sar. The millets are grown in the Mally Sar in the same period under the Kharif season. Table 17.3 reveals the cropping pattern in 2000 under Rabi and Kharif Crops.

The cropping pattern of this village is systematically characterized by mixed cropping. The crop rotation, which is locally known as "Sar System", has been giving good yield of crops. Within a year, the land is given-up fallow for six months. The cropping pattern of this village represents the entire Uttranchal state except the plains of Dun, Dwar, and Terai and Bhaber regions, where the extensive cultivated land extended and the cropping capacity, intensity of crops, crops production and per ha yield of crops is very high.

Horticultural, Herbicultural and Tea Garden Practices

The entire region is suitable for growing fruits, vegetables, herbs, and tea. The production of fruits is limited only on the some patches of Gwaldom-Tharali, Khun-Kothali, and Kalimati-Jangle Chatti, while the climatic conditions are similar in the basin. Among the fruits, apple, orange, nuts, and pears are grown widely. But the quality and quantity are very low. They are produced and utilized locally.

Herbs are naturally grown in the meadows, which are locally known as 'Kharaks'. The main areas where herbs found extensively are Khankharakhet, Bedani Bugyal, Kanpur Garhi, Beni Tal, Shuvtal and Jhal tal. These herbs are locally utilized for medicinal purposes and local people have the positive results using them.

Simultaneously, a tea garden practice is centuries old but presently its existence is very less. Following the creation of Uttranchal state, the government has launched the scheme for establishing tea gardens in the different geographically ideal locations. In the Pindar basin, there are above five suitable belts for tea production. These are included, Gwaldom-Lolti belt, which is still in existence, Shuvtal-Chhaltal belt, Beni tal belt, Kalimati-Janglechatti belt, Khakharakhet-

Table 17.3 : Cropping pattern under Rabi and Kharif crops

	Sar (in ha) 45.6	*Period of sowing and harvesting*			*Sar (in ha) 56.4*	*Period of sowing and harvesting*	
Name of Crop	*First Sar (Mally Sar)*	*Sowing Period*	*Harvesting Period*	*Name of crops*	*Second Sar*	*Sowing*	*Harvesting*
Manduwa (Kodo)	23.1			Paddy	15.03	March	September
Jwar-Bajra (Junoru-Bajuro)	3.5			Koni-Jhangora	10.7	April	September
Urd	8.0	June-July	October-November	Paddy-Koni-jhangora	30.7		
Soybean	7.4						
Tuar	0.5						
Kulthi	0.1						
Bhatt	1.0						
Ramdana	0.5						
Potato	1.5						
	Second Sar (Tally Sar)	Land near and around the settlement					
Wheat	39.2	November-December	May-June	Maize	4.7	June	September
Barley	8.4			Different types of vegetables in rainy season	5.3		
Oil seeds	8.8						
Total	56.4				10.0		

Source : Primary data collected by the author.

Table 17.4 : Similarly the cropping pattern of 2001 under Rabi and Kharif crops are given in the following table :

Crops	*Sar (Area in Ha) 41.6*	*Period of sowing and harvesting*		*Crops*	*Sar (Area in Ha) 45.8*	*Period of sowing and harvesting*	
	First Sar (Mally Sar)	Sowing period	Harvesting period	Name of crops	First Sar (*Mally Sar*)	Sowing period	
Paddy	32.8	March	September	Wheat	32.6	November-December	May-June
Koni-Jhangora	3.4	April	October	Oilseed	6.5		
Paddy-Koni-Jhangora	5.4			Barley			
		Second Sar (Tally Sar)		Land near and around settlements			
Mandua	29.0	June-July	October-November	Maize	5.2	May	October
Jwar-Bajra	7.5			Vegetables	3.8		
Urd	5.2				9.0		
Soybean	5.1						
Tuar	2.7						
Kulthi	4.1						
Bhatt	0.3						
Ramdana	0.5						
Potato	2.0						

Source : Surveyed by author.

Kanpurgarhi belt and many others which are located in middle latitudes along with very gentle slops.

These practices can help to boost the economy of the entire basin. Similarly, they can check the problems of environmental hazards, because, this region is landslide prone.

Cattle Farming

Raising livestock is an integral component of mountain farming systems. In the mixed crop, livestock farming system, there is a dynamic relationship between common property resources, livestock and crops. Livestock depend to a certain extent on fodder and grass grown of common property resources as well as on crops residue; at the same time, the animals return fodder, grass, and crop residue to common property resources and fields in the form of manure, and provide much needed draught power. Livestock are thus integral to the sustainability of hills and mountain farming.

Dairy farming is an integral part of mountain agriculture. In the Uttranchal state of India, small holders who make up and overwhelming majority, have evolved two major systems of livestock management in mixed crop livestock farming system : sedentary and migratory. Cattle and buffaloes are the only dairying species throughout the basin. Use of goats and sheep as dairy animals is extremely rare. The production of cattle over the last three decades has decreased or remained stagnant, while that of buffaloes has recorded a remarkable increase. Cattle make up about 49 per cent and buffaloes 51 per cent of the 61404 dairy animals in the Pindar basin. The number of dairy animals in the Uttranchal state is 1100839. Out of them 52 per cent are cattle while buffaloes are 48 per cent. All the buffaloes in the region are indigenous. This milch species is largely an admixture of improved type of Murrah and Bhadwari. The overall averages composition of indigenous cattle is 99.2 per cent The following table shows per day production of milk and need of fodder in the three different altitudinal village of the similar longitude (Table 17.4) :

Table 17.4 : Per Day Production of Milk and Need of Fodder in the Pindar Basin

Name of Village (elevation)	*Accessibility from the road head*	*Milk Production by cow (indigenous)*	*Milk Production by buffalo (indigenous)*	*Fodder used Oak/others (in kg) Wint./Summ.*	*Crops residue Whole year*
Prethi (1654)	5 km away	50 kg	150 kg	600/700 kg	400 kg
Ali (1200)	km away	20 kg	25 kg	Nil/200 kg	400 kg
Kimoli (1905)	10 km	80 kg	300 kg	800/900 kg	300 kg

Source : Collected by Author.

The above table denotes that per day production of milk and need of fodder in the Pindar basin varies from one place to another according to the height. The production of milk is increased according to increase in the elevation. The milk obtained from buffaloes is much more than cows. Similarly, an altitude and oak fodder play a vital role in the production of milk. The table reveals that comparatively with the increasing altitude, milk production increased. The fodder need is also met at a large scale in the high altitude. The forest mostly meets the fodder need because there are large grassland and forestland in or surrounding of the villages. The excessive dependency on the forest for fodder and fuel need threatens the existence of environment. The growth rate of plants in comparison to per day exploitation is very low. About 32 quintal fodder is being utilized per day in these three villages as the table reflects. Even though, the production of milk is not desirable as it is in the other parts of the nation. Though the climatic conditions are very feasible for production of milk and rearing of high variety animals, they are yet to be introduced. At least in the accessible areas, which are connected by the roads, may be used for introducing cross-bread animals. This process will boost up the economic condition of the villages.

A comparative case study of the three villages located in the different altitudes and their varied accessibility from the road head has carried out by the present author. Table 17.5 reveals the some of the aspects of animal rearing and their economic significance.

Table 17.5 denotes that the village which is located in the high altitude has the high ratio of animal even the number of goats and sheep are more than to the villages that located in comparatively low altitude. Contrarily, the villages have low altitude the number of cows and buffaloes are much while the ratio is less. Again the table shows

Table 17.5 : Aspects of animal rearing and their economic significance

Village Name	*Height*	*Accessibility*	*Cows Number*	*Buffaloes*	*Sheep*	*Goats*	*Oxen*	*Households*	*Animal/Family ratio*
Wan	2100mts.	10 km up from main road head	62	35	135	106	21 pairs	104	Average 4 Animal/family
Meing Gadhera	1010mts.	On the road	116	67	nil	nil	32 pairs	234	Average 1 animal/farmily
Prethi	1600mts.	5 kms away	67	65	nil	nil	16 pairs	105	Comparatively higher than previous one

per day production of milk and need of fodder in the three different altitudinal villages of the similar longitude.

Primitive agriculture is the main occupation of the Himalayan people. Therefore, the economy rests on agriculture and its allied practices. The motivation of government and non-governmental organization towards and application of modern technology in the fields of agriculture in the Pindar basin did not turn and appreciable mode of development due to its remoteness from the means of transportation, rugged terrain an unwillingness of the people towards development. The natural resources are still unexploited. The problems of soil erosion and landslides are the major issue to be solved at the right time. This basin is known for rapid landslides, throughout its course. Consequently, every year, thousand tonnes of soil eroded by the river and heavy losses of lives are to be taken place.

Tourism

Tourism has a vast scope in the entire basin. There are many places of tourist interests. Roopkund, which is known for untold mystery, Bedani Bugyal, Gwaldom and Karanprayag, are the major sites for tourism and its development. The rich cultural heritage in forms of fairs, festivals, and customs can help for the development of tourism. The presence of Great Himalayan Range and its peaks exhibits the potential for tourism.

Ecology and Environment

The Pindar basin consists of very high mountains and steep precipitous slopes. Throughout the area, marginal land farmers have attempted to squeeze out a living from lands not suited for crop production, and have only ruined the entire landscape in the process, but have created hazards for floods and sedimentation in the area down below. These steep mountainous areas have their own characteristics response to erosion. Avalanches, land slides, and land slips, etc. is fairly common. High intensity runoff of base land surface would not usually permit any trace of soil on these hills. However, there are few redeeming features, which make the slopes sufficiently stable to allow some kind of land use. Poor impact of raindrops coerces skeletal soil and fast vegetation recovery is some of the conditions that favour soil protection. The pressure of population, except, in areas where soil is easily workable, is too little. Soil working is restricted to hand digging. Holdings are small, usually confined near the habitation. Poor accessibility of the area would not provide for major technological changes in near future. On the basis of these assumptions, it may not be necessary to think of major land use changes. The strategy for watershed control would mostly confine to improving practices within major land use regions. The survey aspect should therefore aim at locating areas whose potential is not being fully utilized. For example, if marginal land is suited for agriculture, it should ecologically, be treated through reducing its pressure of overgrazing and developing plantation agriculture on it.

Ecological problems dealing with environmental consequently result in low productivity in the Pindar Basin. Natural hazards, i.e., cloud bursting, soil erosion, mainly in rainy season, and unrationalized human interferences like unscientific quarrying and blasting for road and dam construction, over grazing, over lopping, over following, over ploughing, and dam deforestation, consequently, give birth to low fertility rate and low production and productivity. Generally, landslides in the valley occur when hillside or valley side slopes fail due to certain geological, climatic, and biotic factors.

The Pindar river is perennial. During the winter, the river flows far below the danger mark and in summer mostly on the average level and sometimes above the level due to the snow melting in the Himalayan ranges, but during the rainy season, it flows above danger mark. A heavy rainfall occurred in valley on 20th July, 1970, consequently, the two steep slopes slides and blocks the water of the river for more than 15 minutes, and afterwards, flood situation, occurred, swept away human and cattle life and damaged the low lying areas of the valley. The river Alaknanda was influenced by the flood and even the low-lying areas of Shrinagar town were silted. During the blockade people made a good collection of fishes. Beside this, there were series of landslides taken place during the different rainy seasons. The cloud burst of 1972 in Paithani, natural calamity of Shilodi in 1889, calamity of Bainoly, cloudburst of Gadini Gar in 1992, and cloud burst of Musudiar in 1993 were among the other natural calamity they took place in the course of the Pindar river.

A landmark anomaly is found in the basin. Mainly the basin is divided into two climatic zones, namely, 'Danda' a highly mountainous zone and 'Gangar' a low-lying area of the valley. The word 'Danda' is characterized by highly precipitous zone, predominant with humidity in the atmosphere, which is useful in healthy production of crops without irrigation. The effect of drought of 1987-88 was negligible in this zone (Danda). The corps which are grown in this zone are mostly cash crops like ramdana, potato, soybean, beans, kulthi, tore, etc., inhigh elevation of the zone (more than 1500m height), and rice, wheat, barley, manduwa, koni, jhangora, jawar, bajra, corn and various types of pulses grown between the 1200-1500m altitude. Among the millets, jawar, bajra, koni, jhangora, and among pulses; tore and leguminous plants are sown in the steep slopes and between the ridges of the fields. Due to long roots, the plants protect the soil erosion. The term 'Gangarh' points to the low-lying areas below 1200m and is characterized by the low rainfall and a dry hot zone. Irrigation is essential for high production and productivity in these zones. The main means of irrigation are canals and lift irrigation schemes. The zone is more productive than the Danda zone. The main crops are paddy, wheat, barley, pea, lentil (*masur*) and gram.

The land resources of the valley like forests, water, mineral, soil, and physical landforms have been deteriorated frequently due to excessive erosion and frequent occurrence of mass wasting which is a direct consequence of deforestation, overgrazing, overlopping, over-fallowing, over-ploughing, unscientific mining and quarrying, over-blasting for construction of highways and high dams, and the neglect of the conservation meansures. The over lopping is generally taken into illegal practices by the graziers in the whole of the Garhwal (Saxena, 1988).

The heterogeneity in climate and high anomalies in the landforms make a critical variation in the potentiality and capability of crops. Owing to the inaccessibility of the villages, due to lack of means of transportation, the use of fertilizers, insecticides are negligible. The traditional methods of ploughing and small size of fields lead to the low capability of yields.

Conclusion

What appears to be lacking in the approaches to development planning in the Pindar basin is integration between aspects of environment and economic development, on the one hand, and between development of infrastructure and economic activities and among different activities, on the other. Such an integrated approach is feasible only if it is based on area-level planning; decentralized implementation of plans made at the central and state levels has so far proved to be inadequate. No doubt, certain environmental aspects need to be centrally regulated, but development plans need to be prepared and implemented at the decentralized area level within rationally and clearly defined regulations and guidelines in the context of the use of environmentally sensitive resources. This alone can ensure that the development patterns of the region are based on its specific resource endowments and relative advantages and are sustainable both economically and ecologically.

The rough, rugged and uneven landscape, assign of low fertility and low yields of crops, has given birth to a low living standard of people who depend, directly or indirectly on agriculture. This is the fate of the area, which is unable to fulfil the needs of people to meet just a square meal. Hardly, one third of the annual need is met by the local agricultural production. Obviously, inaccessibility of the region, outdated modes of cultivation, near-total absence of fertilizers and insecticides, and new seeds account for the low yield of crops. There is a redeeming feature too. There is a rich fertile linear belt on both sides of the river from Gwaldom to Karanprayag covering more than 95 per cent cultivated land. Easy accessibility, use of fertilizers and insecticides, HYV seeds, sufficient irrigational facilities would definitely raise the production in these river terraces many times higher than anywhere else in the basin. From the poor Danda the legendary term "Ukharh (dary land) still reflects the poor cultivation. The 'Gangarhs' are fortunate enough to have enough patches of 'Talouns' (irrigated land), which provide a scope for experimental farming. However, the developmental agencies have yet to venture into such experiments in the area.

To protect the valley from the geographical hazards following preventive measures should be introduced :

(a) High power explosives should not be used for road construction.

(b) Underground drainage system should be adopted for small channels and streamlets.

(c) To protect roads from erosion, high embankments enforced with iron concrete should be constructed.

(d) Large-scale deforestation, over lopping, over grazing, and bare fallowing should be checked.

(e) Unscientific farming on the valley slops should also be prevented.

(f) Afforestation and fruit gardens should be encouraged on the valley sides and spurs.

(g) The soil and water loss caused by gully formation and fast flow of seasonal and perennial rivers, must be checked and conserved by scientific measures.

REFERENCES

1. Sati, V.P. : (1993) Problems of Agricultural Ecology and their Management in the Pindar Basin', *National Geopgrapher*, Vol. XXVII, No. 2 pp. 143-48.
2. Sati, V.P., *ibid.*
3. Saxena, P.B. (1998) '*Alaknanda Basin : A Modern Approach in Geography*'.
4. *The Hindustan Times*, 12 Nov., 2000.

18

Combating Land Degradation in Fragile Desert Eco-System Using Remote Sensing and G.I.S.

A Case Study of Shekhawati Region, Rajasthan

T.S. Chouhan

Introduction

More than 6.1 billion ha, 47.2 per cent of the earth's land surface, is lying as dryland. Nearly 1 billion ha of this area is naturally hyper-arid desert, with very low biological productivity. The remaining 5.1 billion ha, are made up of arid, semi-arid and dry sub-humid areas. These areas have been degraded since the dawn of civilization. These lands are the habitat and source of livelihood for about a fifth of the world's population. They are the areas experiencing pressure on environment caused by human mismanagement problems that are accentuated by the persistent menace of recurrent droughts.

UNEP (1993) defined desertification as "land degradation in arid, semi-arid and dry subhumid zones arising from various factors, including climatic variation and human impact."

Desertification is the process of land degradation occurring in the arid, semi-arid and dry sub-humid areas of the globe. These areas are referred to as 'the susceptible dry lands' which cover about 40 per cent of the world's land surface. Recent investigations show that they are the home of more than 1 billion people. They are the focus of the United Nations Covention to Combat Desertification (CCD). Desertification directly affects, or puts at risk, the livelihoods of more than one billion people who are directly dependent on land for their survival.

The problems of degradation of productive drylands, like the Savannas of Africa, Great Plains and Pampas of Americas, Steppes of Asia, 'outback' of Australia and margins of Mediterranean sea are very acute. These lands can no longer sustain life.

Drylands are the lands that sustained our transition from hunting to pastoralism and agriculture and still continue to provide much of loud grain, livestock, the habitat to support human population and big game animals. With the increasing insecurity the available land per capita diminishes day by day and desertification continue to convert these lands into less productive or unproductive ones.

Desertification is mainly a man made problem, resulting from excessive exploitation of natural resources, viz. Land, water, vegetation, etc. under any eco-system in the arid, semi-arid and dry sub-humid areas. Thus, the problem of desertification is a continuous one owing to increasing biotic interferences but it could be controlled to a great extent by improved dry land management practices.

Methodology

The present study has been substantiated by extensive fieldwork. The essential data have been collected from a wide range of sources. The remote sensing data available in black and white and false colour composites of IRS-LISS-III of different seasons on 1:1 million and 1:250,000 scale have been used. Survey of India topo-sheets (old/new) for the entire region in both 1:4 miles or 1:250,000 scale, 1:1 mile or 1:50,000 scale and a biolimatic map, have been used as base maps. Additionally, data from reports, maps, pamphlets, research papers, books, monographs, soil survey data from published and unpublished materials have been collected from different agencies. After examining the remote sensing and other data related to physical (climate, soil, land forms and water) biological (flora and fauna) and social (population dynamics, economic activities landuse and productivity) indicators, a few test sites were located on IRS-LISS III imagery. The sites were visited during field survey.

Result and Discussion

Land degradation is a result of natural and anthropogenic land degradation processes like wind erosion/deposition, water erosion, salinisation, physical disintegration. Among natural processes climate and neotechtonic activities play an important role in the land degradation. However, anthropogenic factors, such as cultivation on marginal lands, cutting of shrubs and trees, overgrazing and use of highly saline and sodic waters growing crops are principal causes of degradation/desertification depending upon their degree of inherent vulnerability of different fragile arid ecosystems.

In the study area almost all land degradation processes, particularly anthropogenic are actively operating and resulted into degradation of different landforms. In the north-western, western and south-western parts due to less than 500 mm rainfall and presence of coarse to medium textured Aeolian landforms wind erosion/deposition is the major process of land degradation. Whereas in the central, north-eastern, eastern and south-eastern parts of the region having more than 500 mm rainfall and presence of fine to very fine textured fluvial landforms (ecosystems) and impeded drainage conditions, water erosion and salinity/alkalinity are the major processes of land degradation.

The severity and extent of the land degradation/desertification hazard of area under different categories of Shekhawati region are presented in Table18.1 and Fig. 18.1 and discussed below.

Wind Erosion/Deposition (8439.35 sq. km or 57.57 per cent)

Wind erosion/deposition process involves the displacement and transportation of Aeolian sediments by Aeolian action resulting in soil erosion at one place and its deposition at another place in the form of thick sand sheeting, hummocks and sand dunes of different shape and size. As a result of this 57.57 per cent of the total area has been degraded, and decrease the soil fertility and crop yield.

Slight Wind Erosion/Deposition (3840. 21sq.km or 26.19 per cent)

This hazard has affected 26.19 per cent of the total area and is mainly concentrated in the flat deep sandy alluvial plains in the form of sand sheets (5-20 cm thickness) and fenceline hummocks (0.5-1.5m high) at field boundaries. Such areas occurin Ranoli, Shishu, around Sikar, Govindpura, Sarwari and Mandha villages of Sikar and Data Ramgarh tehsils.

Table-18.1 : Area under different Land Degradation/Desertification Categories in Shekhawati Region *(Area in sq. km.)*

S.No.	*Types of land degradation (Dominant Processes)*	*Area under different categories (sq. km)*	*Percentage to total area*
1.	Wind Erosion/Deposition	8439.35	57.57
	Slight	3840.21 (26.19)	
	Moderate	2917.05 (19.89)	
	Severe	1682.09 (11.49)	
2.	Water Erosion, Wind Erosion/ Deposition and Water Erosion	4831.91	33.00
	Slight	1347.29 (9.19)	
	Moderate	2387.45 (16.28)	
	Severe Water Erosion	226.28 (1.56)	
	Wind Erosion/Deposition and Water Erosion	876.49 (5.97)	
3.	No. Degradation	1382.74	9.43
	Total Area	1466.00	100.00

Note : Figures in parentheses indicate percentage of various degree of land degradation of total area under different categories.
Source : Field Survey and Laboratory Analysis.

Moderate Wind Erosion/Deposition (2917.03 sq. km or 19.89 per cent)

This hazard has degraded 19.89 per cent of the total area that caused massive sand movement in the form of thick sand sheets (50-100cm thickness) and hummocks (1-2 m height) in the fields and its boundaries. The hazard of this type of severity covers the large area of Fatehpur, Sir Madhopur, Lachhmangarh, Sikar and Danta Ramgarh tehsils near Renu, Palsana and Kanwat villages.

Severe Wind Erosion/Deposition (1682.09 sq. km or 11.49 per cent)

This type of hazard has affected 11.49 per cent of the, area and occurs

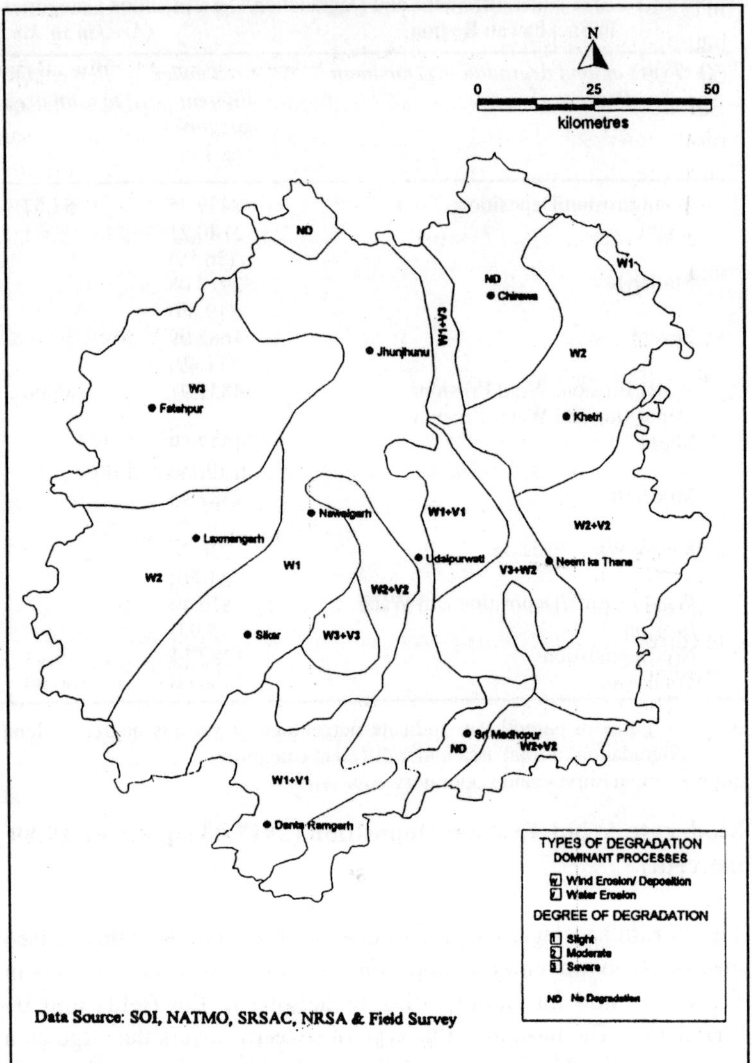

Fig. 18.1 : Shekhawati Regioin : Land Degradation/Desertification Status year, 2001.

in the areas infested by hummocks and dunes of 4 to 60 m height, viz. barchan dunes of 2 to 10 m height and reactivated stabilised dunes of 60 to 80 m height near Bhagot, Lakhawali, Bajor, Mandha, Pipali, Balodi, Dinwa, Thedi, Lawanda and Dantru villages. The increased biotic activities, the dunes have been reactivated resulting in the sand sheeting and deposition of loose sands on the nearby cultivated lands and depletion of their biological productivity in Fatehpur, Lachhmangrah, Sikar Danta Ramgrah and part of Sri Madhopur tehsils near Thimoli Dirwa, Bibipur and Dantru villages.

Water Erosion, Wind Erosion/Deposition and Water Erosion (4837.91 sq.km or 33.00 per cent)

Water erosion in the form of sheet, rill and gully occurs in the north-eastern, eastern, south-eastern and central parts of the region. It affected 4837.91 sq. km. or 33.00 per cent area of the hills, rocky/ gravelly and buried pediments, flat older alluvial plains, younger alluvial plains and river banks mainly attributable to volume, duration, intensity and frequency of rainfall together with variable soil depth, texture, vegetative ground cover and terrain characteristics and presence of perennial or ephemeral drainage channels in the area.

The water erosion on the basis of size and shape of rills and gullies, amount of dissection and loss of sediments has been classified into various degree of severity classes and their salient features are discussed below.

Slight Water Erosion (1347.29 sq. km or 9.19 per cent)

The sheet and rills erosion largely occurs in medium to fine textured flat older alluvial and interdune plains resulted in loss of top soil and exposer of Kanker Pans in Neem-ka-Thana, Sri Madhopur and Danta Ramgarh tehsils near Thimoli, Jugalpura, Shanusar, Bibipura, Gangapur, and Kudan villages that covered 9.19 per cent of total area.

Moderate Water Erosion (2387.45 sq. km or 16.28 per cent)

The severity of moderate water erosion in the form of rills and gullies dominantly occurs in rocky/gravelly and flat buried pediments and gently undulating older alluvial plains and affected 16.28 per cent of the total area near Khandela, Baroct, Raghunathgarh, Gujroki-Dhani and Dudwa villages.

Severe Water Erosion (226.28 sq. km or 1.56 per cent)

Severe water erosion occurred around hills and rocky/gravelly buried pediments and in flat and sandy undulating older alluvial plains which degraded 1.56 per cent of the total area and different sized gullies were formed in Neem-ka-Thana, Sri Madhopur, Danta Ramgarh tehsils near Guda Kalan, Prithvipura, Chankari, Salwari and Samrathpura villages and substrata has been exposed.

Wind Erosion/Deposition and Water Erosion (876.49 sq.km or 5.97 per cent)

Sand deposition and sheetwash as well as gully formation take place simultaneously in 876.49 sq.km (5.97 per cent) area of sandy undulating buried pediments located against hill slopes and affected by moderate degradation near Gauti, Khrol, Mandawa and Todi villages. However, the area degraded by severe wind erosion/ deposition and water erosion hazards lie in the eastern side of Bhergaran-ki-Dhani and in between Kalikhera and Thikaria, Jeronki-Dhani and Bhegeja villages.

The soils of these areas are variable in texture ranges from sandy loam to loam, clay loam and silty clay loam, greyish brown to dark greyish brown, strongly calcareous and dominantly have angular to subangular blocky structure and at places prismatic structure. The substrata is gravelly coated with lime and at places hard lime concretionary layers which restrict the infiltration of surface water resulting in the development of impeded drainage conditions and salinity/alkality problem. The EC of these soils varies from 1100 to 21000 (1:2) Sm^{-1} while pH ranges from 8.6 to 10.2. Natural and Anthropogenic Salinity/Alkalinity occurred in the slightly low lying areas of saline flat older alluvial plans, younger alluvial plains and saline depression *(rann)* of Dams Ramgarh and Aneem-ka-Thana tehsils near Karjkera, Rajpura, Bai, Shyam Khatu, Alodas, Ganora, Rawasa, Raghunathgarh, Khandela and Dunga-ka-Nagal villages.

Anthropogenic salinity/alkanity problem has only developed in pockets due to the use of saline/sodic ground water for raising irrigated crops by unscientific methods of irrigation. This has also resulted in rise of ground water table and development of secondary salinisation in cultivated lands.

In the region considerable area of cultivated lands has been degraded due to irrigation with high residual sodium carbonate (RSC) waters. These degraded lands are located in all the tehsils in pockets.

The soils due to irrigation with the high residual sodium carbonate (RSC) waters, have developed unusual hardness and their infiltration rate has been reduced resulting in poor growth of plants. The crop yield has also declined by 50 per cent within a period of three years.

Thus it could be inferred that out of total 1466.00 sq. km. Area of the region 8439.35 sq. km (57.57%) is affected by wind erosion/ deposition, followed by water erosion wind erosion/deposition and water erosion, 4837.91 sq. km (33%), natural salinity/alkalinity. Rest of the area 1382.74 (9.43) is free from such type of hazards.

Conclusion

Land degradation/desertification is taking place in Shekhawati Region through wind erosion/deposition, wnter erosion, salinity/alkalinity, waterlogging and degradation of vegetation. These processes have been accelerated by increasing anthropogenic, technologenic and zoogenic activities. It has resulted in the depletion of biological productivity of the region. In below 500 mm rainfall zone, wind erosion/deposition is more serious and widespread due to low rainfall and highly vulnerable sandy terrain. Salinity/alkialinity and waterlogging resulted mainly due to faulty methods of irrigation and presence of imprevious strata. Overgrazing and indiscriminate felling of trees resulted in the degradation of vegetative cover and decrease in biomass production In more than 500 mm rainfall zone, water erosion and wind erosion/deposition are the serious problems affecting biological productivity of different ecosystems. In case these problems continue unchecked, large acreage of agricultural land will be affected in future. Based on this integrated information, suitable action plan may be formulated for combating land degradation/desertification.

The present study makes an attempt to analyse the causes and extent of desertification in the arid and semi-arid eco-systems of Rajasthan and suggests measures for combating it. The main objectives of this study are (i) to provide a preliminary synthesis of the available cartographic information on desertification and to present it uniformly on a national and reginal basis; (ii) to locate homogeneous and representative sites for monitoring and developmental programmes and for conservation of resources; (iii) to serve as a framework for more detailed surveys in selected areas. At larger scales, consideration has been given to a greater number of parameters to evaluate vulnerability classes, using more quantiative criteria, particularly on the carrying capacity of land, identifying more degrees of desertification hazards and placing less stress on purely climatic factors; and (iv) to show critical areas. This study will provide information to the planners about the location of possible hazard of a given intensity so that measures can be taken to improve the environment, depending on the intensity of desertification and land degradation processes involved.

The study on the monitoring of the land degradation/desertification has revealed that satellite remote sensing and GIS have great potentials in identification and mapping of land degradation and associated hazards. This integrated information on the land degraded/desertified areas will enable planners and policy makers to evolve and plan suitable measures to reclaim the land degraded/desertified areas and regional development plans.

19

Integrated Water Resources Management and Ecological Concerns

An Overview

R.S. Goel

Introduction

India occupies about 1/50th of the worlds' area, 1/25th of global fresh water but supports 1/6th of global population. It is expected that Indian population will reach a figure of about 150 crores by year 2050. Despite the tremendous economic development and growth of industries in India, the livelihood of about 70 per cent of our population still depends on agriculture, directly or indirectly. Over 170 million people engaged in agricultural practices avail the benefits of irrigation facilities. Availability of water in India is under tremendous stress due to growing population, rapid urbanization, and growing industrial and other demands for maintaining ecology and the increase in per-capita consumption. Deteriorating quality of surface and ground water is further affecting the net availability of water for consumptive uses. The availability of water resources are affecting the environment and, in turn, human well being. Two third of India is arid and semi-arid and is prone to recurring droughts. It is unfortunate that a smear campaign has been launched during last two decades against hydropower and water resources projects by exaggerating the likely or assumed adverse environmental impacts and by suppressing their need and tremendous benefits. It is absolutely necessary that the Indian public is made aware about the fact that twin policy of faster development and better management would have to be vigorously followed. It is essential to make the people aware of the status and issues in water sector on scientific lines. People have to be made an integral part of the water management system. The presentation of facts, analyzed inferences and voicing of unscientific perceptions have to be delinked.

Grave adverse impacts are borne by flora, fauna and domestic cattle and the very life itself fights against nature for its survival. Total absence of precious water during droughts for cattle, wild and aquatic life, forests and plants results in formation and enlargement of deserts. The present dismal share of hydrothermal mix is adversely affecting the optimal utilization of natural as well as financial resources. India has to import the fuel to meet the thermal power needs at a huge cost. Accelerated development of hydropower is highly desirable which would not only help in meeting our energy requirement but will also ensure food security due to increased irrigation from the regulated flows.

Challenges in Integrated Water Resources

Integrated water management is most vital for poverty reduction, environmental sustenance and sustainable economic development in India because water has the potential for both disease causation and prevention. While water is the medium for the transmission of water borne diseases; water is also a primary contributor to the control of infectious diseases through its use for personal and domestic hygiene. Availability of water in India is under tremendous stress due to growing population, rapid urbanization, increase in per-capita consumption, industrial growth and other demands for maintaining ecology. Due to the large temporal variations in river flows in India, storage of water becomes inevitable. To meet the objective of sustainable development, the strategy has to focus on augmentation of available water resources and their optimal utilisation without sacrificing on its quality and with active participation of all the stakeholders.

Concentrated Water Needs Due to Rapid Urbanization

During the last 50 years the share of urban population in the country has increased from 14 per cent to 33 per cent. During the last fifty years the population of India has grown two and half times, but Urban India has grown by nearly five times. India's population has already crossed 1 billion mark and it has been assessed that the urban population may reach 50 per cent of the total population by the middle of this century, as against about 33 per cent at present. Already there is acute shortage of drinking water supply in cities like Bangalore, Delhi, Mumbai, Hyderabad and Chennai and water is being transmitted long distance to cater to the needs. Between years 2000 and 2050 freshwater withdrawals by urban areas will rise from an estimated minimum of about 15 BCM to a projected maximum of about 60 BCM. Options like watershed management, rainwater harvesting, groundwater exploitations, which create spatially distributed resources, are unable to meet these concentrated demands. Supply of safe drinking water to such a large urban population besides meeting the commercial, industrial, cattle and recreational purposes is proving a Herculean task for creating concentrated sources of water.

Water Pollution

Water pollution is a major environmental concern in India. Deteriorating water quality has become a serious problem. The growth of urban megalopolises, increased industrial activity and dependence of the agricultural sector on chemicals and fertilizers has led to the over loading of the carrying capacity of our water bodies to assimilate and decompose wastes. The increasing discharge of domestic and industrial wastes has also led to the contamination of ground water, making it unfit for human consumption at many places. It is estimated that 80 per cent of all diseases and over 1/3rd of deaths are caused by consumption of contaminated water and on an average as much as 1/10th of each person's productive time is sacrificed to water related diseases.

Food Security

National Commission For Integrated Water Resources Development Plan, 1999 has estimated that increase in irrigation facilities alone has contributed to about 52 per cent increase in foodgrains production and improved management, agricultural practices and high yielding varieties have contributed to 48 per cent increase in foodgrains production. Once water is available, crop lands begin to produce more and become less prone to droughts. The threat to sustainability of irrigated agriculture arise due to mismanagement of irrigation water and defective policies that encouraged its injudicious use. The allocation of water for agriculture should be based on the principle to maximize value from a unit of water rather than maximize value from a unit of land.

Electric Power

Hydropower potential of the country has been assessed as 84,044 MW at 60 percent load factor excluding contribution from small schemes. The operation of hydro projects is environment friendly and has longer useful life spans. Hydropower is the cheapest amongst various sources of power supply in the long run since it has no fuel cost component as well as it is almost free from operating cost escalations. The present share of hydrothermal mix in the power kitty of the country is adversely affecting the optimal utilization of natural as well as financial resources. The economic superiority of hydropower has been further enhanced in recent years with steep increases in the prices of imported fossil fuel and rapid exhaustion of exploitable sources of such fuels. Thermal power production requires burning of fossil fuels, seriously affecting the environment adversely.

Droughts

It is estimated that around 263 million people live in drought prone area of about 108 m. ha. which works out to 1/3rd of the total Indian geographical area. Thus, more than 26 per cent of total population of the country face the consequences of recurring droughts, on a wide spectrum of social concerns. During the drought years there is a marked tendency of intensive exploitation of ground water, resulting in abnormal lowering of ground water table thus accentuating the distress. Grave adverse impacts are borne by flora, fauna and domestic cattle and the very life itself fights against nature for its survival. Droughts accentuate problems in cities in the form of mushrooming of slums and pressure on the existing civil amenities thereby adversely affecting urban life.

Floods

Over 40 million hectares of the area of the country experiences periodic floods. The average area affected by floods annually in India is about 7.5 m. ha of which crop area affected is about 3.5 m.ha. Floods have claimed on an average 1529 human lives and 94000 cattles ever year. Apart from loss of life and domestic property, the devastating effects of floods, sense of insecurity and fear in the minds of people living in the flood plains is enormous. The after effects of floods like the agony of survivors, spread of epidemics, non availability of essential commodities and medicines and loss of their dwellings make floods most feared natural disaster being faced by human kind. Large-scale damages to forests, crops and precious plants and deaths of aquatic and wildlife, migratory and native birds in various National Parks, Delta region, low altitude hilly areas and alluvial flood plains of Assam, Arunachal, Uttranchal, U.P., Bihar, Orissa, West Bengal, have always been the matter of serious concern. River Valley Projects moderate the magnitudes as well as frequencies of floods.

Necessitiy of Large Storage Projects in India

Supreme Court Majority Judgement for Narmada Projects has highlighted that against the utilisable storage 690 cu. km. of surface water resources out of 1869 cu. km.; so far storage capacity of all dams in India is only 174 cu. km., which is incidently less than the capacity of Kariba Dam in Zambia/Zimbabwe with capacity of 180.6 cu. km. and only 12 cu. km. more than the Aswan High Dam of Egypt. The impact on environment should be seen in relation to the project as a whole.

It must be noted that future demands of water can be met only by exploiting almost all the 1122 bcm of utilizable water by conventional means. It would thus be essential to utilize the entire 690 bcm of the utilizable flows for which storages of the order of 400 b.cu.m will have to be created. Water demands forecasts show that Rajasthan, Maharashtra, Gujarat, Haryana, Karnataka and Tamilnadu could face heavy water supply shortfalls. 75 out of 99 drought-prone Indian districts have higher water resources potential (with 75 per cent dependability) than their requirement. The water shortages would be far more serious in the water short basins like the Cauvery, Pennar, Sabarmati, Mahi, and Krishna etc. To meet the bulging water requirements, it would be necessary to ensure substantial augmentation of water supplies; requiring sufficient raising of water storage capacities, thus necessitating completion of new large water storage projects. We must realize the basic fact that the medium and small water projects as well as water harvesting schemes cannot substitute the need of large water storages but can at best complement the larger projects. This, too, depends upon the hydrological, geological, topographical and regional limitations.

It is unfortunate that the furious overtones by novelists and pseudo-environmentalists could not have allowed the construction of even the most beneficial projects like Ukai, Bhakra, Hirakud, Ramganga and DVC reservoirs; if these schemes were executed during the present decade. Can we realize the hunger and thirst of 1000 million Indians in the absence of these major water storage projects completed in 1960 and 1970. It is sad that elite novelists, freedom fighters and few activists are spreading fantasies, built around unfounded data, while repeating the same stuff in hundreds of articles and media interviews, speedily followed by battles on railway platforms, *rallies* against water resources projects, threats of *Jal Samarpan* and endless crude methods. India has 1/3rd of its territory drought prone and 1/8th of its territory liable to flooding. About 80 per cent of the surface water of the rivers goes to the sea unutilised while the country reels under the flood-drought-flood syndrome. Social tensions, political instability and street fights are already on the horizon; due to stoppage and slowing down the construction of almost all major dams; ignoring the bulging demands of water and power for municipal uses in metro cities, by few environmental activists and novelists; without professional analysis. Why did America, France, UK, Italy etc. have constructed large and high dams for storage in the range of three to four times of percentage storage in India; even though they are blessed with round the year rains and very little population in sheer contrast to India. The socio-economic development in India is dependant on sustainable conservation of water wealth. We have to act very fast in a decisive manner to accelerate the water resources development due to fast emerging crisis of water for urban use, production of food and power for such large bulging population.

Strategies for Integrated Water Management

Water Conservation

If the projections of future water demand for urban areas as per the National Commission for Integrated Water Resources Development is considered, then the per capita demand works out more than 200 lpcd, which is very ambitious in view of the fact that the fresh water resources in India are depleting and the need of water for various uses is increasing year after year, necessitating conservation of available water and its economical usage. This will, not only help in reducing the total fresh water need of the new as well as augmentation schemes, but also reduce the funds requirement for implementation of such schemes.

Public awareness should be generated through a massive campaign of communication through all available media and by the utility management itself setting an example for conservation. All urban dwellers should be made aware of the source from which water is being brought to the city and from which additional water will have to be brought in the future. They should be aware of the costs involved, not only in financial terms, but also the cost that other communities have to incur in terms of opportunity lost by not using the water. The measure for water conservation should include metering of supplies as a matter of policy and increase in tariff rate on a sliding scale. Use of treated effluents, in place of filtered water for horticulture and large gardens, and fitting of waste—not taps on public stand—posts to avoid wastage of water should be encouraged. Public awareness needs to be created for reducing water consumption, with special emphasis on women's paticipation as they are the major users.

Recycle and Reuse of Water

In almost all major urban centres there is an acute problem of adequate water supply while the sources of augmentation are very few. It is roughly estimated that in urban water supply, 30 to 40 per cent of the municipal water is wasted through the distribution system. In Industrial sector too, there is a scope of economy in use of water. It is estimated by Bureau of Industrial Costs and Prices that 10 to 30 per cent saving in water consumption in industries is possible by recycling, modifications in processing, evaporation control etc. Apart from ensuring leakage control, water conservation strategy in industries should include introduction of appropriate technology to ensure efficient use of cooling and process water and necessary pollution control mechanisms and maximum recycling and reuse. The grey water may be put into various types of treatment such as grease trap, anaerobic filter etc. and the filtered water may be let into wet land, polishing ponds etc. and can be reused for gardening and horticulture etc. The black water may also be put into various types of treatment such as screen, grit removal primary, secondary and tertiary treatment etc. and the treated waste water can be let into wet land for irrigation or for ground water recharge. The municipal wastewater and industrial effluent are being treated upto tertiary level and used for various purposes other than drinking by various industries and cities.

Economic and Legal Instruments

Suitable fiscal concessions and subsidies may be considered by the Central and State Governments to the industries, commercial establishments and any other agencies which adopt/practice waste reuse, recycling and resource recovery. Similarly, in case the Urban Local Bodies on their own would like to take the initiative and set up waste reuse, recycling and resource recovery schemes in their respective areas, similar fiscal concessions and subsidies may also be made available to them by the Central and State Governments. In fact, it may be made mandatory in phases that large industries and commercial establishments must meet a sizeable percentage of their non-potable water requirements from the reclaimed water. Similarly, for irrigating crops, horticulture, watering public lawns/gardens, flushing of sewers, fire-fighting etc. reclaimed water should only be used and to this effect, there is a need for legislation or amendment in the municipal byelaws.

Environmental Impacts — Myths and Realities

Rehabilitation and Resettlement

The crucial problem is that of physical and psychological suffering of the displaced people from the submerged area of reservoirs. Presently full efforts are being made for provisions for more liberal compensation, amenities, land and employment for the displaced population while constructing river valley projects. The rehabilitation of displaced people at some projects like Almatti Dam and Tehri Dam are highly satisfactory. There is a strong need to formulate a strong monitoring group for dealing with rehabilitation of project affected persons. No efforts should be spared to expedite the process of improving the living standards of the tribals by bringing them into mainstream ensuring proper R&R for them.

Experience in the North-Eastern region indicates that the tribal people definitely want water storage projects for bringing in overall economic prosperity to them. Awareness and persuasive approach are needed to tackle this problem. At the same time the rehabilitation packages with scope for more than one option should be made more attractive so that project affected people are induced to accept them. It is expected the National Policy for R&R of project affected persons would soon be adopted to help the State/Project authorities in expediting construction activities. There are many positive aspects noticed in the resettlement and rehabilitation of oustees. In the Upper Krishna Project in Karnataka State, a reputed NGO, Maryada, was associated in preparing the Action Plan for the rehabilitation of PAPs. Similarly, in the Upper Indravati Project, in Orissa State, a voluntary agency viz. M/s Agragamy was entrusted with the work of preparation of rehabilitation master plan. As per the broad assessment made by Central Water Commission through the review of data of 2784 dams, the total project affected persons (PAPs) ranges between 6 to 7 millions. The exaggerated claims by the opponents of large dams blow up this figure up to 70 million by taking the average of the recent few mega dams and multiplying the same by 4291 (total number of dams over 15m height). It has to be borne in mind that most of the high dams (by definition every dam having height of more than 15m is classified as high dam mainly for safety concerns) did not displace persons, firstly due to very thin population in the submergence in earlier dams during construction, secondly very few dams having the height greater than 50m would have the submergence impacts on the upstream habitation. Displacement of human settlements is indeed a painful necessity and must be handled with compassion, fairness and even generosity, aiming at better quality of life for PAPs..

Deforestation

It is said that large dams result in deforestation which is unavoidable in areas to be submerged under water, it would, however, be erroneous to regard Water Resources Development (WRD) projects, as a major cause for deforestation. There are obvious natural limitations on the cattle population that can be sustained on pastures and grass area in a given region. It needs to be stressed that the deforestation on account of water resources projects has been limited to only 4 per cent of the total forest that has been lost during the last 3 decades. Against this loss, the Water Resources Projects have created biomass including tree crops and other plantations several times over the forest lost. Availability of electricity from hydropower projects, in fact, checks demand of fuel and fire wood that greatly contribute to deforestation. Nevertheless, compensatory forestry has become an essential and integral part of all the new water resources projects. In several large projects like Narmada Sagar and Sardar Sarovar Projects, alternative lands for compensatory afforestation have been allocated and the cost of this has been included as part of the project cost.

Sedimentation

Sediment flow in rivers in a natural process and water resources development projects do not contribute to increase in the sediment inflow in the rivers. Provision is made for siltation of reservoirs while planning the projects. The rate of inflow of sediments have been observed to be more than the anticipated values in some of the projects constructed earlier, basically due to the empirical relationships used for assessment of the silt load in the absence of adequate observed silt data. Now, with the advent of modern technology and equipment, rate of siltation can be estimated more accurately. Further, the rate of sedimentation decreases with the age of the project and becomes asymptotic. Moreover soil conservation works, watershed treatment and afforestation are being undertaken in the catchment to reduce the sediment flow.

If water resources projects are not built, silt gets transported unchecked and gets deposited towards the sea thereby gradually flattening the slopes of river in their lower reaches with a consequent tendency to diminish their flood discharging capacity. This, in turn, results in frequent and severe flooding of the lower alluvial tracts. Reservoirs positively control both silt and flood. Several fears have been voiced regarding silting of the Indian reservoirs due to increase of sediment flow in the river. Siltation is a natural process and can be reduced but cannot be eliminated. The reservoirs by themselves have not accelerated the rate of silting. Due to very steep gradients in Himalayan rivers, the run of river and small dams get choked due to boulders as well as coarse sediments. On this account the high dams keeping exclusive provisions for dead storage have an advantage. For example, the Maneri Dam on Bhagirathi (39 m high) and Ichari Dam (60 m high) on the River Tons — Yamuna were designed and filled upto crest by sediments during construction. The storage space for sediment is determined while computing the life of reservoirs, say 50 or 75 or 100 years on the basis of standard reservoir types ranging from lake to gorge. Sediment distribution is assessed to determine new zero elevations for positioning of the outlets for withdrawal of water for various uses. If sub-watershed in the Direct Draining Areas along the rim of the reservoir has very high and high category erodable areas, such areas need to be treated to reduce direct silt inflow into the reservoir and to improve the environment in the vicinity of the reservoir. Geographical diversity precludes precise definition of such watersheds and standardised package for their treatment. Such areas and their treatment will necessarily be project specific.

Flora and Fauna

In case some unique or endangered species of flora and fauna are likely to be threatened by the project, suitable measures are being taken for their rehabilitation. Similarly, if the project interferes with wild life migration, suitable arrangements are made for their habitat. In fact, protection of flora is a contentious issue. However, as the project submerges only a small fraction of forest land, it is difficult to comprehend that the particular endangered species of plants or animals cannot be preserved in the vicinity of the project in the same watershed. Gene banks to preserve species and to regenerate them in favourable conditions, elsewhere, should be possible. Silent Valley project in Kerala, though a promising project planned to develop hydropower, was shelved, as it was projected as affecting prime virgin forest with rare species of plants. On a positive side, Heran reservoir in Gujarat which provided assured water enough for wild life, has actually helped in population growth of wild life and crocodiles, which were on verge of extinction. Pong reservoir is now acting as a resting place for migratory birds and number of rare species of birds have now been sighted in these areas. Water reservoir projects have in general enhanced the natural environment for development of flora and fauna in their vicinity.

Mineral Deposits and Archaeological Heritage

At times mineral deposits, archaeological monuments or shrines are threatened by submergence due to reservoirs. Mineral wealth can be exploited to the possible extent before inundation. Further it is possible to protect the mineral wealth and monuments falling in the shoreline zones by constructing ring bunds etc. Sometimes historical and cultural monuments may fall in the submerged area of a reservoir. The temple of Abu Sibel in Egypt and the Nagarjunakonda in India are living examples of how ancient monuments have been saved and given a greater lease of life. Many temples have been successfully shifted with religious fervour in the Bargi, Srisailam and Sardar Sarovar projects. The Jyotirling temple has been preserved in the planning of the Omkareshwar Dam and improvement in the approach roads and bridges would also be a part of the integrated project itself. Similarly, Dargah at Galiakot, which would have come under submergence of Kadana reservoir on the river Mahi in Gujarat, was protected from submergence by constructing a ring bund. Srisailam, Narayanpur and Almatti reservoirs are shining examples; where historical monuments have been rehabilitated successfully.

Waterlogging and Salinity

The growth of irrigation may subject certain tracts of lands to water-logging, increased soil salinity and alkalinity. Waterlogging can be easily avoided, except for chronic areas due to other topographical reasons, by providing adequate drainage measures right in the beginning of the irrigation project. The change in water table / salinity should also be closely monitored to check any adverse effect on crops and preventive /remedial measures taken. In waterlogged areas, the conjunctive use of surface and ground water, reduction of water allowance and bio-drainage could be useful. Drainage should be integral part of any irrigation project. The Working Group set up by Ministry of Water Resources have estimated areas affected by water-logging, soil salinity and alkalinity in various irrigated commands to be in the order of 2.46 m.ha., 3.06 m.ha. and 0.24 m.ha. respectively. Lack of data restricting long term planning on one side and finance resource crunch on the other hand compelled planners and designers to implement the irrigation projects in phased manner; giving thrust only to the essential components in the first instance so as to accrue the benefits at the earliest and that too with minimum possible investments. Other related issues like efficient drainage system, conjunctive use of surface and ground water etc. are being considered as and when needed. In some projects like Tawa Irrigation Project (M.P.), adopted measures have tremendously removed earlier water logging.

Appreciable seepage from canals, obstruction in the natural/ existing drainage system on account of developmental activities, inadequate support for maintenance, absence of realistic operational plan especially during the initial stages of construction leading to over irrigation in the irrigation projects etc. has gradually led to the change in the ground water table resulting in waterlogging, salinity / alkalinity in a few commands of irrigation projects. State authorities have been taking up necessary measures for controlling the drainage problems depending on their nature, extent and magnitude. Lining of canals, the surface drainage and sub-surface drainage are most widely employed interventions for relieving drainage problems.

Seismic Impacts

It is often claimed that the impoundment of water in a large reservoir triggers earthquakes with their epicenters below or near reservoir. Koyna dam in Maharashtra is very often cited in this connection but there is no conclusive evidence about the correctness. The mighty reservoir like Bhakra and Ramganga, although located in earthquake prone areas did not influence seismicity, since their impoundment. A study of 425 large dams in the world has shown that in case of 15 reservoirs only, the seismic forces were observed to have gone up after construction of reservoirs. In 10 of these 15 cases, the magnitude of earthquakes was less than 5 on Richter scale. It is also said hypothesis that impounding of water in large reservoirs induces seismicity (RIS). There is however little proof of this hypothesis. The failure of large dams due to earthquake is only a fear, which is unnecessarily being projected manifold. Any way, RIS can never be larger than the magnitude of seismicity due to tectonic activities. Indian Meteorological Department, Central Water and Power Research Station and School of Earthquake Engineering, Roorkee, are continuously monitoring the seismic status of the area around selected reservoirs. National Committee on Dam Safety (NCDS) and National Committee on Seismic Design Parameters (NCSDP) for River Valley Projects continuously review and update standards and procedures to ensure earthquake resistant and hydrologically safe construction of large dams. International Symposium on "Earthquake and Dams" held in May, 1987 in China had concluded that "Dams designed by modern techniques and built according to the latest specifications have considerable reserve strength to resist severe earthquakes". They have been shown to resist earthquakes well compared to other manmade structures. There are thousands of dams in the world of over 150 m height but there is hardly any damage to them due to earthquakes. It has been scientifically established that the large and high dams constructed using modern methods of design and construction can safely withstand high intensity earthquakes.

Migration of Fish Species

One of the criticism levelled against the water resources projects is that dams pose barriers across the rivers to affect the migration of fish species like salmon and trout. The provision of fish ladders has successfully taken care of these problems in number of projects such as Farakka, Kosi and Hathnikund barrages. The reduction in the river fisheries can be compensated by rearing fish in the reservoirs. Many new species have taken firm hold in the newly created environment. The striking examples in this regard are fast growing gangetic carp in the reservoirs of Krishna and Cauvery in the Peninsular India and in the Gobind Sagar in north-west Himalayan region. Ukai and Tawa Irrigation Projects can be cited as instances of improvement of fish production where no commercial fisheries existed in the river before impoundment. In Tawa Project the displaced people have been allotted the rights of fishing in the reservoir, yielding substantial income for them.

Positive Environmental Impacts

Health Aspects

Almost half of the world's population suffers from diseases associated with insufficient or contaminated water. Safe water supplies and environmental sanitation are vital for health. Important water borne disease are hepatitis A, cholera, typhoid, polio, diarrhoeal disease and amoebiasis. Important water associated diseases include malaria, filaris, dengu, polio myelitis, heiminthic and infectious hepatitis. Degradation of water quality from industrial and domestic pollution is another major cause of environment-related illness. Water resources projects provide a dependable source of drinking water. People from the irrigated areas may therefore, enjoy better health and sanitation facilities, thus reducing the incidence of disease. General decline in incidence of disease has been reported from some of the selected irrigation commands. It should be noted that the very availability of water leads of improvement in the level of sanitation. The improved economic status also makes people health conscious and capable of availing of requisite health care. The development of area due to irrigation also make available health care units, nearer to irrigated areas. This can be verified from development of villages in Punjab, Haryana and Western UP where many facilities, particularly 'Health Care Units' have become available in villages, alongwith other facilities like roads, electricity etc.

Tourism and Recreation

Recreational facilities like boating, rowing and other facilities for water sports may be created in the river valley projects. Water resources projects lead to social and cultural improvements. Development of gardens and recreation areas have taken place around many reservoirs making the environment more pleasant and providing recreational facilities to the people. The gardens developed in water resources projects like Brindavan garden, Ukai garden, Ramganga garden, Kalindi Kunj, Dhyaneswar Udyan, Jaikwadi Garden, Matalila Garden, Pinjore Garden; Bhakra garden, Mata Tila garden, Jayakwadi garden, Nagarjunasagar garden and Bagalkot New Township garden are great tourist attractions in India. Forests around reservoir created by Periyar dam is a very popular tourist spot to watch varieties of wildlife, while boating in the reservoir. Large reservoirs thus provide additional benefits; as these can be economically developed into tourist attractions for skiing, fishing and recreational facilities.

Improvement in Wildlife

Case Study for Ramganga Dam Project clearly shows that there has been substantial increase in the numbers of tigers, panthers, elephants cheetals, crocodiles and ghariyals (as per countings by Wildlife Preservation Organisation of UP Forest Department, in the famous National Corbett Park. Such increase has been due to continuous availability of green fodder, plentiful availability of clean water throughout the year and environmental security among animals and improved climatic conditions; after construction of the 127.5 m. high Ramganga Multipurpose Dam Project; even though 55 sq. kms. of the Corbett Park has been either submerged or affected by this reservoir. It is also found that rare species of birds flock in and around Ramganga reservoir. This phenomenon of an increase in birds and wildlife has also been observed around Pong, Bhakra, Pandoh, Rihand and Matatila reservoirs. Some of the best tourist places of India like Ukai tourist resort, Periyar wild life sanctuary, Shalimar Garden, Brindavan Garden, Pinjore Garden, Kalindi-Kunj, Matatila Garden, Gyaneshwar Udyan, New Bagalkot Town Garden and Ramganga Udhyan are highly beneficial impacts of water resources projects. Significant increase has also been reported in number of species of wild animals and aquatic life after construction of water resources projects.

Multiplier Socio-Economic Benefits

Large infrastuctural facilities including bridges, metalled roads, telecommunication and postal services, large scale employment opportunities, soil conservation and afforestation in very large areas, assured power and water for agricultural and industrial purposes have resulted in tremendous direct and indirect socio-economic benefits; due to the multiplier chain effect of benefits by most of the water resources projects. Ramganga, Pandoh, Pong, Bhakra, Hirakud and Beas dams and Indira Gandhi Canal Project are shining examples in this regard. The headquarters of Ramganga Project, Kalagarh was being seriously considered for siting the capital of new Uttranchal State due to availability of all facilities and infrastucture by construction of this project.

Assessment and Management of Environmental Impacts

Proper mechanism did not exist upto 1978, for assessing the environmental impacts of the river valley projects in India. CWC and Planning Commission had indicated broad-lines for the preparation of Detailed Project Report and clearance of the Projects. Department of Environment and Forests (DoEF) prepared in 1978 Guidelines for EIA of river valley projects. It was aimed in these Guidelines that each river valley project should be subjected to rigorous assessment of their environmental impacts so that necessary mitigative measures could be incorporated in the project right at its inception stage. These guidelines for site selection as well as incorporation of mitigative measures were prepared and circulated among the project authorities through the Planning Commission. It was made imperative to analyse whether the adoption of environmental measures is going to result in any short or long term social and economic benefits or not. Special emphasis was laid on the health effects, plant genetic resources, aquatic life, waterlogging, salinity of irrigated soils, deforestation and soil conservation for considering the techno-economic viability of the project.

Aspects recommended for consideration in site selection included immediate and long-term impacts on human settlements, flora and fauna in the vicinity, impacts on monuments and mineral resources, ground water levels, etc. The Guidelines include most of the important effects on the environment such as (i) involuntary settlement of project affected persons, (ii) sedimentation, (iii) waterlogging, (iv) salinity,

(v) monuments/shrines, (vi) submergence of flora and fauna, (vii) aquatic life, (viii) health, (ix) climate, (x) water quality, etc.

Under Environment (Protection) Act, 1986, the Government has acquired wide powers for protecting the environment. Some other Acts related to Water and Environment are Water (Prevention and Control) of Pollution Act, 1974 (amended in 1988), Water (Prevention and Control) of Pollution (Cess) Act. 1977 (amended in 1991), Forest Conservation Act, 1980 and Environmental Impact Assessment Notification of 1994 (amended in 1997) and the Ministry of Environment and Forest's Notification of January 1997, constituting the Central Ground Water Authority (CGWA).

The Water (Prevention and Control of Pollution) Act, 1974 seeks to maintain or restore "wholesomeness of Water" and the Central and State Pollution Control Boards have been established under this Act. According to the Water Cess Act, 1997, both Central and State Governments have to provide funds to the Boards for implementing this act. The Forest Conservation Act, 1980 provides for compensatory afforestration to make up for the diversion of forest land to non forest use.

The Environment (Protection), 1986 was enacted in 1986 for the protection and improvement of human environment. The EIA Notification of 1994 has made the environmental clearance mandatory for all the new projects as well as for expansion/modernisation of existing projects covering 29 disciplines including hydropower, major irrigation and flood control projects. Amended Notification of 1997 has made it mandatory to hold environmental public hearing before according environmental clearance to the listed category of projects.

Central Water Commission has published, "Guidelines for Sustainable Water Resources Development and Management", September 1992, to meet the Indian requirement for data collection and analysis, impacts assessment and preparation of management plans. Together with EIA for water resources projects, Environmental Management Plan (EMP) is also essential to ensure sustainable development. Special attention has to be given for rehabilitation and resettlement of affected persons, compensatory afforestation and catchment area treatment of directly draining degraded areas.

Four prestigious publications have been compiled, each containing over 30 invited state of art like articles on well focused subjects from eminent experts, chief executives of water and power development agencies, academicians, apex professional institutions and credible NGOs, under the following titles:

(a) *Hydropower and River Valley Development—Environmental Management, Case Studies and Policy Issues,* Edited by Goel R.S. and Srivastava R.N.

(b) *Environmental Management in Hydropower and River Valley Projects-Techniques of Management, Case Studies,* Policy Issues and Application of Scientific Tools—Edited by Goel R.S.

(c) *Environmental Impacts Assessment of Water Resources Projects — Concerns, Policy Issues, Perceptions and Scientific Analysis* - Edited by Goel R.S.

(d) *Environment Management in Hydro Electric Projects*— Edited by Prasad Kamta and Goel R.S.

First three publications have been published by M/s Oxford and IBH Co. Pvt. Ltd New Delhi and the fourth one by M/s Concept Publishing Company, New Delhi. Above Guidelines and Books have been published for the study and management of environmental impacts in various river valley projects to achieve rapid economic development on a sustained basis; with due care for safeguarding the environment.

Scientific Public Awareness

India may envisage meeting most of the water needs, without causing any large scale distress if the twin policy of faster development and better management has to be vigorously followed even though the supply and demand balance is precarious. Water quality concerns would become serious due to the large scale internal recycling and the reuse of water which would be unavoidable and sustainability would have be ensured through stringent measures. It is essential to make the people aware of the status and issues in water sector, both through the process of creating public awareness through mass programmes and through sharing of information. The environmental aspects and process of planning and operation of water resources projects should also be fairly understood by the experts of different disciplines. Participation of people is a must in the management of water. People have to be made an integral part of the water management system. The community is to be made not only water conscious, but also to be integrated to participate in the planning and management of such projects and pollution prevention programmes.

It is unfortunate that a smear campaign has been launched during last two decades against hydropower and water resources projects by exaggerating the likely or assumed adverse environmental impacts and by suppressing their tremendous benefits. Professional Societies can act as multi-disciplinary fora for national and regional debates, analysis and framing of action plans on water related matters by utilizing their infrastructure, professional expertise, library, publication and documentation services. These societies can serve as rich sources in generating technically sound options with well-defined limitations and assumptions in Indian peculiar situations for taking informed decisions. The services of Professional Societies having vast network, good spread and pool of expertise may be channelised in debating, dissemination and documentation of reports and creating balanced scientific public awareness.

Conclusion

Water requirements of the country would continue to grow partly due to the rise in population and partly as a result of the improvement in the quality of life. It is to be stressed that non-development of water storage projects is not a viable or available option; due to the large temporal variations in river flows in Indian monsoonic climate. We must realize the basic fact that the medium and small water projects as well as water harvesting schemes cannot substitute the need of large water storages but can at best complement the larger projects. This, too, depends upon the hydrological, geological, topographical and regional limitations. The controversy of the large versus small dams is irrelevant. Preserving the quality and the availability of the freshwater resources is the most pressing of the many environmental challenges on the national horizon.

Whole focus on sustainable water development, as it is construed in the present environmental politics, concentrates on what it is not, and then tries to ameliorate the negative impacts. It is essential that environmental activists should be equally concerned for maximisation of positive environmental impacts and socio-economic benefits from the large dams. However, what is more difficult to negotiate is the challenge posed by environmental activism, which if not contained may throw up remedies that may prove to be worse than the disease. The media should set its focus right to set the agenda and making the society conscious through balanced public awareness while seriously considering the presented facts in this article but need to suppress street fights generally activated by self acclaimed environmentalists and novelists without scientific background.

It is evident that the benefits accrued from large river valley projects are so immense that they substantially outweigh the costs of immediate human and environmental disruptions. On the other hand, long-term adverse effects of not utilising the water resources would be catastrophic due to recurrence of floods, droughts and the resulting unemployment, which further increases the backwardness in countries like India. Catchment area treatment and watershed management are development projects in their own right and should be planned and executed as such independently without putting undue financial burden on water resources projects. At best, treatment of direct draining sub-watersheds along the reservoir rim could be charged to the cost of the reservoir project. Mankind is at the center of environment which cannot be protected unless the basic needs of food, drinking water and energy for an all round development are expeditiously met by accelerated vital improvement of environmental condition through irrigation, water supply, hydro and multipurpose river valley projects.

Note- The views in the article may not belong to the Author's Organisation.

REFERENCES

Central Water Commission, (1988) Water Resources of India, New Delhi.

Central Water Commission, (1992) Guidelines for Sustainable Water Resources Development and Management, New Delhi.

Goel, R.S. (Editor), (1993) *Environmental Impacts of Water Resources Development*, M/S Tata McGraw Hill Publishing Company, New Delhi.

Goel, R. S. and Srivastava, R.N (Editors), (2000) *Hydropower and River Valley Development- Environment Management, Case Studies and Policy Issues*, M/s Oxford & IBH Publishing Co. Ltd., New Delhi.

Goel, R.S. (Editor), (2000), *Environment Impacts Assessment of Water Resources*

Projects - Concerns, Policy Issues, Perceptions and Scientific Analysis M/s Oxford & IBH Publishing Co. Pvt. Ltd., ISBN-81-204-1422-5, New Delhi.

Goel, R.S. (Editor), (2000) *Environmental Management in Hydropower and River Valley Projects-Techniques of Management, Policy Issues, Case Studies and Application of Scientific Tools*, M/s Oxford & IBH Publishing Ltd., ISBN-81-204-1423-3, New Delhi.

Goel, R.S. (1999) 'The Good Side to Dams is Rather Unfairly Overlooked in the Vital Matter of Water Resources Management', Editorial Page Article, *The Economic Times*, 27th June,1999,New Delhi.

Goel, R.S. (2000) 'Unquiet Narmada-The Antoganism against River Valley Projects is Unjustified', Editorial Page Article, *The Economic Times*,31st Dece,2000 ,New Delhi

Goel, R.S. (2001) 'River Valley Projects, Dams are Beneficial', *The Times of India*, 11th May, 2001, New Delhi

Goel, R.S. (2000) 'Unquiet Narmada-The Antoganism against River Valley Projects is Unjustified', Editorial Page Article, *The Economic Times*,31st Dece,2000 ,New Delhi

Goel, R.S. and Patel, V.B., (2002) 'Scientific Public Awareness about Ecological Concerns and Integrated Water Resources Management in India', IWRA Regional Symposium on Water for Human Survival, 27-30 November 2002, New Delhi.

Goel, R.S. and Patel, V.B., (2003) 'Ecological Concerns and Challenges in Integrated Water Management in India', 4th International R&D Conference, CBIP, Aurangabad, 29-31st Jan., 2003.

Goel, R.S. and Patel, V.B., (2003) 'Integrated Water Management in India - Environmental Concerns and Prospects', International Seminar on Watershed Development and Drinking Water Supply, 16-18th Oct, 2003, MERGEFIELD State Visakhapatnam.

India Water Partnership, Institute for Human Development, (2000) "India Water Vision 2025- Report of The Vision Development Consultation", New Delhi.

Indian Water Resources Society, (1999) "Theme Paper on Water: Vision 2050", New Delhi.

Indian Water Resources Society, (2002) "Theme Paper on Integrated Water Resources Development and Management", New Delhi.

Ministry of Water Resources, (1999) " Report of the National Commission for Integrated Water Resources Development", New Delhi.

Prasad, K. and Goel, R.S. (Editors), (2000) *Environmental Management in Hydro Electric Projects*, M/s Concept Publishing Company, ISBN-81-7022-870-0, New Delhi.

Supreme Court Judgement, (2000) Narmada Bahao Andolan Vs. Union of India, AIR, SCC, 2000.

Water Management Forum, (2002) 'Theme Paper on Inter-Basin Transfers of Water - Challenges and Opportunities', New Delhi.

SECTION-II

SUSTAINABLE AGRICULTURAL DEVELOPMENT AND FOOD SECURITY

20

Science and Technology for Sustainable Food Security

M.S. Swaminathan

Introduction

Albert Einstein once said "*concern for man himself and his fate must always form the chief interest of all technical endeavours in order that the reaction of our minds shall he a blessing and trot a curve*." It is my conviction that scientific enterprises which are not base on the above principle, will ultimately do more harm than good. Hence, all efforts in science and technology, whether in the public or private sectors, should have public good as their goal.

Private enterprises to be viable have to place economic success as their overriding priority. Their duty is to their shareholders. In contrast, the primary duty of public-funded research institutions, owing their existence to the taxes paid by citizens to government, is to work for the benefit of all citizens of the country.

Public-funded research has become particularly important in the context of the rapid expansion of Trade Related Intellectual Property Rights (TRIPS) regime, following the coming into force of the World Trade Agreement on January 1, 1995. In this paper, I wish to discuss a few implications of the new climate of market-driven economic forces on agriculture and food security.

During the last 25 years many chronically food deficit countries of the Asian region have been able to raise food output above the rate of growth in population. This has been achieved primarily through increased productivity per unit of land and time. Such an improvement in productivity was referred to as *Green Revolution* by William Gadd of the United States of America in 1968. A recent study by the International Food Policy Research Institute, however, indicates that the prospect for reducing malnutrition during the next 25 years among children is bleak. According to the most likely 2020 vision scenario, the proportion of malnourished children will decline from 34 per cent to 20 per cent by 2020. But because of population growth, the actual number will decline only slightly. Thus, there is no time to relax on the food production and distribution front.

Twenty-seven years after the term Green Revolution was coined, we are in a position to draw a balance sheet and chalk out a strategy for the future. Apart from erasing the "begging bowl" image of India the most important gain has been the saving of forests and land, thanks to the productivity improvement associated with high-yielding varieties (HYVs). In 1995, Indian farmers harvested over 60 million tonnes of wheat, as compared to 6 million tonnes at the time of our independence in 1947. Punjab farmers have raised the average yield of wheat to over 40 quintals per hectare. Likewise, Tamil Nadu farmers have raised the average yield of rice to over 50 quintals per hectare. If the yield improvement associated with the Green Revolution in wheat and rice had not taken place,we will need another 70 million hectares to produce the wheat and rice we now harvest. Thus, the productivity improvement associated with the *Green Revolution* is best described as forest or land saving agriculture.

Towards an Ever-Green Revolution

The population of India is growing. The average total facility rate of India in 1992-93 was 3.39. Only Kerala, Tamil Nadu, Goa and Mizoram have so far achieved a demographic transition to low birth and death rates. Besides population increase, improved purchasing power among the poor will enhance the demand for food, since under-nutrition and poverty go together. In contrast, per capita availability of arable land is shrinking. Water use efficiency is still on the whole low and water markets arid disputes are going. In addition to the gradual decline in per capita availability of land and water, various forms of biotic and abiotic stresses are spreading. There is still a widespread mismatch between production and post-harvest technologies. In perishable commodities like fruits, vegetables, flowers, meat and other animal products, this mismatch is often severe, affecting the interests of both producers and consumers. This is why foreign experts frequently refer to the setting in of a fatigue of the Green Revolution. Lester Brown and Hal Kane in their book *Full House* released in 1994 predict that at the current rate of population growth and environmental degradation coupled with an improvement in the consumption capacity of the poor, India will have to import annually over 40 million tonnes of foodgrains by the year 2030. This is four times the quantity we imported in 1996, i.e. before the onset of the Green *Revolution. We should examine this issue seriously and should not allow complacency overtake the farm sector, just because we now have over 35 million tonnes of food grains available with Government.*

Industrial countries are responsible for much of the global environmental problems such as potential changes in temperature, precipitation, sea level and incidents of ultraviolet-radiation. *While further agricultural intensification in industrialised countries will be ecologically* disastrous, *the failure to achieve agricultural intensification acid diversification in predominantly agricultural developing countries will be socially disastrous.* This is because, agriculture including crop and animal husbandry, forestry and agro-forestry, fisheries and agro-industries provide livelihood to over 70 per cent of India's population. The smaller the farm, the greater is the need for higher marketable surplus for increasing income. Eleven million new livelihoods will have to be created every year in our country and these have to come largely from the farm and rural industries sectors. Importing food and other agricultural commodities will hence have the same impact as importing unemployment. Thus what we need now is an environmentally sustainable and socially equitable green revolution or what may be termed an ever-green revolution.

Those who advocate going back to the old methods of farming ignore the fact that just a century ago when the population of undivided India was 281 million, famines claimed 30 million lives between 1870 and 1900. The famine eradication strategy of India comprising the following steps is perhaps the most important achievement of independent India : (a) greater investment in irrigation and technology development and dissemination leading to enhanced production and productivity, (b) better distribution through the public distribution system, (c) adequate grain reserves, (d) purchasing power enhancement through various employment generation and guarantee schemes, and (e) special intervention programmes for children, pregnant and nursing mothers and old and infirm persons.

While famines have been prevented, widespread under-nutrition prevails among the economically underprevileged. Since several non-food factors like environmental sanitation and hygiene also affect food and nutrition security, we need to develop a holistic concept of food security.

Food security implies livelihood security at the level of each household and all members within and involves ensuring both physical and economic access to balanced diet, safe drinking water, environmental sanitation, primary education and basic health care.

Hunger-Free Area Programme and National Food Security Act

To give operational content to such a concept of food security, we should initiate a Hunger-Free Area Programme (HFAP) consisting of the following components:

(a) Ensuring sustainable availability of food by maintaining the growth in food production over population growth through the development and dissemination of ecotechnologies, supported by appropriate packages of services and public policies. Ecotechnology involves the blending of the ecological prudence and technologies of the past, with the best in frontier technologies particularly biotechnology, information technology, space technology, renewable energy technology and management technology. Without ecotechnological empowerment, farm men and women will not be able to produce more food and other agricultural commodities on an environmentally sustainable basis from land, water and energy resources.

(b) Sustain the productivity of the natural resource base, by conserving and improving the ecological foundations essential for continuous advances in crop and animal productivity.

(c) Ensure adequacy of household incomes through promotional social security, such as accessing assets, employment and organisational and marketing empowerment. Agricultural programmes should concurrently aim at more food, more jobs and more income. Integrated attention to farm and non-farm employment and value-addition to primary agricultural commodities will be necessary to enhance income and rural livelihood security.

(d) Provide entitlement to food through protective social security measures to the vulnerable groups, such as employment guarantee and food for nutrition programmes.

(e) Introduce a National Food Security Act with the concurrence of the National Development Council for the purpose of paying integrated attention to :

   (i) Conservation of land, water, forests, biodiversity and the protection of the atmosphere,
   (ii) Enhancing productivity through ecotechnologies,
   (iii) Improving distribution in order to eliminate endemic hunger,
   (iv) Maintenance of adequate food security reserves,
   (v) Strengthening the techno-infrastructure for better post-harvest technology and expanding the coverage of sanitary and phytosanitary measures, and
   (vi) Efficient research, education, extension, marketing systems both to take full advantage of emerging opportunities in international trade and to ensure that research and extension designed to promote public good receive adequate support.

1995 marks the beginning of a new era in global agricultural research due to the onset of (TRIPS). How do we promote research for public good under a fast spreading Intellectual Property Rights Environment (IPR) in order to ensure that our national effort for achieving the ecotechnological empowerment of resource poor farm and rural families does not suffer under an economic environment where profits get precedence over public good? Investment on research designed to promote public good must be stepped up. At the same time, we must promote meal social contracts between the corporate sector and resource poor rural families so that there is value-addition to the time of the poor.

Many of the environmental problems associated with the recent phase of agriculture and coastal aquaculture arise from the excessive and unscientific use of chemical pesticides and mineral fertilisers as well as the abuse of land and groundwater. In my lecture at the Indian Science Congress held in Varanasi in January, 1968. I had warned about potential environmental harm from exploitative agriculture, and I quote, from what I said then :

> "Exploitive agriculture offers great possibilities if carried out in a scientific way, but poses great dangers if carried out with only a immediate profit or production motive. The emerging exploitive farming community in India should become aware of this. Intensive cultivation of land without conservation of soil fertility and soil structure would lead ultimately, to the springing of deserts. Irrigation without arrangements for drainage would result in soils getting alkaline or saline. Indiscriminate use of pesticides, fungicides and herbicides could cause adverse changes in biological balance as well as lead to an increase in the incidence of cancer and other deseases, through the toxic residues present in the grains or other edible parts. Unscientific tapping of underground water will lead to the rapid exhaustion of this wonderful capital resource. The rapid replacement of numerous locally adapted varieties with one or two high-yielding strains in large contiguous areas would result in the spread of serious deseases capable of wiping out entire crops. Therefore, the initiation of exploitive agriculture without a proper understanding of the various consequenses of everyone of the changes introduced into traditional agriculture, and without first building up a proper scientific and training base to sustain it, may only lead us, in the long run, into an era of agricultural disaster rather than one of agricultural prosperity."

Such an action-reaction analysis led to the intensification of efforts in the areas of varietal diversification, development of integrated pest and nutrient management systems and improved soil health care. Consequently, we now have several hundred HYVs in major crop plants and effective methods of pest and soil fertility management. There is every opportunity for pushing ahead with yield improvement without associated ecological harm.

Sustainable green revolution is an ecological and economic necessity in our country of small holdings and 100 million farm families. *Therefore it would he better to coin another term to denote the unsustainable use of yield enhancing technologies. I would like to refer to it as the Greed Revolution.* What we should strive for in both agriculture and aquaculture is the promotion of sustainable green revolution (i.e., improving production through productivity improvement) and the curbing of the greed revolution. Failure to make a distinction between green and greed revolutions often causes confusion in public mind and will do harm to public good.

Food Demand and Availability in the Asia-Pacific Region

Various projections on food demand and availability for the medium-term, up to 2010, indicate that the supply position is fairly satisfactory in the developing countries as a group. Given the untapped yield reservoir existing at current levels of technology, prospective developments in technology, and constraints in organisational pattern and the resource base, it should be able to maintain the balance between demand and supply. In the Asia-Pacific region demand for foodgrains could be met in future decades mostly from production within the region and some imports from outside. In 2000, for an expected population of 3,144 million demand is estimated at 830 million tonnes of food grains and production at 811 million tons, leaving a deficit of 19 million tonnes. In 2010 population of 3,729 million would demand 959 million tonnes of food grains. Taking into account the prospective developments in agriculture (technology) and constraints (resources and investment), the production anticipated is 927 million tonnes which would leave a deficit of 30 million tonnes.

In the long run (2030 or 2060 AD) globally the situation causes concern, primarily arising from the problems of resource degradation, includng a possible change in the climate. But there are some positive indications. Firstly, that if sufficient measures are taken during the period when food availability is ensured, i.e. during the medium-term, then there is the possibility of the problem being overcome. Secondly, towards the middle of the next century, population stabilisation is anticipated for most of the developing countries. World population is expected to stabilise at about 10 billion by then. Additional demand for food on the population score would then come only from enhanced purchasing power. Higher purchasing power leads to the diversification of food habits.

At the aggregate level, globally and regionally as well, food supply was fairly adequate to meet the demand. Despite that, hunger is persisting. A very substantial portion of the population, both at the global level and in the Asia-Pacific region, remained undernourished. In 2010 AD, some 637 million people in the developing countries of the world will remain undernourished, and some 640 million people could be at risk even in the year 2060, if the current trend of jobless econpmic growth persists (Tables 20.1 and 20.2).

Table 20.1 : Demand and Production of Creals *(In million tonnes)*

	Demand				*Production*			
Territory	1969-1971	1979-1981	1988-1990	2010	1969-1971	1979-1981	1988-1990	2010
South Asia	124	161	205	317	121	156	209	307
East Asia	215	327	425	642	211	307	409	320
Asia-Pacific region	339	488	630	959	332	463	618	927
Developing countries	492	709	918	1,460	480	650	845	1,314

* Both for human consumption and animal feed.
Source : FAO (1993).

Table 20.2 : Under-nourshed Population *(In millions)*

	Population				*Proportion*			
Territory	1969-1971	1979-1981	1988-1990	2010	1969-1971	1979-1981	1988-1990	2010
South Asia	254	285	271	202	34	31	24	12
East Asia	497	359	252	70	44	26	16	4
Asia-Pacific region	751	644	523	272	40	28	19	8
Developing countries	941	843	781	637	36	26	20	11

Source: FAO (1993).

Poverty is the primary cause of hunger, and of chronic food insecurity. In the mid-1980s, 72 per cent of the world's poor and 66 per cent of the world's extremely poor lived in the Asia-Pacific region. Of the total population, 32 per cent were poor and 18 per cent extremely poor. Over the years, the proportion of poor has declined in this region. But in absolute numbers, it is staggering, with 560 million poor and 280 million absolutely poor. In 2010, by and large, the food availability will meet the demand in terms of the income generated. Still this region as a whole will be importing food of about 40 million tonnes. Some 272 million people could suffer from chronic malnutrition. This situation can be avoided if the suggestions contained in this report are implimented.

Potentials and Constraints of the Region

Past experiences in food production efforts in the South and South-East Asia region provide signs of both hope and despair. Much of the increase in cereal production in the region arose from enhanced productivity rather than expansion of area. Between 1966-70 and 1976-80, production increased at 3 per cent per annum; area at 0.7 per cent and yield at 2.3 per cent. Between 1979-81, and 1989-91, production increased at the rate of 2.6 per cent per annum, yield at the rate of 2.7 per cent while area declined by 0.1 per cent. Productivity improvements were achieved largely through the use of scarce resources like water, and high-energy externally dependent input like fertilizers which are leading to biotic and abiotic stresses. The growth rate of rice per annum between 1966 and 1974 was 2.1 per cent, between 1974 and 1982 2.6 per cent, while between 1982 and 1988 it was only 1.9 per cent. Figures for wheat for the same periods are 4.1 per cent, 4.4 per cent and 2.1 per cent. Cereal imports into the region still continue (47 million tonnes against the total production of 763 million tonnes in 1989).

The share of agriculture in the Gross Domestic Product (GDP) declined from 40 per cent in 1965 to 23 per cent in 1991, but the dependence on agriculture as a source of employment has not registered commensurate decline (70 per cent in 1970, 64 per cent in 1980 and 61 per cent in 1985). Therefore, the per capita GDP of the rural population is declining. Urban slums are expanding, largely due to the migration of landless families who go to towns and cities in search of jobs.

Food Security Redefined

The persistance of hunger, despite enough food to go around globally, and in most of the regions and nations, brings up the need for redefining the food security concept. Conventionally, food security is defined as the balance of food supply (mainly cereal supply) and effective demand for food. Despite the balance (otherwise termed self-sufficiency) those who cannot afford to buy food are left hungry and undernourished. Food security as conceptualised in this study makes departures from the general perspective, in composition, access and spatial dimension. In composition, food is specified as a basket of commodities consisting of not just calories but a nutritionally balanced diet, as well as a nutrition-effective environment with safe drinking water to minimise wastage arising from waterborne diseases, and basic sanitaion to plug leakages owing to parasitic infestations, and an empowering environment with elementary education to inculcate the entitlement right to food security. In access, it recognises not only physical access or availability, but equally, emphasises economic access to food or recognises entitlement to food. Economic access is reinforced by narrowing the spatial dimension from a geographic or territorial entity to the household, there by equating food security with 'livelihood security. The gender-based infra-household discrimination is warded off by ensuring food security to the individual. Thus food security needs to be redefined as "livelihood security for the households and all members within, which ensures both physical and economic access to balanced diet, safe drinking water, environmental sanitation, primary education and basic health care." Following the World Nutrition Congress Resolution of 1992, food security in its composition is defined with nutrition security and accessibility not only in terms of availability but concurrently with entitlement. Spatially, it moved from a geographic territory to the household and, improving further, to reach down to all individuals within the household.

The Poor and the Food Insecure

Two main groups constitute the undernourished or food-insecure. First those who live in poorly resource-endowed areas, such as degraded regions of the world, the mountainous countries, the tribal areas, etc., where poverty is still pervasive. The second group belongs to the resource-endowed areas, but has little or no asset to fall back upon such as the landless, rural artisans, and even small farmers with extents of lands but not enough to make a living. Iniquitous asset distribution is the prime cause of poverty and food insecurity. Countries in the region which have accomplished greater equity in the asset distribution have also succeeded in making a dent into poverty. And the reverse is true of those countries which have failed in bringing about equity in the ownership of assets, not only land but other forms of productive assets. Degradation of the environmental quality especially of the common property resources has further aggravated the problems of the poor, particularly women. Women and children suffer more and the feminisation of poverty can be overcome only through providing them with ownership rights to land.

The Challenges and Opportunities

South and South-East Asia region faces two basic food security challenges. First, maintaining the availability of food through production from within the region, which is constrained in the immediate future due to conditions of diminishing land and fresh water resources, expanding biotic and abiotic stresses, inadequate investment in rural techno-infrastructure and an inequitable trade environment; and in the long run, due to deceleration in production setting in as a result of ecological factors, such as depleted soil, exhaustion and pollution of ground water and surface water resources, genetic erosion, the accumulation of greenhouse gases in the atmosphere, and the depletion of the ozone layer. Second, expanding the economic access to food, thereby ensuring food security for all at the household level, under conditions of insufficient growth in household income arising from slow growth in diversification in gainful employment opportunities. The untapped potentials both in technology and resources as reflected in the diffrences in field between the best farmers and the average, and between regions, rather than the best farmers and best regions and the research stations the rich ecosystem diversity and the possibilities for intensification and diversification, at the remarkable success met within various countries in dealing with specific aspects of food security through a well-orchestrated, mutually reinforcing package of green technologies, services and public policies, are some of the opportunities with which the Asia-Pacific region is endowed in ensuring food security.

A Pro-active Instead of Business-as-usual Approach

A review of the various scenarios suggests that if the 'business as usual' approach to food sees is pursued, the countries in the region would not be able to resolve the problem of poverty and ensure food security for all even by 2010, as indicated earlier, let alone by the end of this century, as mandated by this project to inquire into the twin basic food security challenges of the present and the decades ahead that of sustaining the availability of food and expanding the economic access to food, thereby ensuring food security for all at the household level—argue for a new paradigm of development which is pro-poor, pro-nature and pro-women by ensuring both economic growth and human development, consistent with the pathways for sustainable development. Recognition of food security as a fundamental right is cardinal to a proactive perspective approach to ending endemic hunger.

This recognition has to be backed by political commitment and public policies, and matched with investment with respect to the sustainable management of resources, building productive capabilities and access to income and entitlement to food.

Strategies

Reflecting the holistic concept of food security and the perspective approach, a set of strategies for attaining food security has been designed based on four integral elements, viz. (i) ensuring sustainable availability of food through maintaining the growth in food production over the population growth, as has been possible during the past decades; (ii) sustaining the productivity of the resource base over the period by keeping the economic and environmental costs at the minimum; (iii) ensuring adequacy of household incomes through promotional social security measures such as accessing assets, employment and empowerment; (iv) provide entitlement to food through protective social security measures to the vulnerable groups and unreached; and (v) enlarging support for national and international agricultural research designed to promote public good.

Sustainability of Food Availability

Sustainability of food availability is sought to be achieved through a set of strategies consisting of (a) changes in the planning process from commodity-based to resource-based planning, including investment in research, delivery systems, support services and infrastructure; (b) systems perspective and integrated use and management of biophysical (land and water) and human resources to maximise income and employment; and (c) interventions that ensure economic viability, environmental compatibility and social equity to sustain the production capabilities of the resource base and minimise the environmental and economic costs through resource-friendly land use systems and policies, integrated management systems for nutrients, energy and pest control and conservation of genetic resources for biodiversity.

Following from the above propositions, the need to have policies for the sustained use of resources emerges. Land use policies, particularly those which will place restraint on the diversion of prime agricultural land irrevocably to other uses, are of major concern in the countries of the region, and user-friendly water use and management, including pricing polies to increase efficiency and to conserve these finite resources.

Minimising Economic and Environmental Costs

Restructuring Research and Technology Development : Minimising the environmental costs as well as the economic costs are necessary to sustain the productive capabilities of the resource base, on the one hand, and to make food available at affordable prices and incentive enough to farmers to continue production, on the other. To accomplish both increasing knowledge has to be substituted for biophysical resources within biological production limits. Substitution of biophysical resources with knowledge requires, on the one hand, reorienting and restructuring research and technology development by (a) maximising the productivity of the resources and inputs and production systems, in addition to maximising the productivity of

Table 20.3 : Population Doubling Time

Year	*Population (millions)*	*Life expectancy (years)*	*Population doubling time (years)*
10000 BC	6	20	8,000
0 AD	250	22	1,900
1750 AD	771	27	1,000
1950 AD	2,530	35	116
1990 An	5,292	55	40

Source : Keyfitz aced Lindahl-Kiessling (1994).

Table 20.4 : Global Population Growth

First billion	*Millions of years*
Second billion	123 years
Third billion	33 years
Fourth billion	14 years
Fifth billion	13 years
Sixth billion	11 years

Table 20.5 : Population Size and Availablity of Renewable Resources

	1990 (millions)	*2010 (millions)*	*Per capita change*
Population	5,290	7,030	-
Fish catch (tonnes)	85	102	-10
Irrigated land (hectares)	233	277	-12
Cropland (hectares)	1,444	1,516	- 21
Forests (hectares)	3,413	3,165	- 30

Source : FAO (1992,1993).

commodities to meet the needs of the resource-poor; (b) expanding location-specific research to adapt not only to the variations in biophysical endowments but also to socio-economic conditions; (c) blendig traditional technologies with frontier technologies to integrate the ecological and social strengths of the farmer with the production potential, cost-effectiveness and consumer appeal of the latter; (d) fostering participatory mode in research and training to ensure the active involvement of all the interest groups especially of the farmers, evolving a clear division of responsibilities and commitments between the research systems at the international, regional, national and sub-national levels so that all the systems compliment each other rather than be competitive. The international systems could focus on some of the basic problems of commodities and take a supportive role in training personnel and providing materials while the national and sub-national systems could concentrate on systems research and resource management and on the other augment resource support commensurate with the expectations from research and technology development.

Restructuring Technology Delivery and Support Systems

Following the restructuring of technology generation and development, technology delivery also requires reorientation to meet the multiple needs, including intensive training in skills, of a production system rather than specific commodities, and restructuring to minimise the proliferation of agencies at the same time to research the services and support at the grassroots. A single-window systems at the grassroots level for technology delivery and support services is proposed.

Social Mobilisation and Group Action

Fostering social mobilisation for the adoption of eco-friendly and other technologies which ensure sustainability of production, and group action to enable the resource-poor to have access to the kind of support services they need and which are available with the government, and credit with the financial institutions, is an integral part of the proposed strategy.

Affirmative Actions and Empowerment of Women

Strategies for achieving food security proposes a decisive affirmative, pro-women bias to development, because in Asian countries women are primarily the purveyors of food and, in addition, substantial contributors to farm work. Affirmative, pro-women orientation calls for action on three fronts. Firstly, in designing developing programmes which would help increase the income earning capability of women; secondly, in ensuring that the incomes earned will be available to them for use; and lastly, in reducing the drudgery of work in which women are involved. A monitoring mechanism comparable to the 'environmental impact assessment', should be made mandatory to assess the impact on women for every intervention on food security, in order to take timely corrective measures. The successes achieved with the *Grameen Bank* movement in Bangladesh and with Self-Employed Women's Association (SEWA) in India in empowering women through mobilising credit for income-generating activities are illustrative.

Ensuring Adequate Household Income: Strengthening the Asset Base

Poverty is the prime cause of food insecurity. That lack of an adequate asset base has much to do with poverty is well recognised. Not only that in general the poor are in areas which are poorely endowed, but a significant proportion of the people is poor in well-endowed areas with poor distribution of assets. Providing the population, especially the food insecure, with sources of entitlement based on production, trade, own labour or inherited asset is a means for ensuring adequacy of household incomes to access food. The strategies proposed include: (a) Redistribution of rural assets, especially land; (b) Enabling the acquisition of productive assets such as livestock, fishponds, cottage enterprises, etc. (India's Integrated Rural Development Programme is illustrative); (c) Facilitating access to the usufruct of common property resource, including the granting of exclusive rights to enjoyment of usufruct from common property resources; and enable them to develop these resources by extending capital support, evolving a management system and fostering group action; (d) Sharing knowledge through skill development and support services for translating knowledge into income and employment-generating activities; and (e) Facilitating access to credit through financial institutions owned and managed by the resource-poor (Grameen Bank Movement in Bangladesh).

Crop Intensification and Farming Systems Diversification

Farm families with small holdings can earn higher income only through crop intensification and farming systems diversification. Agro-forestry, crop livestock and crop-fish combinations will have to be promoted on ecologically sustainable lines. Productivity of farming systems should be measured by net-income per unit of land, water and energy, and not merely by yield per hectare.

Promotion of Rural Employment and Diversification

Employment generation in the off-farm sector to enable the resource-poor to have direct access to income is suggested as an integral strategy for ensuring food security, promotion of employment through rural industrialisation, exploring the opportunities for value-addition by integrating production with processing and marketing, and direct investment by the State for employment generation in programmes such as 'food for work' in building infrastructure and community assets, etc.

Providing Entitlement to Food

Food security for those vulnerable groups who are unable to transact in the market are to be assured through a comprehensive social security system complemented by an institutionalised public distribution system that would take care of the impacts of production cycles, meeting special needs of vulnerable groups and improving resilience in the face of contingencies.

Surveillance and Food Reserves

Food accounts for a substantial part of the income of the resource poor and food-insecure. A reasonably stable price regime is very vital for sustaining production as well as protecting the vulnerable groups from the fluctuations in food prices. A surveillance system to monitor the supply, demand and prices of food, and a food reserve system (sub-national, national and regional levels) as a basic instrument to influence food prices would help maintenance of stable price regimes which would protect the vulnerable from fluctuations in price and at the same time provide adequate incentive for the food producers to maintain supply.

Trade Related Intellectual Property Rights (TRIPS)

In the fast spreading IPR environment, private profit, rather than public good, could become the prime motivating factor for decisions relating to the investment of scientific and financial resources. It is important that under such an environment research and development (R&D) efforts relating to sustainable food and nutrition security do not take a back seat. Ecological agriculture needs a major investment of resources. Investment in research is excluded from the calculation of subsidies under the World Trade Agreement Developing nations will therefore be well advised to step up R&D investment directed towards public good.

Agenda for Action

Against the advocacy of a proactive perspective approach to ensure food security as conceived in this study in the South and South-East Asia region by the year 2010 A.D. which coincides with the implementation of Agenda 21 of United Nations Conference on Environment and Development held at Rio de Janeiro in 1992, the strategies proposed and the arguments marshalled in support, are translated into an agenda for action. While recognising the complexities of, and the interlinkages of a variety of factors in, ensuring food security, such as food production, technology, environment, economy, population growth and the socio-political system, the focus in this study is zeroed in on issues of food production, resource base, technology, employment and support-led security and the related enabling policy environment as the world community collectively and individual nations on their own, have initiated action on other related but vital areas such as population. The agenda for action is presented first arguing the need for public action, then by outlining the enabling macro-policy environment, and finally laying out the action proposals.

Public Action

The recognition of food security as a fundamental right of the citizenry makes it imperative that the governments in this region uphold this right and commit themselves to its realisation. A crucial commitment is fostering a democratic political system, an active opposition, and free and indipendent media as freedom from hunger and under-nutrition is seldom achieved without constant public pressure through an alert and politically-conscious citizenry for timely State intervention. While recognising the importance of the market and respecting its role in ensuring food security through efficient resource allocation and pricing, distortions in the market and limitations of the market on account of the inequitable purchasing power, and lack of capacity for the resource-poor to transact in the market, rendering them food-insecure, cannot be discounted. Public action is needed in :

- integrating the provision of safe drinking water, basic sanitation, housing and elementary education as part of food security;
- achieving mass literacy by making school education compulsory and by abolishing child labour;
- strengthening the national system of agricultural research through continued public investment in research, technology delivery and support systems;
- fostering promotional security in the form of State-supported employment programmes and employment-enhancing rural growth;
- making food available through universal public-supported systems at prices affordable to those identified as poor;
- instituting programmes of social security which involve income and welfare support to those who are unable to transact in the market;
- building and maintaining a well-orchestrated food reserve system at local, national and regional levels to overcome shortages and price fluctuations;
- facilitating the democratic transition leading to low birth and death rates, by providing the enabling environment where children are born by choice;
- institutionalising a healthy, democratically functioning system of local self-government, indipendent judiciary and fostering freedom for the mass media to enable the citizenry to be alert and conscious; and
- above all, bringing about institutional changes that would lead to the transformation of backward agrarian relations, and the elimination of the worst forms of oppression and discrimination against women and socially depressed populations.

Macro-Policy Environment

A conducive macro-policy environment would seek to accomplish the goal of food security by limiting the population to carrying capacicty, providing food at affordable prices to consumers while keeping incentives to food producers, and ensuring incomes and support to those who lack resources to access food. The elements of such a macro-policy environment are :

(a) Institutinalisation of the food security strategy by providing a legal personality through legislative bodies encompassed in a 'Food Security Act';
(b) Economic policy interventions in the direction of price including food reserves, credit and investment in infrastructure to sustain production, employment, public distribution and disaster prevention to ensure entitlement to food;
(c) Land and water use and management policies including the pricing of public supported infrastructure and services, armed with legislative measures, to sustain the productivity and to pevent the erosion of these finite resources;
(d) A grain-saving livestock policy to improve the availability of animal products at affordable prices and to increase the income of the resource-poor with emphasis on non-grain resources of animal feed and more energy-efficient sources of animal protein; and
(e) A coastal aquaculture policy for the sustainable utilisation of the coastal systems by ensuring environmental compatibility and social equity.

Fighting the Fatigue of the Green Revolution

The actions expected to resolve the twin prime concerns of food security, viz. availability of food and the economic access to food, are :

- Restructuring of development planning, including investment in techno-infrastructure on the basis of ecosystems and identifying potential opportunities to optimise the use of both biophysical and human resources through intensification;

- Creation of a revolving fund for the rehabilitation of the fragile ecosystems such as mountain regions, tribal habitations, etc., and for enabling the inhabiting population to overcome the problems of survival during the period of transformation into sustainable resource management systems;
- Evolving a new vision for the agricultural research systems, both national and international, that would lead to a new paradigm of agricultural development based on a job-led economic growth rooted in the principles of ecology, equity and energy conservation by developing technologies which are environmentally sound, socially equitable both in gender and economic terms, and employment incentive. New forms of linkages based on the need to promote public good and not just private profit and new forms of partnership between the international and national research systems sharing responsibilities and commitments are to be established. While permitting no let-up in commodity focused research opening a second front of research concerns of production of food from less land with less water and energy is necessary to maintain the profitability to farmers; at the same time producing food at affordable prices to the resource poor is part of that vision.
- Ensuring resource support to agricultural research by allocating a minimum proportion of the agricultural GDP.
- Development of farming systems for sustainable livelihood security of the resource-poor households through maximising income and employment by increasing the productivity of resources;
- Providing technology delivery and support services through the single-window system;
- Empowerment of women by assisting in the generation of sustainable income and employment through skill development and fostering group action in creating support institutions a mandatory monitoring mechanism to assess the fact on women for every food security intervention;
- Creation of an appropriate socio-political climate for the redistribution of the rural assets through land reforms and asset-creating mechanisms;
- Redistribution of the knowledge asset for skill development

and income-earning capacity by translating technology into income and employment opportunities through technological empowerment;

- Fostering self-employment, especially in -the service sector through group action and creating the instrumentalities for providing investment support;
- Generation of employment in the rural areas through rural industrialisation based on local resources and value-addition by linking production with processing and marketing;
- Fostering agribusiness consortium of small farmers through a new social contract between corporate industry and the resource-poor families;
- Institutionalisation of a support-led purity system, including a public distribution system for those who are unable to transact in the market and for other vulnerable groups;
- Maintaining surveillance and building food reserves to ensure food supplies at affordable prices and to provide incentives for the producers;
- Preparedness for disaster prevention and management to overcome the hardships arising from natural and man-made calamities;
- Strengthening regional food security through fostering the maintenance of regional food reserves to assist the participating countries in overcoming food shortages and abnormal price increases arising from contingent situation like floods and other natural calamities through exchange of food;
- Promoting a pro-developing national trade regime for commodities with appropriate incentives and disincentives that facilitate the shifting of the present surplus food production base from the developed nations to the developing countries, so as to enable the developing countries to find export opportunities in the markets of developed countries, and that secure support for the resource-poor to alleviate the hardships arising from economic restructuring; and
- Initiating co-ordinated action at regional and global levels to ensure that the fast expanding trade related intellectual property environment does not ignore the needs to foster the application of science and technology for national, regional and international public good.

Conclusion

The period needed far adding a billion persons to the human population is now 11 years (Tables 20.3 and 20.4). At least 200 million tonnes of extra food grains will be needed for an additional billion individuals. In addition, several other agricultural commodities such as milk, fibre, fodder and fuelwood will be needed. While needs are expanding, the per capita availability of cropland, water and forests is shrinking (Table 20.5). The sustainable food security challenge can be met only by conserving genetic wealth and utilising it efficiently. This is why all women and men who make contributions to this conservation-utilisation chain need support, recognition and reward. Development which is not equitable will not be sustainable in the long-term. We should lose no further time in integrating the principles of equity and employment with those of economics and ecology in the development of food and agricultural policies.

The path ahead in the farm sector of developing countries involves accomplishing three major goals in an integrated and mutually reinforcing manner. These are: productivity improvement, natural resources conservation and poverty eradication. For all these purposes, national, regional and international research designed for public good is a basic necessity.

REFERENCES

Agarwal, Bina (1994), *A Field of One's Own Gender and Land Rights in South Asia,* Cambridge University Press, New York, U.S.A.

Food and Agriculture Organisation of the United Nations (1992), *Production Yearbook 1991,* Vol. 45, United Nations, Rome.

Food and Agriculture Organization of the United Nations (1993), *athan, e fit and erstin, Lindahl issling 199 , The World Population dabate rgenc* of *the Problem in Sir Francis Graham Smith Ed 199 Population: The Complex Realit* The Royal Society, London, pp. 21-51.

Swaminathan, M.S. (1968), presidential Address at the Agricultural Sciences section, Proceedings of *the Fift -Fifth Indian science Congress, Varanasi*

Swaminathan, M.S. (d.) (1993), *Wheat Revolution :* A *Dialogue,* Macmillan India td., Madras.

21

Food Security
An Overview

Tony Binns

WHAT IS FOOD SECURITY?

Some Definitions

1. "Availability at all times of adequate world supplies of basic foodstuffs..., to sustain a steady expansion of food consumption.....and to offset fluctuations in production and prices" (UN, 1975).
2. "A condition in which the probability of a country's citizens falling below a minimal level of food consumption is low" (Reutlinger and Knapp, 1980).
3. "Everyone has enough to eat at any time—enough for life, health and growth of the young, and for productive effort" (Kracht, 1981).
4. "Ensuring that all people at all times have both physical and economic access to the basic food they need" (FAO, 1983).
5. "Access by all people at all times to enough food for an active, healthy life" (World Bank, 1986).
6. "Always having enough to eat" (Zipperer, 1987).
7. "A country and people are food secure when their food system operates efficiently in such a way as to remove the fear that there will not be enough to eat...In particular, food security will be achieved when the poor and vulnerable, particularly women and children and those living in marginal areas, have secure access to the food they want" (Maxwell, 1988).

8. "The absence of hunger and malnutrition" (Kennes, 1990)
9. "A situation in which all individuals in a population possess the resources to assure access to enough food for an active and healthy life" (Weber and Jayne, 1991).
10. "The viability of the household as a productive and reproductive unit (not) threatened by food shortage" (Frankenbergerand Goldstein, 1991).

Food Security : Changing the Focus

From the Global and the National, to the Hosuehold and the Individual (1975-85)

1974-World Food Conference

Pre-occupation of African governments with national food self-sufficiency.

Sen (1981)—emphasised the question of *Access* ('Food entitlement') and *Control* over resources.

Different scales of analysis–interlinked :

- Global
- National
- Community
- Household
- Individual

From a 'Food First' perspective, to a 'Livelihood' perspective (from 1985)

Stimulated by observations of the African famine, 1984-85.

Conventional view of food security—sees food as a primary need.

But food (especially short-term nutritional intake) is only one of the objectives people pursue.

Importance of Coping/Adaptive strategies.

Oshaug (1985) : 3 types of household :

- *'Enduring' households*—maintain household food security on a continuous basis.

- *'Resilient' households*—suffer shocks, but recover quickly.
- *'Fragile' households*—become increasingly insecure in response to successive shocks.

Sensitivity—measure of extent of change following a shock. Interaction between resilience and sensitivity—analysis of food security over time.

Livelihood Security—most food insecure households show high sensitivity and low resilience.

From Objective Indicators to Subjective Perception (1990s)

Poverty literature distinguishes between *'conditions'* of deprivation and *'feelings'* of deprivation.

Conventional approaches to food security—rely on objective, 'target' levels of consumption, e.g. below 80 per cent of WHO average required daily calorie intake.

But, nutritinal adequacy is problematic—dependent on age, gender, health, size, workload, environment, behaviour.

What is an 'Optimum' state of nutritional health—Whose knowledge counts?

Quantitative measures ignore important qualitative issues; technical food quality, consistency with local food habits, cultural acceptability, autonomy, self-determination.

Food Entitlement—not just about *quantity,* but also *quality* (removing the fear that there will not be enough to eat, and securing access to the type of food that people want).

Importance of *perceptions* of food problems.

Diversity of coping strategies.

Key Words :

- Flexibility
- Adaptability
- Diversification
- Resilience

'Modern' and 'Post-Modern' Approaches in Development

	Modern	*Post-modern*
Underlying reality	Simple, uniform	Complex, diverse
Objectives	Growth	Development
	Preoccupation with macro	Preoccupation with micro
Research approach	Measure	Listen
	Survey	Participatory Rural Appraisal
	Reductionist	Holistic
	Deduction	Induction
	Abstract models	Complex reality
	Aggregate	Disaggregate
Planning approach	Plan	Enable
	Model	Interact
	Top-down	Bottom-up
	Centralize	Decentralize
Implementation	Blue-print	Process
	Role culture	Task culture
	Standardization	Flexibility, innovation

Source : S. Maxwell (1996), Food Security a post-modern perspective, *Food Policy,* 21(2), pp. 155-70.

Post-Modernism and Food Security Policy

- *Post-Modernism is critical of over-arching theory and meta-narratives*

Recognise diversity of causes of food insecurity, situations and strategies

- *Post-Modern Food security policy gives 'privilege' to the livelihoods of individuals and communities*

Priority given to providing households and individuals with choices which contribute to self-determination and autonomy in developing livelihood strategies.

- *Post-Modern food security policy emphasises diversity of ideas and sources of ideas*

Good ideas can be borrowed from many different fields—applying to policy the principles of 'collage'.

Case Study : Issues Concerning Food Security in Southern Africa

- Food Security situation has deteriorated in Southern Africa : number of food-insecure peoople in the region (excluding South Africa) almost doubled in 1980s from 22 million people in 1979-81 to 39 million in 1990-92, while their share of the population grew from 37 per cent to 46 per cent (FAO, 1996)
- On average, 200 calories per person per day available in Southern Africa region during 1990-92. This is the same as in 1970s, but almost 10 per cent lower than in the early 1980s.
- Considerable variation in daily per capita calorie consumption.

Average daily per capita calorie consumption in Southern Africa, 1971-72 and 1991-92 (selected countries)

Country	*1971-72*	*1991-92*
Angola	2141	1832
Malawi	2402	1884
Mozambique	1944	1696
Namibia	2185	2200
South Africa	2848	2769
Swawziland	2335	2484

Source : FAO, 1994.

- A continuation of past trends would result in dramatic decreases in calorie consumption to very low levels in Angola, Malawi, Mozambique and Zambia.
- Excluding South Africa, the 10 remaining countries of Southern Africa have more than 4.3 million malnourished pre-school children—1 out of 4 children under 5 is malnourished. Mozambique and Tanzania each account for more than a million of these children.
- That every second person is food insecure and every fourth child is malnourished, implies that hunger is and will remain a primary challenge in Southern Africa.

A continuation of recent food production trends will lead to a rapidly increasing gap between production and requirements : Botswana, Mozambique, Namibia—imported 66 per cent of total cereal consumption in 1991-92. Mozambique, Zimbabwe, Malawi, Angola—largest receivers of food aid.

Understanding Indigenous Livelihood Systems : Ensuring Food Security

Key features of indigenous livelihood systems :

- Resilience
- Flexibility
- Adaptability
- Diversification

Key factors of production

- Land : (quantity/availability, quality—slope, fertility, pests etc.)
- Labour : (quantity/availability, quality—age, gender, health, efficiency)
- Capital : (livestock, trees, buildings, infrastructures, boats, nets tools, savings)

Diversification and coping strategies

- Non-farm work and income (rural/urban)
- Division of labour (men/women/young/old)
- Different crops and varieties of crops, different types of livestock
- Combinations of production systems : farming, fishing, pastoralism
- Use of tree produce, wild foods
- Sale of produce or livestock
- Exchange and reiprocal transactions
- Sustaining the environment : Preventing erosion (terracing, bunds, basins), managing pasture and water resources (wells, irrigation)

- Knowledge of pests.

Cycles of vulnerability

- Seasonal patterns—'hungry season'.
- Life-cycle—changing ratio of workers to dependents.
- 'Shocks'–

  Work shocks : quantity/availability of work changes
  Output shocks : fall in quantity or price
  Food shocks : lack of food in markets, price rises
  Asset shocks : fall in assets, death of livestock, theft debt.
- Illness/AIDS schoks : affects production and purchasing power.

Future Food Security : An Action Plan

1. *Primary focus*—supplying vulnerable people and households with secure access to food. Individual and household needs should take precedence over national food self-sufficiency and self-reliance.
2. *Overwhelming need for peace and physical security*—war is the greatest cause of famine.
3. *Need for poverty*—reducing economic growth to generate secure livelihoods (with buffers against destitution) and buoyant national economies to provide jobs and, where necessary, purchase food.
4. *In agricultural sector*—need for generally labour-intensive strategies which emphasise job creation and better incomes for poorest groups.
5. *Balance between food and cash crops*—emphasise long-term comparative advantage rather than self-sufficiency for its own sake.
6. Efficient food marketing to store and distribute food, at reasonable prices, to all parts of the country in all seasons and in all years. Private sector involvement, but state to control relief buffer stocks.
7. Establish more effective safety nets by strengthening

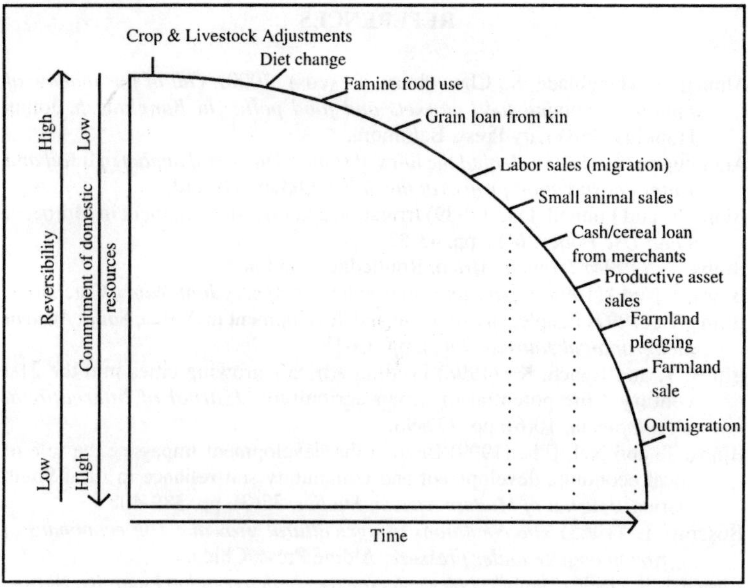

Source : Adapted from Watts (1983) by Frankenberger and Goldstein, Office of Arid Lands Studies, The University of Arizona.

Fig. 21.1 : Respones to Household Food Shortage.

community institutions, targeted food and nutrition interventions and improved famine preparedness and response, especially at local level.

8. Famine preparedness is critical in many countries, involving not just early warning, but also improving capacity to respond. Early decisions on importation, small, locally-based stocks for relief distribution, programmes for vulnerable group feeding and rehabilitation.
9. Need for strong international support—through more favourable trading environment, debt relief and larger and better-focused aid flows.
10. Food security planning should follow a 'process' rather than a 'blue-print appraochdecentralization, bias to action over planning, encourage risk-taking and innovation, multi-disciplinary and multi-sectoral planning teams.

(Source : adapted from Maxwell, 1996)

REFERENCES

Ahmed, R, Haggblade, S., Chowdhury, T., (eds) (2000), *Out of the shadow of famine : evolving food markets and food policy in Bangladesh*, Johns Hopkins University Press, Baltimore.

Atampugre, N., (1993), *Behind the lines of stone : The social impact of a soil and water conservation project in the Sahel,* Oxfam, Oxford.

Binns, T. and Funnell, D.C. (1989) Irrigation and rural development in Morocco, *Land Use Policy,* 6(1), pp. 43-52.

Binns, T., (1994) *Tropical Africa,* Routledge, London.

Binns, T., (ed.) (1995) *People and environment in Africa, John Wiley,* Chichester.

Binns, T., (1997) People, environment and development in Africa, *South African Geographical Journal.* 79(1), pp. 13-18.

Binns, T. and Lynch, K., (1998) Feeding Africa's growing cities into the 21st century : the potential of urban agriculture, *Journal of International Development,* 10(6), pp. 777-93.

Binns, T. and Nel, E.L. (1999) Beyond the development impasse : the role of local economic development and community self-reliance in rural South Africa, *Journal of Modern African Studies,* 37(3), pp. 389-408.

Boscrup, E. (1965) *The conditions of agricultural growth : The economics of agrarian change under pressure,* Aldine Press, Chicago.

Brown, L.R. (1996) *Though choices : facing the challenge of food scarcity,* Norton, New York.

Chambers, R. (1983) *Rural development : Putting the last first,* Longman, London.

Chambers, R. (1988) Poverty in India : concepts, research and reality, *IDS Discussion Paper,* 241, Institute of Development Studies, Brighton.

Chambers, R. Paccy, A and Thrupp, L.A. (1989) *Farmer first : Farmer innovation and agricultural research,* Intermediate Technology Publications, London.

Chambers, R. (1993) *Challenging the professions : Frontiers for rural development,* Intermediate Technology Publications, London.

Chambers, R. (1997) *Whose reality counts? : Putting the first last,* Intermediate Technology Publication, London.

Conway, G. (1997) *The doubly green revolution : Food for all in the Twenty-First Century,* Penguin, London.

Devereux, S. and Maxwell, S. (eds.) (2001) *Food security in sub-Saharan Africa,* ITDG Publishing, London.

Dyson, T. (1996) *Population and food : Global trends and future prospects,* Routledge, London.

Ehrlich, P.R. (1970) *The population bomb,* Sierra Club, New York.

FAO, (1999) *The state of food insecurity in the world. Food insecurity : When people must live with hunger and fear starvation,* FAO, Rome.

Garrett, J.L. and Ruel, M.T. (eds.), (2000) Achieving urban food and nutrition security in the developing world, Focus 3, 2020 Vision, International Food Policy Research Institute, Washington.

Geheb, K. Bins, T. (1997) 'Fishing Farmers' or 'Farming Fishermen'? The quest for household income and nutritional security on the Kenyan shores of Lake Victoria, *African Affairs,* 96, pp. 73-93.

Haddad, L. Kennedy, E. and Sullivan, J. (1994), Choice of indicators for food security and nutrition monitoring, *Food Policy,* 19(3), pp. 328-43.

Haddad, L. (ed.) (1997) *Achieving food security in southern Africa,* International Food Policy Research Institute, Washington.

Khan, M.M. (1994) Market-based early warning indicators of famine for the pastoral household of the Sahel, *World Development,* 22(2), pp. 189-99.

Leach, M. and Mearns, R. (eds), (1996) *The lie of the land : Challenging received wisdom of the African environment,* James Currey, London.

Leach, M. Mearns, R. and Scoones, I. (1999) Environmental entitlements : dynamics and institutions in community-based natural resource management, *World Development,* 27(2), pp. 225-47.

Lynch, K., Binns, T. and Olofin, E. (2001) Urban agriculture under threat : the land security question in Kano, Nigeria, *Cities,* 18(3), pp. 159-71.

McCalla, A.F. and Revoredo, C.L. (2001) Prospects for global food security : a critical appraisal of past projections and predictions, *Discussion Paper* 35, 2020 Vision, International Food Policy Research Institute (IFPRI), Washington.

Maxwell, S. (ed.), (1991) *To cure all hunger : Food policy and food security in Sudan,* Intermediate Technology Publication, London.

Maxwell, S. (1996) Food Security : a post-modern perspective, *Food Policy,* 21(2), pp. 155-70.

Maxwell, S. and Fernando, A. (1989) Cash crops in developing countries : the issues, the facts, the policies, *World Development,* 17(11), pp. 1677-1708.

Maxwell, S. and Frankenberger, T.R. (1992) *Household food security : Concepts, Indicators, Measurements. A Technical Review,* UNICEF/IFAD, New York.

Meadows, D.H., Meadows, D.L., Randers, J. and Behrens, W.W. (1972) *The limits to growth : a report for the Club of Rome's project on the predicament of mankind*, Universe Books, New York.

Mortimore, M. (1989) *Adapting to drought : Farmers, famines and desertification in West Africa,* Cambridge University Press, Cambridge.

Mortimore, M. (1998) *Roots in the African dust : Sustaining the drylands,* Cambridge University Press, Cambridge.

Motteux, N., Binns, T. Nel, E.L. and Rowntree, K. (1999) Empowerment for development : taking participatory appraisal further in rural South Africa, *Development in Practice,* 9(3), pp. 261-73.

Nel, E., Hill, T. and Binns, T. (1997) Development from below in the 'new' South Africa : the case of Hertzog, Eastern Cape, *Geographical Journal,* 163(1), pp. 57-64.

Oshaug, A. (1985) The composite concept of food security, in W.B. Eide *et al.* (eds.), Introducing nutritional considerations into rural development programmes with focus on agriculture : A theoretical contribution, *Development of methodology for the evaluation of nutritional impact of development programmes,* Report No. 1, Institute of Nutrition Research, Unniversity of Oslo.

Potter, R.B., Binns, T., Elliott, J.A. and Smith, D. (1999) *Geographies of Development,* Longman, London.

Reij, C. Scoones, I. and Toulmin, C. (eds.), (1996) *Sustaining the soil : Indigenous soil and water conservation in Africa,* Earthscan, London.

Richards, P. (1985) *Indigenous agricultural revolution,* Hutchinson, London.

Richards, P. (1986) *Coping with hunger : Hazard and experiment in an African rice-farming system,* Allen and Unwin, London.

Scoones, I. and Thompson, J. (eds.), (1994) *Beyond farmer first : Rural people's knowledge, agricultural research and extension practice,* Intermediate Technology Publications, London.

Scoones, I. (ed.), (1995) *Living with uncertainty : New Directions in pastoral development in Africa,* Intermediate Technology Publications, London.

Scoones, I. *et al.*, (1996) *Hazards and opportunities : Farming livelihoods in dryland Africa, lessons from Zimbabwe,* Zed Books, London.

Sen, A. (1981) *Poverty and Famines : An essay on entitlement and deprivation,* Clarendon Press, Oxford.

Tiffen, M., Mortimore, M. and Gichuki, F. (1994) *More people, less erosion : Environmental recovery in Kenya,* John Wiley, Chichester.

Walker, P. (1989) *Famine early warning systems : Victims and Destitution,* Earthescan, London.

World Food Summit, (1996) *Rome declaration on world food security and world food summit plan of action,* World Food Summit, Rome.

22

Organic Farming and Agribusiness for Food Security in India

Jagdish Singh

Food Security Scenario

According to Pilot Analysis of Global Ecosystems (PAGE) : Agroecosystems, "Our current global population, currently about 6 billion people, is expected to increase by more than one quarter over the next two decades," "We must find ways to increase food production to sustain growing populations in developing countries. But this challenge must be accomplished without major increases in the amount of new land under cultivation, which would further threaten forests and biodiversity, and without resorting to *unsustainable farming* practices." (*sic*).

Stanley Wood, one of the co-authors of the report, stressed that since agricultural land dominates the earth's populated landscapes, we need it to do more than produce more food. "We also rely on agricultural land to provide other goods and services, including clean water and habitat for threatened species."

Agricultural lands could produce more food and help to prevent global warming b returning more carbon to the soils. "Unfortunately, many current agricultural practices actually contribute to global warming."

The *report* also reveals that :

- Soil degradation, including nutrient depletion, erosion, and salinization, is widespread.

- Twenty to 30 per cent of the world's forests areas have been converted to agriculture, resulting in extensive species and habitat loss. Agriculture is encroaching on many national parks and other protected areas.
- Agriculture consumes 70 per cent of the fresh water withdrawn annually by humans. Irrigation is draining more water than is being replenished by rainfall, causing water tables to fall. Moreover, many water sources are being polluted by excessive use of fertilizers and pesticides.

"We must not continue to take nutrients out ofthe soil faster than we replace them. We must not continue to deplete water resources faster than they can be replenished."

The Indian Context

Food security constitutes the basic building block of human welfare; and therefore, has to be treated not just as a matter of availability of and accessibility to food including potable water, but as "a right to nourishment" (Thiel, R.E., 2002) India, in the post-Green Revolution scenario, confronts the contradictions of ample food availability but limited accessibility; falling prices of food grains making their cultivation unviable for the producer unless propped up by high MSP while taking it beyond the reach of a large chunk of the consumers unless sold at the heavily subsidised prices. The following statement succinctly describes this contradiction." The last two years of poor agricultural growth have witnessed the growth of a food grain mountain on the hands of Food Corporation of India. The current year's anticipated record output of 209 million tonnes ....will push up the height of this mountain further still. After dumping 3 million tonnes of grain into the market at prices below the 'economic cost' and another 3 million tonnes in the states' distribution network at a fraction of the 'economic cost', and after exporting as much grain as the global market would take at the subsidised prices, mostly for use as animal feed, FCIs foodstocks amounted to around 60 million tonnes in January 2002...Essentially the Government has priced food out of the reach of the poor.... .The rise in the prices since the mid-90s has been sharper than the rise in the consumer price index or the wholesale price index...

It is all because procurement is confined.... to just four states that account for 90 per cent of all the grain procured that hunger has not spread throughout the country...The net per capita availability of cereals had dipped below the level achieved in 1984, even before it further fell sharply in 2001..... the net per capita availabilty of cereals has come down sharply and an increasing part of the grain available in the country is impounded in Government godowns, it is inevitable that the consumption by the people, the poorest among them should go down ...Even the farmers are losers in the long run. The high domestic prices keep them out of the international market"(Need for a Paradigm Shift; Editorial, EPW Vol. XXXII, OM No. 10, March 9-15, 2002).

The waves of globalisation and liberalisation have further accentuated the contradiction. They are affecting different regional economies differently. It is the task of geographers to decipher the contours of such changes and indicate the directions and scopes of adjustments at the local level. India, where 60 per cent of the active population finds employment in the agricultural sector which contributes one-fourth of the national income, confronts fresh challenges and threats on the food security front in the post-WTO era with the national Govrnment's policy actively pursuing the objectives of liberalisation in the agriculture sector as well, in the second generation reforms.

Similarly, India has grown to be the largest producer of milk in the world. The availability of milk is increasing but potable water is growing scarce. Indian agriculture is threatened by cheaper imports in the wake of removal of quantitative restrictions and lowering import tariffs while her products are uncompetitive in the world market hedged by high levels of subsidies.The chemical inputs based Green Revolution is petering out even without fully covering all parts of the country; while the water is getting highly toxic all over due to increasingly heavy doses of chemical fertilisers, insecticides, pesticides and heavy withdrawal of the groundwater.

Why Organic Farming

In the light of the above scenario, the Indian agriculture, must undergo radical transformation with a view to increase the accessibility to food by enhancing the purchasing power of the masses not directly dependent on agriculture; not just distributing food at the heavily subsidised prices, on the one hand, and ensuring the availability at sustainable basis by ensuring renunerative prices to the farmers, on the other."Achieving food security necessitates increasing food production as well as employment opportunities. Agriculture plays a dual role in the abolition of hunger; it produces the food and it can also produce a great many of the jobs needed by the households to buy food....Raising production and productivity in this sector can immediately place additional purchasing power in the hands of the rural poor, who will in turn, utilise the additional income for purchasing more food, clothing and other basic consumer goods that will create more jobs and higher incomes for countless others. The increased agricultural produce becomes raw material for a wide range of agro-based industries and services that stimulates formation of new enterprises and create down stream jobs, as well as products for further processing, domestic sale or export" (Swaminathan,1995 p.4). Under such a backdrop *organic farming*, with or without the combination of transgenic seeds holds promise for the "right to nourishment" of the masses; directly by ensuring remunerative return to the overwhelmingly marginal and small farmers; and indirectly through a well knit chain of agro-processing industries spread equitably over the space economy to generate gainful employment for the non-agricultural workers, especially the deprived women.

Organic farming based on biofertlizers and Integrated Pest Management as it is, reduces the cost of production while improving the quality of food produced. It also calls for judicious utilisation of soil nutrients and soil moisture by raising crops through inter-culture and allows for maximum diversification by raising interdependent and complementary crops, more so, on small and marginal farms. It adopts a comprehensive approach, i.e. an ecological plus a socio-economic and a technical factoring of modern agricultural systems (Duckham & Masifield, 1970). "Stress is placed upon the principle of developing and facilitating farming systems that take into account ecological conditions and interactions, and the need to balance the relationship between productivity and stability of the system" (Troughton,1991). "Agricultural wastes, intercrops and crop rotations can take care of the inadequacy in any particular organic matter. The farm has to be viewed as a whole system where each part and activity

Some Negative Aspects of Modern Agriculture

Prevailing Agricultural Policies and Objectives	
Miximise Productivity & manipulate distribution for profit & political influence (agro-power), Increase farm size, particularly with respect to non-physical economies of scale.	Vertically integrate, specialize & simplify. Short-terms economic policies encorage use of finite resource & damage to environment & human health.

Physical Base	**Soil Management**	**Food & Nutrition**
Emphasis on high capability land; mining of resource. "Improvements of drainage & irrigation.	Most food wastes not retunred to land physically & chemically manipulated; results in soil pollution, salinization, erosion & declining levels of organic matter, soil biota, fertility.	Reduced quality (deficient &/or toxic) Remval of food wastes from agro-eco-system by export
Buildings & Machinery	**Plant Production**	
Proliferation & increase in size of machines Genetic-engineering	Based on few species & varieties (often hybrids) selected on non-nutritional basis; simple planting design (monoculture). Crops often unable to compete with weeds & susceptible to pests; dependent on herbicides, pesticides & synthetic fertilizers.	Crops often harvested unripe, Transported, stored, processed & prepared (nutrients lost and/or toxins added at each stage).
	Animal Production	
Battery housing; artificial environments; automation; Feedlots; massive waste accumulation	Characteristics similar to pant production; few varieties; emphasis on uniformity of product. Dependent on dietary supplements, hormones, antibiotics & pesticides. Stressed by crowding; cruelty.	Emphasis on animal protein; saturated fats.

Resource inputs	**Environmental Quality**	**Socio-Economic Aspects**	**Human Health**
Dependent on finite fossil fuel energy & other non-renewable resources (exchanged for food). Destruction of renewable resources. Linear nutrient flows replace natural cycles.	Responsive to technological not biological constraints. Waste overload & contamination by synthetic, toxic chemicals; results in declining environmental quality. Loss of wildlife habitats & certain species. Loss of prime land to urban & exurban uses.	Centralization of wealth & power. Corporate intervention, absentee Landlords. Declining farm population. Farmer dependence. Rural community decay.	Increase in nutritionnally & environmentally related diseases (diabetes, Cancer, heart disease etc.) Over consumption; Junk foods.

Source : Modified by M.J. Troughton from Hill, S. B.& J.A. *Ramsay; Limitations of the Energy Approach in Defining Priorities in Agriculture.*

Alternative Agriculture

Eco-Agricultural Policies and Objectives

Based on renewable resources.
Holistic approach to problem solving using multidisiplinary preventive methods.
Environmentally supportive, minimally disruptive, using, yet respecting the laws 'of nature'—part of survival ethic.

'Permanent' production strategies to meet individual nutritional needs while maintaining ecosystem stability.
Regional self-sufficient, decentralised food system, supportive of :

1. self-sufficient homesteads with small surplus,
2. large farms, redesigned along ecological lines, and
3. urban food production.

Physical Base

Embodying concept of 'stewardship'. Geared to climatic knowledge & influences. Deintensification & long rotations.

Building & Machinery

Energy efficient technology with low environmental impact.
Machinery for managing mixed crop operations including small-scale, fuel & animal powered equipment.
Composter; seed energizers; Solar greenhouses & barns.
Smaller scale, less crowded livestock housing.

Soil Management

Conservation or improvement of soil fertility; application of microbial innoculants, conpost, sewage, green manure and other organic materials. Utilizing biological & chemical indicators in balancing all plant nutrients. Minimal, low impact tillage; mulching; drip irrigation.

Plant Production

Expanded gene pool (new species & varieties). Breeding, selection & management to minimize environmental impact, dependence on synthetic energy & chemicals, & to optimize nutritional quality.
Complex planting design; strip & mixed cropping; rotations. Preventive disease & pest control (resistant varieties, nutritional & habitat management).
"Herbal : & indigenous mixture for pasture. Seed & foliar appliation of naturally occurring plant nturients, hormones & microorganism.

Animal Production

As above alternative sources & improved quality of feed; mixed pasture management systems.
Use of early indicators in disease prevention & treatment; humane handling of livestock.
New sources of animal protein (e.g. wild games, fish ponds, insects).

Food & Nutrition

Improvement of food quality to meet individual nutritional needs (improved labelling standards).

Minimal processing to permit storage & distribution.

Promotion of vegetable proteins; more direct use of legumes, oil seeds.
Organic foods as normal rather than expensive 'niche market' items.
Reduction in consumption of animal proteins; emphasis on lean, pasture fed red meat.

Resource inputs

Dependent on solar & renewable energy, managed on a regional basis for permanence without pollution; con-

Environmental Quality

No waste overload through controlled production, recycling & monitoring of environmental quality.
No synthetic organic chemicals manipulation of divesity & succession

Socio-Economic Aspects

Meta-economic approach, sensitive to non-profit criteria (externalities ecological & human values distinguishing between real & manipulated needs)
Redistribution of wealth & power, deurbnization, decentralization, direct &

Human Health

Environmental/nutritional models of health & disease.
Congnisant of soil-food health relation-

(*Contd.*)

serving non-renewable resources. Supportive or cyclical nutrient flows through optimal management of wastes.	to provide stability. Conservational of rural landscape & wildlife habitats. Rejuvenation of damaged & non-productive land.	co-operative marketing. Use of cybermetics in long-term planning. Human capital intensive, employing ecologically appropriate technologies. Sensitive to work quality. Supportive or rural people.	ships & importance of identifying optimal diet for each individual.

complements the other. If interrelationships and complementarities within the farm are taken advantage of, farming can be both cheap and more productive on sustainable basis... It is ideal for peasant farms run mainly on family labour and supervision though it does not eschew hired labour. A policy thrust in favour of natural (organic) farming, therefore, would be instrumental in promoting equity in agricultural sector and providing fuller employment because of its being labour intensive, in addition to promoting a communitarian ethos. Thus it is not just a technology but also a value system and a way of life... Because of its potential for wider spread both across farmers and regions natural fanning can raise the overall productivity of agriculture much more than chemical farming"(Nadlami,1991).The details of various advantageous attributes of organic farming set against Chemical monocultural fanning are outlined in the Chart. Organic farming may prove its effectiveness further by the correct fusion with transgenic seeds of non-food crops, like Bt cotton, (it has just been cleared by GEAC for widespread cultivation by the harried farmers) by enhancing the yield without resort to the costly pesticides and thus promote a chain of agro-industries that will go a long way to enhance employment opportunities in rural areas and thus contribute to the enhanced accessibility, the need of the hour, in the interest of food security.

Misplaced Apprehension

Apprehensions were expressed earlier when isolated experiments were made in organic farming in different parts of India by adventurous and sagacious farmers (including Narain Reddy, the pioneer in India), regarding the productivity of organic farming to meet the food requirements of the country. A recent indepth analysis based on case studies (Prakash, *et al.*,1997) reach the conclusion that of the three forms of agriculture; viz. Growth oriented, relying heavily on external inputs; Stable agriculture with local varieties and internal inputs, and Ecologically benign agriculture (read organic farming)," ecological farming out-performs the remaining two alternatives with respect to most of the criteria; i.e. productivity, equity and sustainability. The cost effectiveness is very high here...... the growth oriented path relies heavily on petro-based external inputs.....giving rise to various externalities in terms of pollution, degradation, and depletion of natural resource base. As the growth oriented path resorts mostly to monocropping, stability as well as resilience are also expected to be very poor here..... On the other hand, ecological farming reflects the interests of the 'Farm agroecosystem' where 'private benefit' is the main consideration. To this end, various forms of diversification meet the profitability objective very well." Now that the country has huge buffer stocks she is in a position to take the leap without the danger of losing the bottomline in the initial transitional years of reduced productivity. The various experiments being carried out individually also reassure that although the individual crop yields may be small in comparison to the Chemical farming, the total yield is certainly more and beneficial to the producer in the long run. Also, it helps bring about the necessary diversification in agriculture to meet the food requirements of the consumer which show a distinct shift in food preferences. According to the latest Economic Survey, vegetables, fruits, milk, fish and meat account for an increasing share of the consumption basket."It is relevant to mention that new agri-enviromnental policy measures with focus on either improving water quality (including those to reduce fertiliser and pesticide use) or promoting organic agriculture have, been introduced in OECD countries" (Thamrajlakshmi, 20).

As regards management of water for irrigation, organic farming is more amenable to irrigation systems dependent on local water harvesting as the finely calibrated crop and animal husbandry combination allows for utilisation of the residual soil moisture for raising a second complementary crop after the primary *kharif* crop like rice. It is also well suited to irrigation technologies like sprinkler or drip irrigation that econmise on water use under the condition of growing water scarcity in India (for details, Singh & Singh, 2001).

Agribusiness : The Way to Food Accessibility

As indicated earlier there has to be greater stress on accessibility to food by enlarging the employment opportunities for the deprived sections of the society; the landless poor and the women. To this end, it would be easier and more cost effective to promote small scale agribusiness strewn over the rural landscape utilising the revolution in infommation technology and resorting to better spatial organisation (Singh, 1979). The World Bank and other international donor and catalytic agencies have again veered round the concept of Integrated Rural development requiring better access to resources, education, medicine and other social services and above all creating a development promoting atmospere and institutions (Thiel, 2002). The renewed interest of the bank (World Bank, 2000 & 2001) in Comprehensive Rural Development includes poverty reduction, food security at the household as well as national and global level and widely dispersed growth (as against the increasing polarisation in the space economy under globalisation propelled growth) among its objectives. The key strategies visualised are intensive agricultural production systems, management of natural resources and forestry and water conservation and management; all integral features of organic farming.

"The Prosperity 2000 strategy for India prepared by the Study Group of International Commission on Peace and Food (ICPF), (Swaminathan, 1995) and incorporated by the Government of India in the Eighth Plan emphasises the role of agribusiness in generating additional employment opportunities for 100 million people of India. This strategy utilises agriculture as an engine of growth by accelerating the development of commercial agriculture, agro industry and agro exports. It calls for a shift in thinking concerning agriculture from production for survival and subsistence, to production for maximum and sustainable profit, from emphasis on meeting minimum needs to realising maximum potentials, from commodity based to resource based planning. It seeks to tap the country's competitive advantage in labour intensive agricultural crops and allied industries to double agricultural production (raising the annual growth rate to 4 per cent *versus* 2.3 per cent in 80s). Creating new jobs through this low cost strategy can be accomplished entirely with the country's own resources, though foreign firms will find investment in India's agro-industrial sector very attractive. Initially about half of the new jobs will be generated on farms by raising productivity through methods to *improve farm management of micro nutrients and water, expanding the total irrigated area by more complete utilisation of the substantial additional capacity* that has already been created giving greater emphasis on *labour intensive commercial crops, reclaiming wasteland for forestry and fodder, and increasing subsidiary incomes from animal husbandry and poultry..... About ten million jobs will be created in down stream agro-industries* located in rural areas—in sugar mills, cotton and textile mills, processing units for fruites, fish and silk and in marketing and distribution—thus serving as a counter magnet to urban migration. Growth of agro-industry will stimulate demand for industrial machinery and services. The multiplier effect of sky-rocketing rural demand will stimulate demand in a broad range of consumer industries and create an estimated forty-five million rural and urban jobs in industry and services" *(*Swaminathan, *ibid. Box).*

The agro-processing industries which are coming up fast throughout the rural landscape benefiting not only the farmers by ensuring remunerative prices to their products, especially the fruits and other horticultural ones, but also providing employment to the landless and the women need all possible assistance and encouragement from the state. The Goverment of India has already recognised this need in its New Agricultural Policy by proposing to set up 20 agro-export processing zones for fruits like mango at Lucknow, Litchi at Muzaffarpur etc.The new EXIM policy announced for a five-year period (2002-07) with a view to increase exports to $80 billion (1 % of the world) has laid particular stress on agro exports which cannot be realised unless adequate infrastructure particularly power supply, roads, telecommunication network, agro-clinics, are put in place. The rural road connectivity programme and improving telecommunications are steps in the right direction. Much, however, remains to be done quickly and adequately. The most neglected aspect has been the proper maintenance of infrastructure already created.

The greatest hurdle in their way is the escalated and discriminatory import tariff imposed by the developed countries on different processed food products, according to the value addition stages of processing. (Windfuhr, 2002). This must be fought out in the W.T.O. in favour of the developing countries' exports of processed and ready-to-use products. "India had hoped that with the prospect of dismantling domestic support and export subsidies by the developed countries-market access for Indian agricultural products would expand. With a negligible 0.7 per cent share in world imports and 0.6 per cent share in exports we are not visible in the area of international trade... We have enabled better market access to the rest of the world by dismantling QRs. We have to harmonise the quality of our products to internationally accepted standards.... Information dissemination to farmers on higher standards has to be promoted, financial assistance extended and training imparted to them on methods to attain these standards". "State intervention in agriculture must shift from aimless subsidisation of all inputs to massive investment in research and development and extension, rural roads, rural power, and support for the development of an agro-processing industry. Crop diversification and integration of value addition chains into mainstream crop husbandry are required to wean the rich farmers of north-west India away from state subsidy" (Thamrajlakshmi, *op.cit.*).

The potential for labour absorption is high in agro based industries, small and medium industries and the rural services sector. "Interdependencies in the food and labour markets are important for the development process. A strategy combining promotion of agricultural growth, productive non-farm employment and high levels of social development would be needed for labour-intensive growth in rural areas," with a view to generate self-reliant employment. The organically produced raw and processed food has better chances to compete in the quality conscious *niche* markets, provided the discriminatory tariff policies are abondoned by the developed countries."Investment in Indian agriculture has been declining for quite some years. In the context of international trade, there is an added urgency to reverse this trend and increase investment in research integrated market development, storage and warehousing facilities, road development, creation of facilities for efficient and quicker transport and development of scientific systems of standard setting and grading. Public expenditure on research and technology, infrastructure creation and rural development will raise our AMS without attracting reduction commitments" (Thamrajlakshmi, *Ibid.*).

Conclusion

For such a strategy of food security laying equal emphasis on the availability as well as accessibility to materialise, there has to be greater concern for growing diversity of specialised crops consistent with the hierarchy of agro-ecological regions in *integrated farming set up* which attains a finely calibrated combination ofthree A's, i.e. agricultural crops, aquaculture, animal husbandary; plus horticulture—agroforestry; thus generating year round employment for the agricultural workers on farms as well as at value addition procesing centres along with remunerative returns to the farmers. Conserving and optimising the use of preferably locally harvested water for irrigation, adopting IPM practices and avoiding altogether the chemical inputs most be encouraged by the state even by extending subsidies. Also, the WTO dispensation has to be favourably altered by forcing the implementation of the promise to abolish subsidies on agriculture and export by the developed world to make it a level playing field and insist on the *right of labour to reside* before the *"right of capital to reside."*

REFERENCES

Duckham, A.N. & Masifield, G.B. (1970) *Farming Systems* of *the World,* Chatto & Wilson, London.

Nadkarni, M.V. (1991)The Crisis of Increasing Costs in Agriculture : Is There A Way Out? in Nadkarni, M.V. *et al.* (eds.) *India* : *The Emerging Challenge,* New Delhi.

Prakash, T. *et al.* (1997) Sustainable Development—Experiences of Indian Agriculture in Agarwal, A. (Ed.) *The Challenge of Balance,* CSE, New Delhi.

Singh, J. (1979) *Central Places and Spatial Organisation in a Backward Economy*: *Gorakhpur Region*, Uttar Bharat Bhoogol Parishad, Gorakhpur.

Singh, J., & Singh, B.K (2001) Irrigation Technology and Food Security, in R.B. Singh (ed.) *Land Resources, Technology and Poverty*, Tara Book Agency, Varanasi.

Swaminathan, M.S., Agriculture, Food Security and Employment : Changing Times, Uncommon Practices, *Nature and Resources,* Vol.l, No. 1, 1995.

Thamarajlakshmi, R. Doha Declaration and Agriculture in Developing Countries, *EPW*, Vol. XXXVII, No. 1, Jan. 5-11, 2002.

Thiel, R.E. (ed.) Laendliche Entwicklung nict nach Marktkriterion, E+Z43, (3) 2002.

Troughton, M.J. (1991) Ecological Assessment of Modern Agriculture in Limits to Rural Land Use G.M.R.A. van Oort *et al.* (editors) Pudoc, Wageningen.

Windfuhr, M. (2002) *Everything but Farms*, E+Z, 42(3).

World Bank (2000) *From Vision to Action*; update (concept paper) Washington DC.

World Bank (2001) *Reaching the Rural Poor An Updated Strategy for Rural Development.* Washington DC.

23

A Plan for Mitigating Drought and Floods

Watershed Development through Employment Guarantee is the Key

H. M. Desarda

"I will give you a talisman. Whenever you are in doubt, or when the self becomes too much with you, apply the following test. Recall the face of the poorest and the weakest man whom you may have seen, and ask yourself, if the step you contemplate is going to be of any use to him. Will he gain anything by it? Will it restore him to a control over his won life and destiny? In other words, will it lead to Swaraj for the hungry and spiritually starving millions? Then you will find your doubts and your self melting away."

— Mahatma Gandhi

Immediate Task

Recurrent drought and floods coupled with mass poverty, chronic unemployment and pervasive undernutrition is the major challenge before India. However, instead of *indulging* in the usual rhetoric of truant rains and doling out funds in *adhoc* and arbitrary manner for drought-relief the Union Government should urgently *launch* a nation-wide Watershed Development Programme through the statutory *employment guarantee,* with the active involvement of state governments, panchayats and the people, to eradicate the scourge of *man-made* famine, floods, hunger and poverty permanently.

The Perspective

First and foremost, we should make a clear distinction between the *meteorological, agricultural* and *hydrological* drought. A comprehensive crash programme for turning the drought adversity into an opportunity, to utilize the available food and labour-force for effectively drought-proofing the nation end eradicating the mass poverty and hunger permanently, is presented for the consideration of policy-makers at the highest level and all those concerned about finding a solution to the problem of famine, floods, unemployment and poverty. Undeniably, this is the national challenge before India today.

In our opinion, *water* and *employment is* the key to build India in accordance with the vision and values of founding fathers and mothers, which are succinctly enunciated in the Constitution of India, as the Fundamental Rights and the Directive Principles of State Policy. 'The objective of our freedom movement is to wipe every tear from every eye', said Gandhiji. And he gave us the *talisman,* cited above. Very aptly Jawaharlal Nehru observed, 'so long as there are tears, we shall not relent.' Indeed, the watershed development we are advocating is the significant substantial step—the talisman—to solve the major national problem we face even after the 55 years of the Independence and 50 years of the Planned Development. Unequivocally, it is the tragic consequence of the *growth* and *governance model* we are pursuing in the name of 'national development'!

The author has formulated the following proposals after visiting large number of states and interacting with social scientists, land and water experts and social activists and the policy planners in the Thar desert, Indo-Gangetic plains, Brahmaputra valley and peninsular India. Agreeably, the scientific and societal approach to achieve this objective is: *micro-watershed development.* Monsoon management and community-based rainwater harvesting is the major thrust of this strategy. Quintessentially, we should follow a socioecological-ethical pathway to equitable and sustainable development.

The quantum of water India receives through the precipitation is 4000 billion cubic metre (BCM), of this annual average runoff is 1953 BCM. Of course, there is a marked *spatial* and *temporal* variation in precipitation and water availability. We are familiar with paradox of the floods and droughts. The flood ravages in Brahmaputra valley and Gangetic plains; and droughts in the western and peninsular arid and semi-arid regions pose a big challenge.

After the independence we have built more than 4000 large dams entailing an investment of Rs. 5000 billion at the current cost norms. Despite this, day by day the water-scarcity is becoming more and more acute. Surely, the present supply-side approach is not a solution to the 'scarcity' problem we are facing.

The need of the hour is to come out of the trap of the cult of gigantism and technoengineering growth mania. Indisputably, this is the root-cause of India a water-rich country becoming a water-short nation. As a consequence, we are fast heading towards water-wars between regions/states, sectors, groups and persons. Despite the Supreme Court's direction the tug of war and tensions we are witnessing between the two riparian states, Tamil Nadu and Karnataka, over sharing of the Cauvery waters should awaken the nation. The prevalence of whole gamut of the highly ingenious indigenous rainwater harvesting systems—which are succinctly documented in the CSE publication Dying Wisdom—are fast falling into disuse, needs to be revived and rebuilt. If areas of the low and scanty rains like the Thar desert were able to manage their meagre water resources wisely, there should not be any difculty in providing water to each and every Indian, anywhere in our subcontinent. The global experience tells us that there are local solutions to ward off the water scarcity. In fact, water is not only what we see as rains and storage on surface but the vast link from the ocean to rains, snowmelts, dew and all allied water-entities. Agreeably, the soil moisture is most strategic part of the water regime. Instructively, over 40 per cent of our total rainwater is stored in soil-profile. Hence, we have to carefully grasp the complex character of the hydrological cycle. Above all, the role of the vital eco-entities like the crop lands, grazing lands, forests and entire range of the biotic and biomass system which helps to augment, renew and recycle the vital life source—water, that is the synonym for the life! Well, water is used to extinguish fire but when water itself catches the fire the situation becomes volatile, violent and threatens national peace and progress. The chief cause of this is the avarice and greed, which is making us crazy about the grid of the water and electricity. As Gandhiji's famous remark reminds us : 'There is enough for everyone's need but not for anybody's greed'. As long as we fell to make this as our conscious national value and vision, we will not find a lasting solution to any problem including the most pivotal one like water. With the aforesaid diagnosis, perception and perspective, albeit very short and succinct, we proceed to present the broad contours and components of the action-agenda, including the planning and policy proposals and the active involvement of the people.

The Plan : Conceptual Framework and Operational Strategy

A master plan for initiating and streamlining the watershed works throughout the country is outlined below. The conceptual framework and the operational strategy of the *plan* are detailed below.

Bid Good-bye to Gigantism

As a first step we need to move beyond the present large dam-centric irrigation planning and policy of water resource development, which has distorted the water resource planning and draining the exchequer excessively. Secondly, Drought-prone and Flood-prone area programmes and scattered projects of watershed development should become a part of broad-based *national movement* for rainwater harvesting. The major focus of this should be integrated development of the land, water and biomass resources. For this we have to adopt a holistic approach.

It has now become amply clear that *in situ* rainwater harvesting is the cost-effective pathway. As such, the prevailing perception about storage needs to be thoroughly reviewed and radically restructured. We should always remember that the protective irrigation for drought-proofing is different from the perennial irrigation. The large dams and mega storages with a vast network of canals are required for the water-guzzling crops like the sugarcane and the crop varieties of the Green Revolution oriented agricultural development. And also, for the water supplies to the metropolises and water-squandering industries and the Westernised lifestyles. Hence, we have to overhaul the present water policy and planning paradigm. This is the primary pre-requisite to initiate the rational, resource-conserving and pro-people water resource planning. We have to clearly grasp the ecological and ethical challenge we are facing and the genesis and consequences of this.

Resource Literacy and Empowerment

In the framework of long-term perspective the watershed works should be *dovetailed with* the employment programme which is statutorily guaranteed and *linked* to 'food for watershed works scheme'. The edifice of this should be community empowerment through employment and entitlement. For this systematic public awareness campaign should be launched. The major plank of which will be resource-literacy, care and conservation of prime natural resources—land, water, flora, fauna and biodiversity. Concurrently, a training of the *panchayat* members and those who wish to participate as *barefoot planners* should be initiated at every village and micro-watershed level. Clearly and carefully the spirit of voluntarism and professionalism has to be blended beyond the NGO-cult. Exceptions apart, the present government/non-government alliance generally *excludes* the community and creates an army of middle-men in the delivery system, who are often outfits of the vested interests. Therefore, a plea for *power* to the people through the Panchayat, Gramsabha and watershed community. Given the fact that the programme needs to be launched in thousands of villages simultaneously, there is a pressing need for forging a community institutional arrangement, which the panchayat-raj system alone can provide across the length and breadth of the country. Notwithstanding the present *politicking* prevalent at the panchayats there is no *alternative* to this. We have to accept it as a part of democratic-schooling. For this there is an urgent need to redefine and recast 'the growth and development' as concepts and categories. Boldly, we have to *de-mystify* and *decentralise* the meaning and process of development and make it pro-people, in particular pro-women, and pro-nature.

Utilising the Labour force

The major favourable factor in the wake of drought is labour *availability,* which must be seen as an opportunity for creating *durable social assets.* Moreover, we should remember that the currency notes are meaningless sans labour resource and social capital in terms of community participation. This is the *core* around which the programme should be designed and implemented. Hitherto this has been missing element and that is why nearly 5000 billion rupees spent on irrigation projects, DPAP and such other programmes have failed to rescue the agriculture from 'vagaries of monsoon'. With this clear perspective, an alternative approach should be properly worked-out. This means a *paradigm shift* in the prevalent drought-mitigation and also macro-economic policies of the growth and globalisation.

The Dimension

This scheme should aim at an employment of 20 million people for 200 days at a wage of rupees 50 in cash and two and half kilogram of grains per day. This would entail an expenditure of Rs. 20,000 crore in cash plus 10 million tonnes of food worth Rs. 5,000 crore, at the minimum procurement price. Thus, a sum total of Rs. 25,000 crore investment would generate a work of 4,000 million person-days in a year. Meticulously and meaningfully it should be used for creation of durable social assets. Broadly this includes the works of soil conservation, moisture retention, afforestation, groundwater recharge, water harvesting structures etc. Retarding the runoff and enhancing the recharge is the only option to augment the water availability. The noteworthy aspect of this is that the returns to the investment are the most positive and quick. As such, it is most viable economic programme on the benefit-cost criteria. Certainly, it is not a dole distribution or charity programme but a very sound employment and production programme, which fulfil the considerations of economic efficiency, social equity and ecological sustainability.

Ridge to Valley

Going by the current state of (degradation) watersheds it would require on average around 100 person days of physical work per hectare to carry-out the watershed works. That means around 20 million hectares (mh) can be treated and tackled in a year. True, as per the local or watershedwise availability of the number of labourers it would have to be phased out over 2 years. In any case, once initiated the entire watershed must be completed in a period of 3 years. The work should be carried out strictly *sequentially from* the "ridge to valley" in each and every micro-watershed within the area of sub-watershed in that river basin. This alone can ensure that size, scale and sequence of impounding water is appropriate. Regrettably, at present it is highly topsy-turvy and this is the chief cause of water-scarcity and imbalance in water availability in the country. Hence, we have to adopt an alternative approach, which is integrated and holistic.

Need to stop Degradation

According to the National Commission of Agriculture (1976) 175 mh of land in the country is degraded and is in need of urgent treatment and improvement. Even at the scale suggested here, it would take 9 years to cover the watershed works to arrest wind, water and biotic-pressure-related erosion and degradation. The erosion, degradation and damage has been going on unabated for years causing an annual soil-loss of 12 billion tonnes. Indeed, it is a colossal national loss of precious natural resource. As such, the task of conserving the resource is most challenging. The well-being of the present generation as well as posterity depends on addressing this *number one* national problem. Surely this calls for a National Movement with the clarion call, 'Mitti Bachao, Desh Bachao'—Save Soil to Save the Nation. Undoubtedly, the *life* and *livelihood of* people directly depend on health of the soil and natural resources. In short, the overriding national objective should be to minimise the soil-loss. That would maximise the moisture and mitigate the water-scarcity and the crop failure. Ideally we should plan for a zero soil loss. Curiously, we face, at the national level, a paradox of floods and drought simultaneously. Not only the traditional flood prone areas in Brahmaputra and Gangetic valleys are more intensely ravaged by the flash floods but even arid and semi-arid regions are experiencing floods. The chief cause of this is: indiscriminate and ruthless deforestation in the catchment areas and mountain regions. The silt carried by Brahmputra is 735 million metric ton every year. The amount of soil loss on account of the destruction of the vegetative cover-trees and gasses—in each and every riverbasin, rivulets and watercourse is causing tremendous soil loss and as a consequence it reduces the recharge. The co-relation between rainfall, runoff and recharge, the three fundamentals of water, is diverged and disturbed. It has direct bearing on the water availability, in each micro, milli and mega watershed in any river basin. Unless and until we take care of this, the water-scarcity can never be tackled. We have to see it in totality and deal with the basic issues. Unequivocally, this is the paramount task before the nation today.

Ecological Fundamentals

It should be carefully noted that the formation of an *inch of soil* takes several hundred years. Further, we have to reckon with the fundamental fact: margin of the life on earth squarely depends on the nine inches of the top soil, just that, which we are so callously squandering through the mind-boggling erosion that is going on unabated for several centuries. Therefore, we have to put our house in order to protect the soil resources without any laxity or excuse. After all, the protection of the *biophysical foundations* of growth is the most pressing challenge before the country today. The threat posed to the life-support-system is most alarming and hence, we have to put our acts together to protect the same from the indiscriminate deterioration and degradation of the resources on which hinges the well-being of the present and future generations.

Diabolically, our planners, policy-makers, the industrial tycoons and the army of economists and experts who are the protagonist of the market and are in the 'service' of the state and *noveau-riche* see the solution to every problem in globalization. Forgetting altogether the pressing problems of the mass poverty, illiteracy, ill-health, hunger and ubiquitous unemployment, day in and day out they keep on chanting the mantra: our *macro-economic fundamentals are sound!* The degradation, denudation and decline of natural resources is not a matter of serious concern and consideration to them. Needless to say that it is not only a myopic view but tantamount to criminal attitude to the eco-system. What a pity that our ruling elite and the moneyed classes care so little about the state of our planet.

As a perceptive critique of water resources management in India observes,"Water resource management strategy in India has so far been guided by the philosophy that water is strictly a fixed resource and we cannot really destroy it on any significant scale. We have successfully destroyed much of our water resources in our streams and rivers, fields and tanks, wells and ponds. Water is not available automatically as a renewable resource because the ecological processes that maintain and give stability to the water cycle govern its renewability. Disrupted water cycles can turn water from an abundant renewable resource into a vanishing non-renewable resource." Precisely, this is what has happened during the past 50 years of the water resource planning in India. We have forgotten altogether the *causality* of the renewal of the resources and the result is: pervasive water scarcity. Therefore, we have to seriously *rethink* and restructure the present approach of the water resource planning and overhaul the project model.

Alternative Approach

Whether it is the large irrigation and power projects or watershed works or any other programme of economic and social development, it cannot succeed just by pumping—in more and more money. Obviously "More of the same" is no answer at all. To avoid the past mistakes the need of the hour is social motivation, transparency and accountability. That alone can break the crooked chain of corruption, which has turned development projects and programmes into a *kick-back* industry! That is tote real name of the game! This chromic disease cannot be cured without decentralisation of political power and devolution of the economic resources. All in all, the centre stage should be accorded to the Gramsabha and the Panchayat institutions. In this respect the recent experience of state like Madhya Pradesh is worth examining and emulating. At the national level the Prime Minister and in the states the chief minister should personally monitor the watershed development programme. That will accord it the due priority and put it on high national action agenda. The main focus should be the community mobilisation, social participation and empowerment and the role of the government has to be facilitative in terms of providing institutional framework and finance and allow the community to execute works, keeping in view the location-specificity and drawing upon collective wisdom of the community. Hence, the plea for adopting socio-ecological pathway.

Bhoomi, Jal Aur Jangal Vikas Sena

In addition to the availability of *funds* the most crucial factor is effective organisation of the labour-force to do the actual work. For this an institutional framework on the pattern of the National Literacy Mission is necessary. As such, there should be a National Watershed Development Authority under the chairmanship of Prime Minister and full-time executive vice-chairman, who is committed to the cause. At state level Chief Minister should head the State Watershed Development Authority, with much more functional and field-oriented inter-disciplinary team to guide, supervise and strengthen primary watershed works. It should also look-after the production and maintenance activities. Further, we suggest a creation of special task-force named as "Bhoomi, Jal Aur Jangal Vikas Sena"— a land army to carry out the watershed works in a planned and effective way. In fact, it would give the labour-force, particularly the landless labour and poor peasants a sense of security and also provide a steady supply of labour to complete the work in orderly and scientific manner, with due regard to the location-specificity of the drought-prone areas in different agro-climatic zones; topography and terrain.

Administrative Streamlining

Besides, the political priority, an appropriate administrative mechanism has to be evolved. We suggest an appointment of a senior (secretary rank) IAS officer as "Watershed Development Commissioner" in each district. Given the vital importance of the programme in solving the problem of mass poverty, it would be quite appropriate to accord it the highest place in the administrative hierarchy; and in implementing policies of the growth and governance of the nation. Moreover, these functionaries should be actively involved in the work of resource-mapping and location specific adoptions in *land-use planning* and watershed plus activities like diversification of farm production and value addition through processing, and crafts of wide variety. Remember, the twin pillars of India's development are : *Kastkari and dastkari*—cultivation and crafts in all its ramification. Innovative revival and renaissance of the indigenous knowledge, skills, resource-literacy and symbiosis of the agriculture and industry alone can ensure the livelihood security and social harmony. Well, the idea is not to create any additional bureaucratic structure but to refurbish the set-up so that functionally it becomes more effective and accountable.

Consciously, it will have to work with the community in a spirit of partnership. To make it truly participatory. The total transparency should be the guiding principle.

Power to the People

The plea for the administrative streamlining and overhauling made in the preceding paragraph is likely to create misgiving that we are pleading for the expansion and strengthening of the bureaucracy. Absolutely not. The purpose is the opposite of this, to devolve the resources to the community. The control and the command would be expressly at the hands of the village and watershed community. Hence, it is not at all the *top-dawn* strengthening but the total transparency and the accountability to the Grantsabha and day to day execution, supervision, monitoring of the work that should rest with the watershed community. The composition of this should clearly provide for the representation of the major interest groups like landless, women, small farmers, artisans and some elders, who can exercise the balancing influence. In short, the finance and functions are devolved and decentralised without any ambiguity. That should take care of the community being at the centre-stage and bureaucracy and technocracy will work in close co-operation with the cross-sections of the stake-holders. The spirit of functioning as a true public servant should be the prime criteria in choosing the personnel from the cadre.

Towards Meaningful Education

Given the present trend of 'jobless growth" large majority of youths do not have any interest in so called 'higher education'. Sadly, the universities and colleges have become dens of ignorance, nuisance and parasitism of the worst kind. Hence, it would be worthwhile to declare "examination holiday" for a year or two and involve the students and the teachers along with the drop-out youths (who have a tendency to shun the physical work and are ashamed to soil their hands) in the national watershed development movement. Certainly, it would be an unique opportunity to engage them physically and intellectually in the task of nation-building on the land. Indeed, it would make them really educated and not just literate. Pervasive parasitism, which has become the hallmark of the elite, an urbane and so-called educated member of the civil society needs to be consciously changed. Therefore, they should welcome this opportunity to work with community and learn the value of *Shrama Pratishtha*, dignity of labour, which will help them to overcome their alienation from nature and society. By actively participating in this unique venture they would learn the true meaning of life, nature and productive work—a much needed 'watershed' in education-for-life!

We Have the Resources

Notwithstanding many disturbing and diversionary trends in our development journey, we have the financial, technical and administrative skills as well as physical resources to fund such a strategic *nation-building* programme. Notably, the outlay of Rs.25,000 crate it requires is just one per cent of India's National Income today. Thoughtfully, we should be able to garner this amount by levying a surcharge on the income-tax and/or through a special professional-tax for the purpose. Well, if we cannot muster and mobilise bare one per cent of nation's GDP for such a vital task then all tall-talk of planning and policy is utterly futile. The most critical missing factor is moral capital. Obviously through the social mobilisation and community action we have to consciously build this vital moral capital. If we fail to act on this forthwith, the acute suffering of masses will lead to anarchy, violence and disorder. We have to see the writings on the wall and act with a sense of national urgency—here and now!

Movement for National Reconstruction

In order to create wider social support and involve civil-society in this national programme, a district level advisory committee of social visionaries, artists, literary figures and social analysts/activists should be created to guide, supervise and evaluate the programme. Social commitment coupled with political will, administrative skill and social mobilisation in the spirit of the Freedom Movement is necessary to ensure the success of this pivotal programme. In short, we have to launch a national movement for watershed development through the statutory employment guarantee. Finally, a time has come to take it from a stage of project to movement, so that it replicates fast and ensure quality and efficiency. Ultimately, we should strive to create a *Jal Sanskriti* (water culture) which promotes health and happiness of everyone. To educate the public opinion and create societal awareness we have suggested the wider resource-literacy campaign, focusing in particular on *Jal Sakshrata* (water literacy), which would emphasize the care, conservation and availability augmentation simultaneously.

All in all, the objective of creating employment for 50 million persons during the Tenth Plan in non-formal and non-industrial sector can be made viable, productive and meaningful (not just the salaried jobs!) only if it is property linked to the watershed development programme and watershed plus activities of the value addition through the diversification of agriculture and allied activities. Thus, it can be very well interwoven with the planned programmes of agricultural and rural delopment in the 10th and subsequent plans. Hence, this plea for adopting the plan of the watershed development outlined above as the *nodal* programme for national regeneration with an elan of *peoples' movement* for community based rainwater harvesting. This alone can ensure *sustainable* and *equitable* development for every Indian and make India a truly developed and civilized society, free from the greed and violence.

24

Sustainable Livelihood and Lifestyle in India

Issues and Challenges

Ali Mohammad

The sustainable livelihood security has been defined by the Brundtland Commission in 1987 that sustainable livelihood maintaining adequate stocks and flows of food and cash to meet basic needs by maintaining or enhancing of resource productivity on a long term basis. Sustainable livelihood security may be gained in many ways. Through ownership of land, livestock, through rights to grazing, fishing, hunting or food gathering, through stable employment with adequate remuneration or through varied activities.

For such livelihood equity indicates the degree to which the poor benefits from development in absolute terms. The need to reduce rural poverty and to bring about greater equity is one of the most pressing economic and social issues in developing countries. In the context of sustainable development, measures towards poverty alleviation like land redistribution and tenancy reforms, improved access of the poor to technologies, extension of training facilities, benefits from food distribution scheme etc. will also go a long way in ameliorating environmental damage.

Measurement of Sustainability

It is a very difficult task to get most of what society demands from renewable resources without unacceptable damage to the production potential. The appropriate method and measurement largely should be reviewed in such a way: deprivation from food and basic resource and other natural products available earlier but looses it capability at present because of the lack of management to increase the greater productive potential from natural resources for sustainable development.

At the very outset, I would like to make it clear that how far the idea of sustainable livelihood and lifestyle is very much applicable in Indian context? Firstly, the concept of sustainability for sustainable society must be made clear otherwise the philosophy of sustainable development may lead a society to a wrong direction. For the sake of sustainable development the aspect of environment and natural resource may be misunderstood and actual potential may not be used optimally or properly. The over and under-use of the resources may not lead to the stage of sustainable development and chances of loosing its long-term potential may be faced because of the phenomenon of the degradation of the various resources such as land, soil, water, forest and overall environment. Consequently, the sustainability of livelihood and lifestyle may also be degraded and much cherished aim of the developing countries, i.e. to eradicate poverty may not be achieved. Therefore, the concept of 'Resource', 'Environment' and 'Sustainability' may not be understood in Western perception.

The existence of each phenomenon related to sustainable approach should be analyzed geographically and in a regional perspectives and treatment of each phenomenon should also vary spatially. For example the land as a resource, either with sustainable or unsuitable combination may not be treated uniformally with variations in space and location. The land resource combination should not be treated in isolation otherwise the objective of optimal use of land resource may not be achieved.

The land degradation phenomenon is prevalent everywhere. It is being perceived as one of the main impediments all over the country, specially in advanced agricultural pockets. Therefore, these regions must be identified because of its immense importance. The mountain regions of the country even at a low level of performance should also be identified on top priority to reduce land degradation in order to achieve sustainable livelihood.

The agriculturally developed pockets such as Punjab, Haryana, western U.P. and large number of western Ghat districts as well as newly emerged high level productivity districts of southern India are facing the problem of soil nutrients degradation of a particular type of nutrients because of the absence of crop rotation practices. In order to maintain existing level of livelihood and gains the farmers have increased the infrastructural inputs from 10 per cent to 40 per cent between 1971 to 1995. How far such a model of agricultural development will continue for a longer time far the sake of sustainable livelihood and lifestyle? The failure of such a model of agricultural development is experienced and recorded in about seven districts of Uttar Pradesh adjacent to Haryana, i.e. the practice to grow surplus food grains by devoting higher share of the cultivated area in Punjab and Haryana is not being adopted even in the adjoining part of this state (UP). Hence 70 per cent of cultivated area is being devoted to non-food grains and only 30 per cent is devoted to food grain crops. It shows that for a better sustainable livelihood grow more food grain became unsuccessful because of the achievement of food self-sufficient at national level and surplus food grain production in some particular regions. It shows that at national level grow more food grain is not at all an issue but market surplus of commercial crops are gaining more importance for a better attainment of sustainable livelihood. However, such type of problems should be tackled at regional level in an integrated manner. The integration between small and marginal farmers with the large size of landholders in adoption of higher remunerative value crop should be observed objectively so that the problem may not be aggravated. The situation may take a very dangerous turn in many pockets of the country where more than 80 per cent landholders belong to the category of small and marginal farmers. The tendency emerging recently is not very healthy for sustainable livelihood as the small and marginal landholders are leasing out their lands to the big size of farmers to cultivate their lands because of the adoption of agricultural support technology with the large size of landholders.

The problem of sustainable livelihood and lifestyle, if taken into consideration at various regional levels the problem can well be understood such as:

1. Potential of sustainable livelihood in mountainous areas.
2. Sustainable livelihood in drought prone areas.
3. Sustainable livelihood in desert and unirrigated and dry farming areas.
4. Culturable wasteland and land reclamation and sustainable livelihood and lifestyle.

5. Sustainable lifestyle and livelihood in a high food grain fluctuation situation.
6. Major issue of livelihood lies with the small and marginal farmers.
7. Common property resources and sustainable livelihood.

There are many other sectors and sections of the society where the study and analysis of sustainable livelihood is needed but among all these sections and regions some of the regions should be tackled on priority basis.

The problem of sustainable livelihood in mountainous regions, covering more than 21 per cent of total area of the country with diversified resources and potentials to produce diversified crops, are significantly different with the thickly populated regions such as Gangetic plain area. Slope, elevations and soil erosion is a predominant feature playing a negative role. Therefore, even steep slopes and marginal lands are being used for cultivation and grazing purposes. Secondly, lesser familiarity with the slopes is identified because of the lack of technological inputs as compared to the developed countries. The future of livelihood is very bleak and in danger because of the lesser management and lack of appropriate technology. The resources are diversified. Hence an integrated planning can provide better opportunities of livelihood even more than the people living in any other regions of the country such as coastal areas, plateau regions, deserts and Gangetic plain regions.

Culturable Wasteland and Sustainable Livelihood

It has been experienced from an extensive survey of a large number of villages that with the increase of population, landlessness and poverty the wastelands of all types are being used as common poverty resources. The survival of the poorest of the poor segment of village population is dependent for their livelihood on such type of land. These lands are characterized as no one resource use it up as you can. No doubt it is a major resource but not made as a resource. It is observed that culturable wasteland resources are of crucial importance to the survival and livelihood to the poor. The intensity of importance varies with the size of this resource at village level. This resource is also accessible for the whole community of the village for pastures, community forests, community threshing grounds, waste dumping grounds, drainage, village ponds, tanks etc. A wide variety of essential items are also collected for the use as well as for income to the poor households such as food, fuel, fodder, fibre, small timber, manure, bamboo, medicinal herbs, oils, material for the construction of house, resin, gum, honey, spices etc. in the absence of property rights a number of problems are being faced by the existence of such type of resource. Firstly, the dispute is very common feature if it is being claimed and owned by someone. Secondly, an indiscriminate use of these resource has created a situation of land degradation.

The issue of the existence of such type of resource is also very important and can be taken for a meaningful discussion in the case of regions where a considerable amount of culturable wasteland exists. Due to the lack of irrigation and low rainfall, high elevations and hills the region is under high culturable wasteland, which will remain as an important component of the village social and economic life.

The following issues should be included for discussion for a meaningful discussion and solution.

(i) An inventory of various types of culturable wastelands should be prepared at regional and village level,
(ii) Identify the use of such type of land as a common property resources,
(iii) How long will it remain as a crucial resource for the poorest of the poor of the village society,
(iv) Mostly the women and the children are involved in collecting the materials from such type of land considered of a very low level,
(v) It is a resource of sustainable value and benefits to the poor during the lean season,
(vi) Degradation of such land is not desirable, and
(vii) Distribution without reclaiming the land among the rural community is depriving the poorest from such type of resource benefits.

It is estimated that culturable wasteland add about 5 billion US $ a year to the incomes of poor rural households in India or about 12 per cent to incomes of poor rural households. This is about two and half times of the total World Bank lending to India in 1996 and more than twice the amount of official development assistance in the same year. Therefore, culturable wasteland is of major importance to the poor. The area under culturable wasteland is about 64 million hectare whereas the cultivated area is about 124 million hectares. Its distribution is not uniform throughout the country as it vary from region to region based on the elevations, slopes, rainfall amount, inaccessible areas, deserts and rugged topographical areas. In many of the regions culturable wastelands are more than the cultivated area such as in the Himalayan region, which constitute about more than 64 per cent of total geographical area. In deserts it is about 78 per cent, while in Uttar Pradesh it varies from 5 to 20 per cent. It is also evident through literature available that no concerted efforts have been done to study such regions.

An extensive survey of a group of villages located between Ghazipur and Ballia district of eastern UP was conducted in which a group of 30 villages was included (Annex I). It was observed that about 13 per cent of total area is under culturable wasteland though this is not the most ideal region to study such type of problem but was studied for some other purposes. But even though this traditional resource is still playing a vital role for few poor households for grazing, fuel, grasses, playgrounds etc.

Sustainable Development and Issues of Small and Marginal Farmers : A Critical Issue

The issue of the development of small and marginal farmers lies to increase the productivity and employment of this section of population. Among the rural poor, the small and marginal farmers constitute a sizeable portion in a country like India in particular and in all the developing countries in general. According to the studies conducted by FAO which has shown that small farms constitute about 65 per cent on an average of the total farms in India and contribute around 30-35 per cent to the total agricultural output (Randhawa and Sundaram, FAO, 1990). The largest 10 per cent of farms control about 50 per cent of land. The cultivable land per capita decline in India from 0.48 hectare 1951 to 0.26 hectare in 1981 and it is expected that it will be 0.11 hectare in 2000 (Swaminathan, 1990).

The size and number of' small and marginal Farmers vary from one region to another region in India such as in Punjab it is 54 per cent of the total number of farm households, in Jammu and Kashmir 92.0 per cent, Arunanchal Pradesh 35 per cent, Bihar 87 per cent, Uttar Pradesh 86 per cent and in India as a whole it is 73 per cent.

The issue for the sustainable livelihood and development of small and marginal farmers is most significant because the land share is declining at a very alarming rate and if it continues the small and marginal farmers are not going to sustain their livelihood from the land left with them. The strategies so far identified, i.e. to increase the size of landholding by distributing the surplus or culturable wasteland or to increase the productivity by introducing the high value commercial cash crops seems to be laudable but how far applicable needs a concerted planning otherwise a major chunk of rural population seems to be displaced and no alternative is left to tackle this problem. If it is not planned properly and the nation is not well prepared it can lead to a major chaos and peace of some of the regions may be highly disturbed which may take a turn to plan and prepare the country to follow a different type of strategy for a better and sustainable livelihood.

This Estimate is Rough and Probably Low

Tony Beck and Madan G. Ghosh (10, 2000), *Economic and Political Weakly*, Vol. 35, No.3, January 15-21, 2000 have studied in detail common property resources and the poor.

ANNEXURE I

Akrampur, Mubarakpur, Nurpur, Rampur, Fatehpur, Bazidpur; Chaurahi, Borsia, Usmanchak, Rasulpur, Hainsi, Nizampur, Nasirpur, Para, Nonehra, Susundi, Ismailpur, Khalilpur, Arzanipur, Huserpur, Miranpur, Aslatpur, Sultanpur, Firozpur, Jakrauli, Alwalpur, Alamchak, Adilabad, Yusufpur, Bahadurpur.

Land Use in Selected Villages in Uttar Pradesh

Districts	*Villages*
1. Agra	1. Khanpur 2. Nagla Punu 3. Thapi 4. Thansera 5. Mohammadpur Bcharipur 6. Gidraun
2. Aligarh	1. Daheli 2. Kondra 3. Dharampur 4. Nagla Jhau 5. Rangpur 6. Pipli. 7. Manzur Garhi 8. Alampursubkaramuafi 9. Sipchrabhanpur 10. Sikandarpur Chherat 11. Barheti 12. Chherat Sudhal 13. Palamazrakastalibais 14. Chandaukha 15. Sumeradaryapur 16. Jawan Sikandarpur 17. Nagaula 18. Gurshikran 19. Nahaukhera 20. Pikhlauni 21. Ikri
3. Allahabad	1. Sarai 2. Glioripur 3. Mahera 4. Sinkikhurd 5. Bhagwatpur 6. Dandi 7. Malakpanchambha 8. Ghughapur 9. Chipaundha 10. Patulki 11. Kaitha
4. Almorah	1. Chidu 2. Phattania 3. Palyura 4. Kotila 5. Akheti 6. Jaitha 7. Jali Khan
5. Azamgarh	1. Katghar Sadar 2.Patna Ahiyai 3. Laphia 4. Lakchmipur
6. Bahraich	1. Bankatawa 2. Sahshrawan 3. Tawakkalpur 4. Roshan Kola 5. Purettindusing 6. Lohti
7. Balia	1. Bhuarkhund 2. Bauria
8. Banda	1. Ragauli 2.Jariha 3.Khajuri 4.Khichril 5.Badhauli 6. Sidhalla 7. Gadhiwa 8. Guraulishukul 9. Amraiya 10. Barwar 11. Gauhanikhurd 12. Chatkan 13. Jharri
9. Barabanki	1. Bhatkhera 2. Chakkodar 3. Bakshupur
10. Bareilly	1.Dhaunr Satuia 2. Sahodar Nagla 3. Mastipur 4. Waris Patti 5. Rahmanpur
11. Basti	1. Bamhnikhurd 2. Pikaura 3. Bankata 4. Siswari 5. Khoria 6. Kashipur 7. Madhwapur
12. Budaun	1. Salimpur 2.Mullanpur 3.Asdharmai 4. Junawai 5. Bichoula 6. Kayanpur 7. Mugarratatei 8. Bliurkhitaura
13. Bulandshahr	1. Jatpura 2 Chainsa 3. Mohammadpur Gujar
14. Deoria	1. Chauria 2. Gopalpur Baraicha 3. Gobrainkhas 4. Faridpur
15. Etah	1. Ghilaua 2. Nagladuria 3. Mankai 4. Mirzapur
16. Fatehpur	1. Khairabad 2. Sudesra 3. Saleempur 4. Saraiabhaiyya 5. Dasauli

17. Ghazipur	1. Kalyani
18. Gonda	1. Sitlapur 2. Saddupur 3. Chauka Kalan 4. Materia 5. Kharhatia 6. Qiyamjote 7. Khanpur
19. Gorakhpur	1. Raipur 2. Urhwatia 3. Rachchapar 4. Bankatia 5. Dubauli
20. Hamirpur	1. Todarpur 2. Siyawari 3. Kishanpur 4.Dhundhaiyan
21. Jalaun	1. Narchlia 2.Banupura
22. Jaunpur	1. Madapur 2.Sarai Kewat
23. Jhansi	1. Padri 2. Basai 3. Kemari 4. Rasoi 5. Parsai
24. Khiri	1. Roshan Nagar 2. Lilauti 3. Mundia Mahadeo 4. Prem Nagar 5. Kamlapur
25. Mainpuri	1. Bikrampur 2.Budsui 3. Kulipur
26. Mathura	1. Lalgarhi 2. Rampur 3. Ekdanta 4. Azampur 5. Naglaqazi 6. Daulatpur
27. Meerut	1. Malmazra 2. Dudhli 3. Meghrajpur 4. Kuri
28. Muzaffar Nagar	1. Dhandaoli 2. Salajudi
29. Nainital	1. Bishguli 2. Bagaur 3. Ucholi 4. Mallibamori 5. Chorpani 6. Chaumehla 7. Thotupura 8. Raipurkhurd
30. Pilibhit	1. Nibadandi 2. Jarabuzurj 3. Alampur
31. Pithauragarh	1. Rilkot 2. Majgaon
32. Rampur	1. Nandgaon 2. Ahmadabad 3. Rasula 4. Mohabbat Nagar
33. Saharanpur	1. Bhaupur 2.Asrakheri 3. Amrupur 4. Padli Garunt 5. Ladpur Kalan
34. Shahjahanpur	1. Mohammadpur Azampur 2.Paindapur 3. Dhaka 4. Shamspur 5. Moazzam Nagar 6. Gutetiuttar 7. Shahzadpur
35. Sitapur	1. Bhagwanpurmafi 2. Unera 3. Narainpur 4. Gauria
36. Tehri Garhwal	1. Paraugi 2. Dibnoo 3. Gaphar 4. Ghorpur 5. Pathoeegaon 6. Reechak 7. Simalikatal 8. Khadolitalli 9. Kot 10. Dobri 11. Joshayna 12. Museti-Bangla 13. Matheliwalli 14. Kandhyal Sera 15. Mandeupur 16. Anchmalli
37. Varanasi	1. Sarai Takqi 2. Chhapra Tarukha 3. Kethkahani

25

Regional Patterns of Food Security in India

Noor Mahammad

Introduction

Food security has become a matter of great concern to academicians, politicians, planners and decision makers all over the world during tile last two and half decades due to the unprecedent growth of population. The present world population of 6 billions is likely to reach 8 billions at the end of first quarter of this century. Each year there will be approximately 90 million additional mouths to feed and bodies to clothe, and about 90 pet cent of this increase would be in the low income countries. According to an estimate about 800 million people already suffer from hunger, and approximately 1.3 billion people subsist on an income of less than one US dollar per day. The World Bank has estimated that 1115 million people were below the poverty line and out of these 630 million suffered from extreme poverty. It further stated that 80 per cent of the world's pool live in rural areas where bulk of the people earn their living from farming (Chopra *et al.*,1998). But the status of farming on the whole is not encouraging. Recent FAO assessment indicates that world agricultural production including both crops and livestock, is going down in comparison to 1980s, the growth rate of food grains output in 1990s was almost half, i. e. 1.8 per cent against 3.54 per cent (Murlidharan, 2001). Among the developing countries as a whole, agricultural production of 1997 has experienced the lowest increase since 1979, and has barely kept pace with population growth, while in developed countries overall agricultural output slowed down in 1997 and it is not expected to be able to meet the anticipated consumption requirements, and the global stocks are to be drawn down (Luppin and Menzol,1999). Therefore, food security has been a persistent issue raised in several national and international seminars and conferences organised during the last two decades. It has also found a prominent place in the research themes of scholars specially the geographers, economists and sociologists (Minhas,1976; Sen,1981; Radhakrishna,1991; Tendulkar, *et al.*,1993; George,1996; and Suryanarayana,1996).

The World Food Conference of 1974 held at Rome, which was organised in the wake of 1972-74 world food crisis, made a call for ending hunger by1984 which led to the enunciation by FAO of the concept of food security defined as physical and economic access to food to all people at all times. In 1976 Minhas in his presidential address to 36th Annual Conference of the Indian Society of Agricultural Economics has cautioned the politicians, people and the academicians about the issue of food security in India. The UN Conference on Desertification held at Nairobi in 1977 alerted political leaders and the public to the serious threats to both food and livelihood security arising from the growing deterioration in the biological potential of the soil as a result of anthropogenic pressure and unsustainable exploitation. The 1992 Rio de Jeneiro Conference on Environment and Development emphasised the need to ensure food security at all levels within the framework of sustainable development. In the International Conference on Population and Development held at Cairo in 1994 the linkages between population growth and food production and the need to evolve global measures to satisfy the ever growing food needs was the focul theme of discussion. In 1995 the World Summit on Social Development held at Copenhagen made a strong commitment to compaign against hunger through its emphasis on poverty eradication. Similarly the World Food Summit held at Rome in 1996 gave a new impetus to fight for food security by focusing attention on food issues. Recently in 2001 a National Consultation with the focal theme Towards Hunger Free India was organised in Delhi with the objective of evolving an action plan and policies that would ensure sustainable food security at the national level and the elimination of poverty induced chronic hunger at the household level by the end of the Tenth Five Year Plan. Two documents, viz. "Food Security Atlas of Rural India" and "Enabling Development—Food

Assistance in South Asia"—have been released. Food Security Atlas of Rural India (Swaminathan, 2001) highlights the food security as the story of poverty, discrimination and neglect of livelihoods unhygienic living, lack of literacy, basic amenities and health care which is the indicator of failed global and national governance. Three dimensions of food insecurity, viz, availability, access and absorption have been highlighted. Food absorption the ability of an individual to assimilate and absorb the food consumed depends on the status of health.

Concept of Food Security

The concept of food security is very complex, multidimensional and complicated and is debated since last three decades. It has been visualised at international, national, regional, household and even at individual levels. The World Food Conference of 1974 was mainly concerned with global food security and it has realised that world food crisis was a common responsibility of all nations and that international approaches were required to achieve world food security (Sarris and Taylor,1976). Fluctuations in food supplies over time and unstable food prices are the basic problems of food security which require international and national solutions such as inter-regional grain reserves of buffer stocks, grain insurance and others (Johnson, 1976, and Vades, 1981). Food security is generally understood to imply arrangements whereby the people would be assured of a minimum level of food grains supply in year of normal as well as poor harvest, which implies the physical supply of a minimum level food grains during all periods including those of harvest failures (Reutlinger, 1977). Subsequently, it was realised that physical availability alone would not ensure economic access to food for all population but satisfactory production levels and stability of supplies should be matched by reduction in poverty and an increase in the effective demand to ensure economic and physical access for the poor (FAO,1987). As such the concept of food security at the individual and household levels was taken up (Frankenberger and McCarton, 1998). Hence, according to FAO (1984) the basic concept of food security implied that "all people at all times have both physical and economic access to the "basic food they need" but the World Bank

(1986) has modified it indicating that "food security is access by all people at all times to enough food for an active and healthy life". Its essential elements are availability of food and the ability to acquire it. The World Bank has made a distinction between chronic and transitory food insecurities. The former refers to persistent inadequate diet caused by the inability to acquire food either by purchase or by produce of their own, while the latter refers to a temporary decline in the household's access to enough food due to instability in food prices, food production and household income. However, the most comprehensive and perhaps, largely acceptable definition came out of the World Food Summit at Rome in 1996. Food security exists when all people, at all times, have physical and economic access to sufficient, safe and nutrition food to meet their dietary needs and food preferences for an active and healthy life. This definition emphasizes three critical dimensions of food security: (i) physical supply of desired food in sufficient quantity; (ii) an economic access indicating close link of purchasing power to food security; and (iii) stability in supply which include access to global food market (Bhattacharya, 2001). A more comprehensive explanation of food security includes livelihood security for each household ensuring both physical and economic access to balanced diet, safe drinking water, environmental sanitation, primary education and basic health care (George,1999). Finally, I would like to quote Chung *et al.* (1997) who have summarised the diverse determinants of food security status in a general conceptual framework focusing on the links between resources commanded by the household, levels of farm and non-farm production, household income, household and individual consumption and individual nutrition. The above review of the studies and the explanation given clearly reveals that the concept of food security is complex and multidimensional.

Indicators of Food Security

As the concept of food security is complex and multidimensional, and its various variables are interlinked with one another, it is difcult to evolve a single indicator for it. Chung, *et al.* (1997) have taken a number of socio-economic indictors such as owning poor quality land or no land, holding distress sales of livestock productive and valued

A Conceptual Framework of Food Security and Generic Indicator Categories

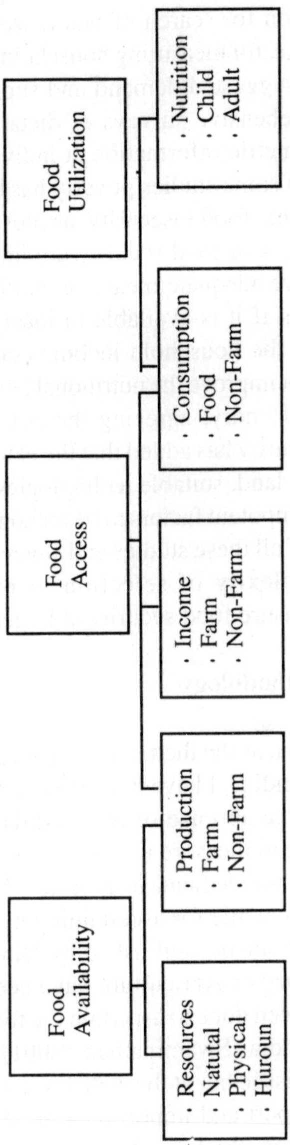

assets, taking loans, substitution of inferior quality staple food and vegetables, migration for search of work, working of pregnant and lactating mothers, etc. for measuring household food insecurity. Barun *et al.* (1992) have suggested demand and supply of food at country level, direct comprehensive surveys of dietary intake at household level and authropometric information at individual level to measure the food security. In some studies poverty has been taken as the main determinant of chronic food insecurity, as most of the qualitative and quantitative indicators of food security are linked to poverty issues. The poor do not have adequate means or entitlements to secure their access to food, even if it is available in local and regional markets (Sen,1981). Hence, the household income contributes to the access of food but it may not improve the nutritional level of all the individuals of the households. Hendry, agreeing the poverty as most effective indicator of food security has added that the extent of access to gainful employment, arable land, suitable technologies and other productive resources are other impotent factors to be reckoned within this context. The above review of all these studies and suggested indicators clearly indicates the complexity of selection of precise indicator and methodology to measure food security at larger scale.

Data Base and Methodology

Ladies and gentlemen, as the theme of my paper is "Regional Patterns of Food Security in India", I have to confine my study to demand and supply indicator in the absence of required data at district level for the nation as a whole to measure it more rationally and scientifically. In oider to ascertain the demand of food for the inhabitants, census data have been taken, while for ascertaining its supply the data from Indian Agricultural Statistics published by Directorate of Economic and Statistics, Ministry of Agriculture have been tapped. As far as I know, no study has been done to ascertain the food security/insecurity at district level in India. Swaminathan (2001) has made a maiden attempt to study this issue but at the state level. Realising the problem of getting precise export and import data of all agricultural crops at district level, a direct method developed by the writer for measuring food security or insecurity is used which is based on the computation of net food available in terms of kilocalofes and monetary value minus the consumption by the weighted total population and agricultural population; with allowances for wastage, seed and feed. This approach has been adopted because the direct estimate of production and consumption is the only reliable way to ascertain food security in a unit area, and to compare one area with another.

The trennium average for the years 1994-95, 1995-96 and 1996-97 have been taken of the projected total population and agricultural population with the simple reason that the latest agricultural data at district level for the whole country is available up to 1996-97. As this average data consists of the persons belonging to different demographic and socio-economic condition, but the requirement of food either in weight or calories varies according to age, sex, occupation, body size, income, culture and climate. Hence, Singh's scale of consumption unit, which is almost a modified form of Aykroyd's co-effecient has not been taken here firstly because it does not give a co-efficient for children under 4 years of age, although substantial percentage of this age group of population depend on diets, and secondly, he has used the same coefficient for adult female and children of 12 to 13 years of age (Chakraborty, 1970). According to Singh's scale each unit of the population is equal to 0.773 consumption unit or 1000 persons are equal to 773 consumption unit. Therefore, the total population and the agricultural population of each district have been multiplied by the co-efficient of 0.773 in order to get the consumption unit of the total population and the agricultural population. Chakraborty has tested it in several districts of India selected on random basis and has found it reasonably sound enough to be used as consumption unit.

Food availability has bean worked out at district level by taking trennium average data of agricultural produce for the years 1994-95, 1995-96 and 1996-97. There are many sources of leakage, e.g. losses in transport and storage, destruction by insects and pests, kitchen waste, use of seeds etc., between the production and consumption of agricultural produce but the precise data for all these aspects are not available. Besides these, certain percentage of the net sown area is given to fodder crops in India. Therefore, Chakravarty (1970), after examining various estimates of losses on these accounts given by different scholars and institutions has suggested the standard deduction of 10 per cent exclusive of seed requirement. Similarly after evaluating various estimates of seed requirements ranging from 3.5 per cent to 14.0 per cent, he has recommended 6.8 per cent of the total production for seed requirements. Hence, the total deduction for wastage, cattle feed and seed comes out to be 16.8 per cent of the total gross production. The co-efficient of production is, therefore, 0.832 (100-16.8 = 83.2). The availability of food, in the present study has been worked out at two levels, i.e. availability of food in terms of monetary value which ascertains the access to food. For the former all the crops such as rice, wheat, bajra, ragi, jawar, maize, barley, gram, peas, other pulses, vegetables and fruits, sugar cane etc. which could be converted into calories have been considered. As the nature of all the crops varies, they cannot be aggregated in terms of weight. Therefore, they have been converted into calorific value with help of the conversion table and have been added together to get the food production in calories for each of the district. The resulting figures have been multiplied by the co-efficient of production, i.e. 0.832 in order to get the net food available in terms of calories.

But in India agriculture is highly diversified and the crops ranging from opium to rice are grown in different parts of the country. The nature and character of the crops grown is as such that they cannot be added or aggregated together. Hence, various methods such as weight, food grain equivalance, standard nutrition unit, calories and monetary value have been suggested (Singh, 1974, Kendal; 1939, and Moonis, 1992). In the present study monetary value, although it also has the problem of selection of prices due to its seasonal and annual fluctuation has been taken up as the other methods have several limitations (Mohammad, 1992). Thus the total agricultural produce have been converted into monetary value by multiplying the average production of crops with their respective state level average prices. The result, thus obtained, has been multiplied by the co-efficient of production, i.e. 0.832 in order to get net availability of agricultural produce in monetary value which ensures access to food within household, local, regional or national markets.

Regional Patterns of Food Security

The regional patterns of food security is ascertained for the total population and agricultural production in terms of kilocalories and monetary value. The results have been presented in tables and in the form of choropleth maps in which the principle of contiguity has been followed in all the maps. However, each one is discussed separately under the following headings :

1. Distribution of Total Consumption Units.
2. Regional Pattern of Food Availability in Kilocalories.
3. Food Security of Total Population in Kilocalories.
4. Regional Pattern of Food Availability in Monetary Value.
5. Food Security of Total Population in Monetary Value.
6. Distribution of Agricultural Consumption Units.
7. Food Security of Agricultural Population in Kilocalories.
8. Food Security of Agricultural Population in Monetary Value.

Distribution of Total Consumption Units

The trennium average of the total projected population of each district has been calculated and is multiplied with the co-efficient of consumption. The result, thus obtained, has been arranged in descending order and the districts have been classified into four categories (Table 25.1).

Table 25.1 : Distribution of Total Consumption Unit

Category	*Consumption unit range*	*No. of district*	*% age to total*
Very High	> 20,00,000	146	34.27
High	20,00,000-15,00,000	77	18.07
Moderate	15,00,000-10,00,000	87	20.42
Low	< 10,00,000	126	29.57
Total		426	100.00

The pattern of distribution of total consumption unit is highly varied and complex. Almost 52 per cent of the districts have large pressure of population. Out of it 34 per cent of the total districts have very high pressure of population. This region is mostly concentrated in the fertile Indo-Ganges plain, eastern coastal areas and the western districts of Maharashtra and Karnataka. Substantial number of districts of Kerala are also covered into it. The region of high pressure of population covering about 18 per cent of the total districts is widely scattered and is by and large found in the peripheral areas of very high population pressure. The region of moderate and low pressure of population cover about half of the percentage of the total districts and are found either in mountainous, plateau or hilly areas or in the desert regions (Fig. 25.1).

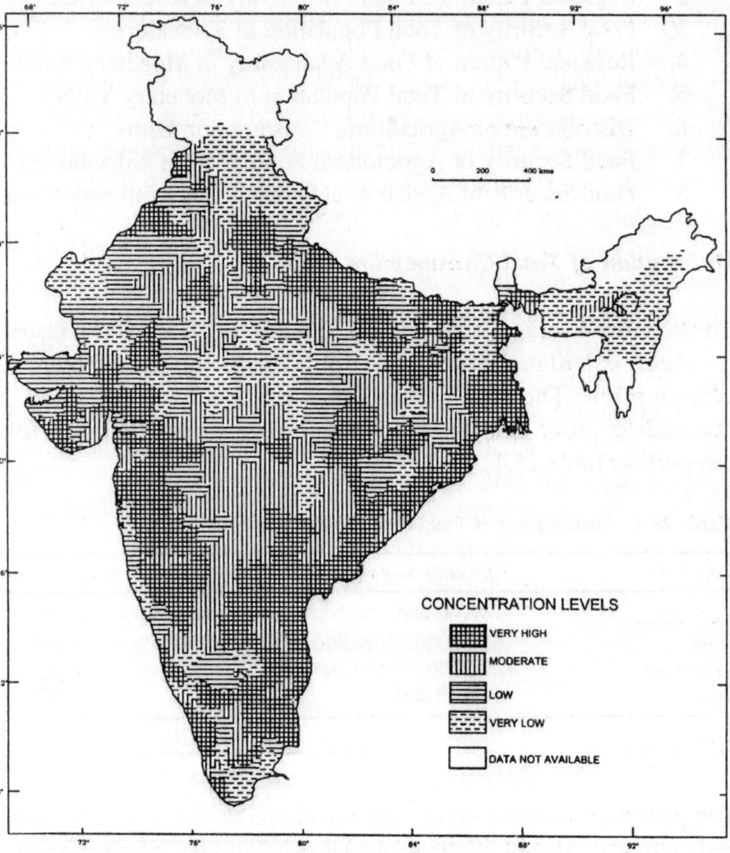

Fig. 25.1 : Distribution of Total Consumption Units.

Regional Patterns of Food Availability in Kilocalories

All the food crops including cereals, small millets, pulses, fruits and vegetables put together account more than 90 per cent of the total cropped area in the country and they also constitute almost equal percentage of the total food consumption in India. Hence, they all have been considered, in whatsoever form their data were available, to work out the food availability. Due to their varied nature to be aggregated, they have been converted into kilocalories and added together for each district. The resulting figure is multiplied by production co-efficient of 0.832 and the product, thus obtained, is divided by cultivated land to get per hectare of production in kilocalories. The districts have been grouped into 4 categories on this basis (Table 25.2).

Table 25.2 : Food availability in kilocalories

Category	*Calorific value*	*No. of districts*	*%age to total*
High	>9000	103	24.18
Moderate	9000-7000	135	31.69
Low	7000-5000	59	13.85
Very Low	<5000	129	30.28
Total		426	100.00

The regions of high food availability constituting a little less than one-fourth of total districts is concentrated in the large zones lying in north-western part of Indo-Ganges plain and the eastern coastal region (Fig. 25.2). The third zone covers whole of West Bengal and is separated from the coastal zone by the district of Baleshwar. The basic causes of high availability of calories in these regions are high productivity and higher concentration on the production of food crops. The region of moderate concentration covering about 32 per cent of the total districts, by and large, emanates from the zones of high availability and are found in 8 pockets. The largest zone covers the predominantly rice and wheat producing areas of eastern Uttar Pradesh, Bihar and northern part of Jharkhand. The other region is located in the southern half of the coastal areas of the Western Ghats, the state of Kerala, and the western coastal districts of Karnataka which are the areas of intensive rice cultivation. Yet another prominent region is found in eastern Gujarat and few northern districts of Maharashtra. The other regions are scattered in central, eastern and northern parts of the country. These are also the areas of food crop production. The regions of low availability are mainly found in the peripheral areas of high and moderate food availability. The largest chunk of very low

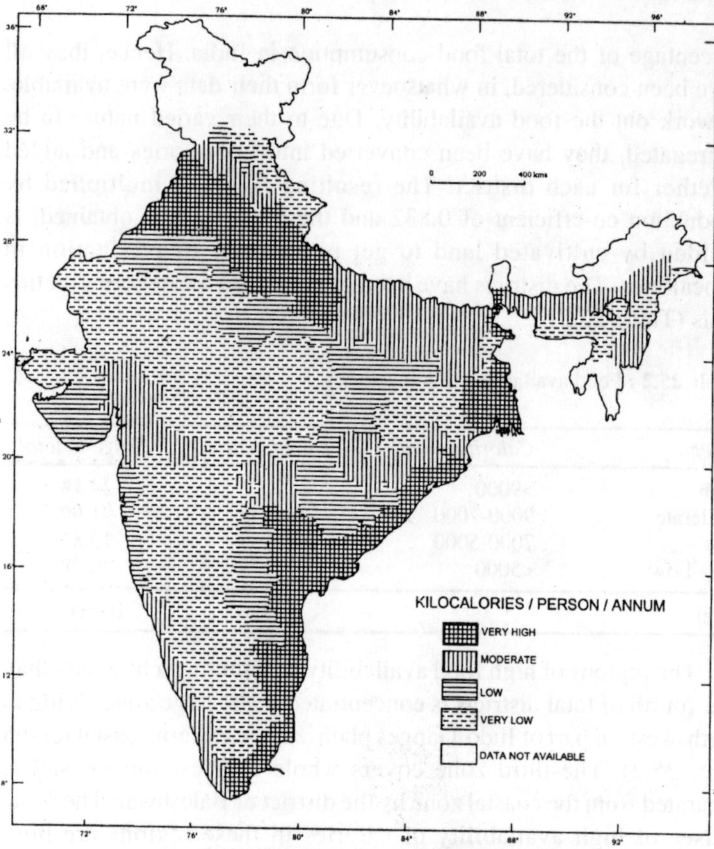

Fig. 25.2 : Food Availability Patterns in Kilocalories.

availability of food which together covers about 30 per cent of the districts, is a contiguous belt running from north to south and covering arid region and Aravali range in Rajasthan, Vindhyan and Satpura ranges in Madhya Pradesh, and the Western Ghats and the projected plateaus in the east. This is the hungry belly of India having low production of food crops due to adverse environmental conditions specially in case of the relief, climate, and soil.

Food Security for Total Population Units in Kilocalories

The distributional patterns of total consumption unit and the food availability in calorific value is highly variable. Hence, in order to ascertain the food security, total food availability has been divided by total consumption unit of each district. The resulting figures have been arranged in descending order and the districts have been grouped into four categories (Table 25.3).

Table 25.3 : Food Security/Insecurity for Total Population in Kilocalories

Category	*Calorific Value*	*No. of district*	*%age to total*
Secure	> 2830	124	29.12
Moderately Secure	2830-2430	99	23.24
Insecure	2430-2030	86	20.18
Severely Insecure	< 2030	117	27.46
Total		426	100.00

Regional patterns of food security becomes complex and very few compact zones emerge (Fig.25.3). The food security zone covers

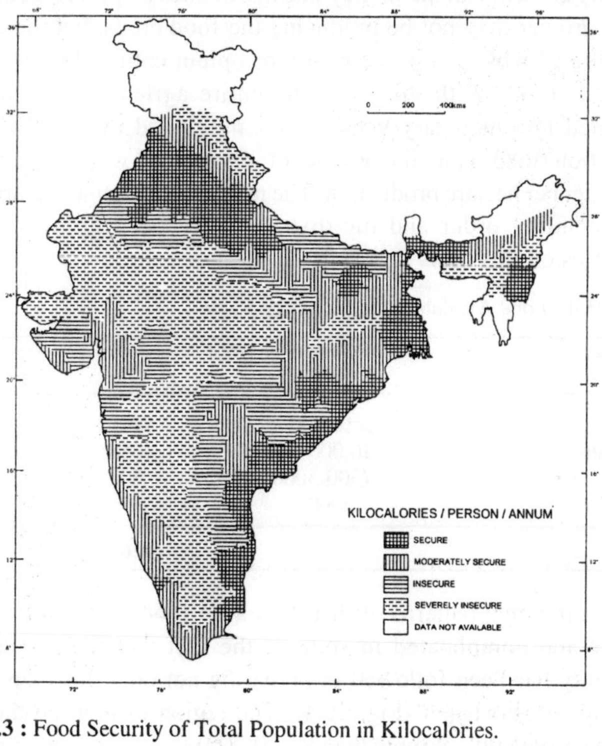

Fig. 25.3 : Food Security of Total Population in Kilocalories.

about 29 per cent of the total districts and it, by and large, coincides with the zone of high food availability. Twenty-one districts have been added from the zone of moderate availability. The region of moderate security covers about 23 per cent of the districts and is found in 7 subzones scattered in different parts of the country. The region of food insecurity covering about one-fifth of the total districts also has a scattered pattern of its distribution found in 4 pockets. The hungry belly has got slightly shrunk covering 27 per cent of total districts.

Regional Patterns of Food Availability in Monetary Value

In the present day world of globalisation and liberalisation it is not the physical availability of food but the adequate means or entitlements to secure the access to food which is even more important, and it basically comes with the money one has at ones disposal. A cotton or opium grower may not be producing the food crops, but the money which he gets by selling the cotton or opium entitles him to secure access to food. With this view the entire agricultural produce is converted into monetary value and is multiplied by co-efficient of production (0.832) and the product of it is divided with the cultivated area to get per hectare production. The result, thus obtained, is arranged in descending order and the districts have been grouped into 4 categories on this basis (Table 25.4).

Table 25.4 : Food Availability to Monetary Value

Category	*Value range (Rs.)*	*No. of districts*	*%age to total*
High	> 10,000	114	26.76
Moderate	10,000-7500	165	38.74
Low	7500-5000	60	14.08
Very Low	< 5000	87	20.42
Total		426	100.00

The regional patterns of distribution of food availability is very diverse and complicated in spite of the fact that the principle of contiguity has been followed in preparing not only this map but all the maps of this paper (Fig. 25.4). The region of high productivity with the production of more than Rs.10,000 per bectare includes more than one-fourth of the total number of districts and is found in four

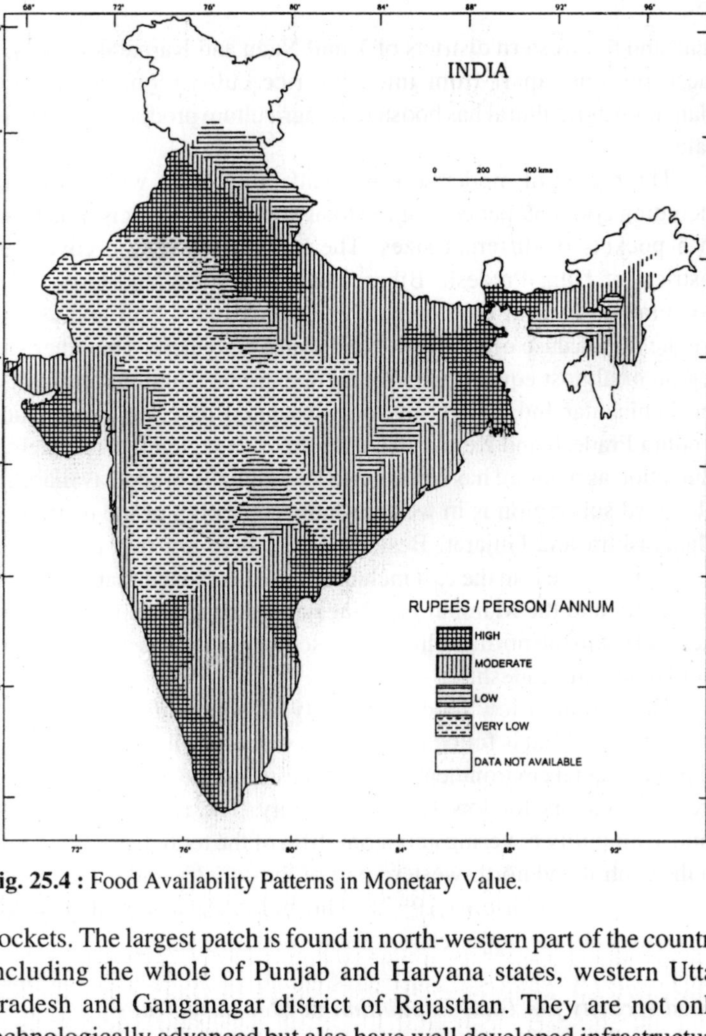

Fig. 25.4 : Food Availability Patterns in Monetary Value.

pockets. The largest patch is found in north-western part of the country including the whole of Punjab and Haryana states, western Uttar Pradesh and Ganganagar district of Rajasthan. They are not only technologically advanced but also have well developed infrastructure and sound economic base, besides being the area of Green Revolution. The other region is found in West Bengal, south-eastern Bihar Plain and few districts of the Brahmaputra valley of Assam. Intensive rice cultivation, tea plantation and horticulture have contributed to high level of food availability in this region. The third region covers the coastal belt of Andhra Pradesh, while the fourth region covers the southern part of the Western Ghats including the whole of Kerala state and the western districts of Tamil Nadu and Karnataka states. In these regions, apart from intensive rice cultivation, commercial plantation agricultural has boosted the agriculture production in money value.

The region of moderate food availability has a wider coverage including about 39 per cent of the total number of districts. It is found in 5 pockets of different sizes. The biggest sub-region covers the districts of Uttar Pradesh, Bihar, Madhya Pradesh and Orissa. The favourable physical environment especially relief, soil and climate are the main cause of moderate food availability here. The other sub-region of almost equal size is found in extreme south-eastern part of the Peninsular India covering the substantial parts of Tamil Nadu, Andhra Pradesh and Kerala states. In this sub-region the commercial plantation agriculture has contributed substantially to food availability. The third sub-region is in western India consisting of the districts of Maharashtra and Gujarat. Besides these, there are two more small sub-regions: one is in the east including the eastern Brahmaputra valley in Assam, and the Manipur and Nagaland states, while the other is a narrow belt in the north including the southern districts of Uttaranchal and Himachal Pradesh.

The region of low food availability is found only in 14 per cent of the districts, but is found in 5 pockets scattered all over the country. The detrimental environmental condition and nature of the crops grown are main factors for low food availability. The region of very low food availability covering about one-fifth of the total districts is found in the central and north-western part of the country and is the hungry belly of India (Moonis,1992). The harsh detrimental physical conditions are the major cause of low food availability.

Food security for Total Population in Monetary Value

There are studies to indicate the standard requirement of calories or food grain per head per annum but there is none, as far as I know, to indicate it in terms of monetary value. Therefore, based on the standard requirement of the food grains and the dietary survey conducted by the writer (Mohammad,1978), an amount of Rs 1150 per annum has been considered to be reasonable enough to support one person at the current level of prices. Hence, the total food available in money value in each district is divided by the total consumption unit in order to get its per head per annum availability. Thus the districts with availability of Rs. 1150 or more have been put in the food secured regions, while the remaining with less than the standard requirements have been kept in food insecure region. These two broad categories have further been divided into two categories each (Table 25.5).

Table 25.5 : Food Security/Insecurity for Total Population in Monetary Value

Category	*Availability range (Rs.)*	*No. of districts*	*%age to total*
Secure	> 1500	174	40.85
Moderately Secure	1500-1150	124	29.10
Insecure	1150-800	48	11.27
Severely Insecure	< 800	80	18.78
Total		426	100.00

The distributional patterns of food security is very complex due to the differential patterns of total consumption unit (Fig. 25.5). However, there are 70 per cent districts where food is secure, while in remaining 30 per cent it is insecure. The food secure regions covering about two-fifths of the total districts are found in 10 sub-regions of small and large sizes. There are three large sub-regions which, by and large, correspond with regions of high and moderate food availability. The remaining sub-regions are of small size and are found in western and eastern parts of the country. The region of moderate security covering about 29 per cent of the districts are the continuation of the region of food security. The largest sub-region runs from north-west to south-east covering the districts of Rajasthan, Uttar Pradesh and Bihar. The region of high food availability has gone to moderate food security due to high density of population consumption unit. The other major sub-regions found in central eastern Peninsular India, covers the districts of Madhya Pradesh, Andhra Pradesh and Orissa. The moderately insecure region, which covers only 11 per cent of the districts, is found in scattered pockets all over the country. The main chunk of the severly insecure region, extending from east to west is found in arid region of Rajasthan, and in the Vindhyan range of Madhya Pradesh. Low food availability and comparatively large consumption unit are the main causes of severe food insecurity.

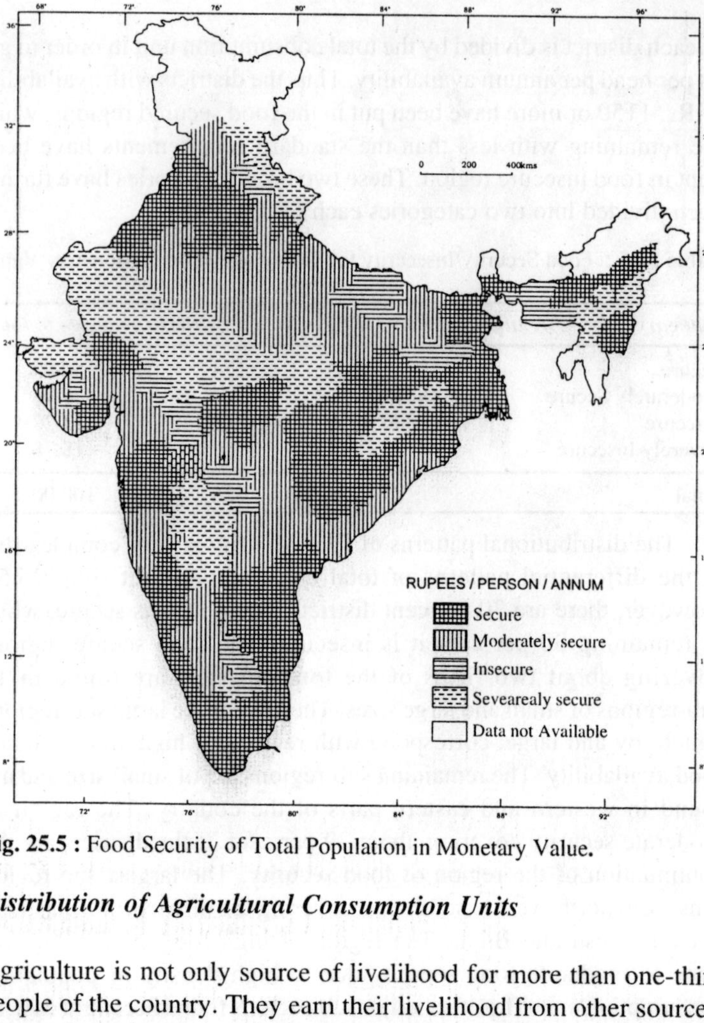

Fig. 25.5 : Food Security of Total Population in Monetary Value.

Distribution of Agricultural Consumption Units

Agriculture is not only source of livelihood for more than one-third people of the country. They earn their livelihood from other sources. Hence, in order to get more precise picture of food security, only agricultural population has been considered separately. Like total population, agricultural population is also multiplied by coefficient of population, i.e. 0.773 in order to get total consumption unit of agricultural population. The result, thus obtained, has been arranged in descending order and districts have been grouped into 4 categories (Table 25.6).

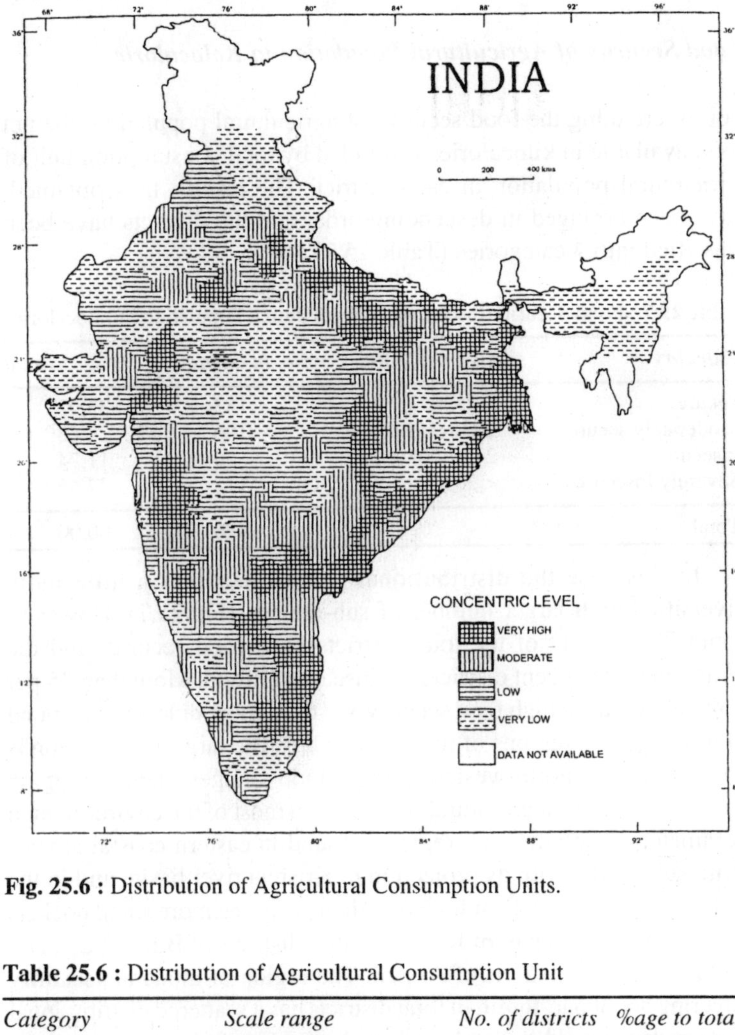

Fig. 25.6 : Distribution of Agricultural Consumption Units.

Table 25.6 : Distribution of Agricultural Consumption Unit

Category	*Scale range*	*No. of districts*	*%age to total*
Very High	> 20,00,000	61	14.31
High	20,00,000-15,00,000	70	16.43
Moderate	15,00,000-10,00,000	114	26.76
Low	< 10,00,000	181	42.48
Total		426	1011.00

Food Security of Agricultural Population in Kolocalories

For ascertaining the food security of agricultural population, the net food available in kilocalories is divided by total consumption unit of agricultural population in each district. The results, thus obtained, have been arranged in descending order and the districts have been classified into 4 categories (Table 25.7).

Table 25.7 : Food Security/Insecurity of Agricultural Population in kilocalories

Category	*Calorific value*	*No. of districts*	*%age to total*
Secure	> 2830	153	35.92
Moderately secure	2830-2430	106	24.88
Insecure	2430-2030	50	11.74
Severely Insecure	< 2030	117	27.46
Total		426	100.00

In this case the distributional pattern becomes a little more diversified with larger number of sub-regions (Fig.25.7). However, about 70 per cent of the total districts have food security and the remaining 30 per cent districts are insecure of food. More than 35 per cent districts have high food security and they, more or less, correspond to food security regions of monetary value. The largest sub-region is concentrated is north-western part of Indo-Ganges plain—a region well known for its agricultural suitability in most of the environmental parameters. The other sub-region is found in eastern coastal area of Peninsular India with its projection in Krishna river basin, and in the state of West Bengal. Besides these, there are three more small pockets located in northern part of Kerala, central districts of Bihar, Nagaland and few eastern districts of Assam. The region of moderate security covering about one-fourth of total districts has a scattered distributional pattern. It is found in 8 sub-regions located in different parts of the country. The region of food insecurity also has the similar distributional pattern with pockets of varying areal strength. The region of severe food insecurity covering about 27 per cent of the total districts, is found in 4 sub-regions. The largest contiguous belt in this region runs from the arid parts of Rajasthan to the Nilgiri hills of Peninsular India, covering the area of Aravali ranges, Vindhayan, Satpura and the highlands of Western Glats together with eastward projected hills and plateaus. The other sub-region consists of mountainous districts of

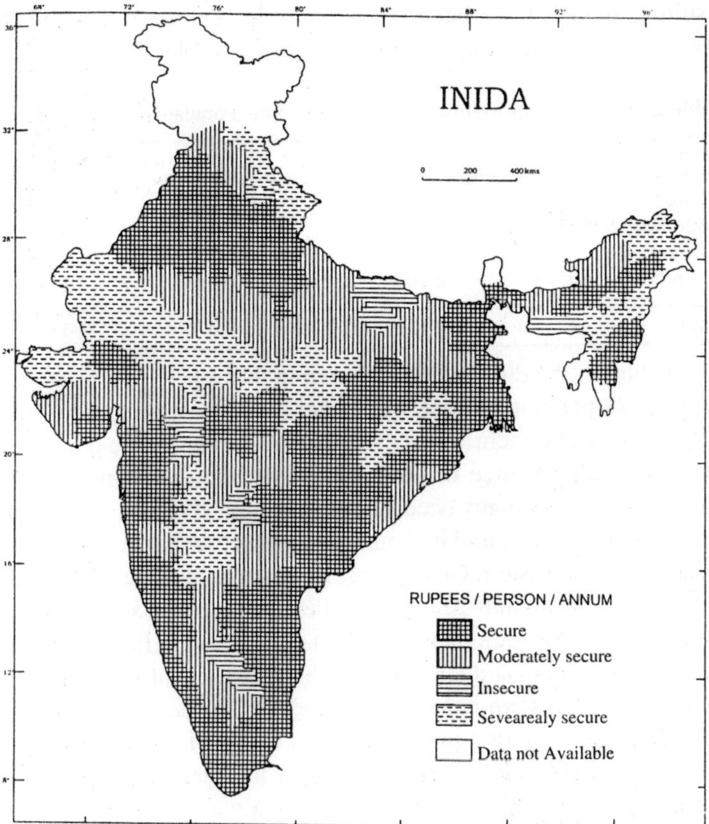

Fig. 25.7 : Food Security of Agricultural Population Monetary Value.

Himachal Pradesh and Uttaranchal, Rajmahal hills and Chotanagpur plateau and the district of Meghalaya and Nagaland. In all these regions the detrimental physical condition coupled with unfavourable techno-economic and socio-institutional factors are basically responsible for severe food insecurity.

Food Security for Agricultural Population in Monetary Value

The availability of agricultural produce in money value has been divided with the total consumption unit of agricultural population to get the food security of agricultural population of each district. The resulting figures have been arranged in descending order and the districts have been grouped into 4 categories (Table 25.8)

Table 25.8 : Food Security/Insecurity of Agricultural Population in Money Value

Category	*Value range (Rs.)*	*No. of districts*	*%age to total*
Secure	> 1500	203	47.66
Moderately Secure	1500-1150	132	30.19
Insecure	1150-800 -	28	6.57
Severely Insecure	< 800	63	14.78
Total		426	100.00

The picture of the distributional pattern in this regard is encouraging as about 78 per cent districts fall under the food security zone. The basic reason is the exclusion of more than 30 per cent of the population which has other source of sustenance covering secondary, tertiary and some of the primany occupations. The largest region with 48 per cent districts in it is found in Peninsular India covering the districts of Western Ghats, Eastern Ghats, West Bengal and agriculturally fertile districts of Orissa and Madhya Pradesh. The other sub-region of considerable areal strength is located in north-western India including the states of Punjab and Haryana, northern districts of Rajasthan and the districts of western-central Uttar Pradesh. Besides these, there are 5 smaller regions found in Brahmaputra valley of Assam, Manipur, Gujarat and Central India. The moderately food secure region covers about 31 per cent of the districts and has a wider distribution with larger concentration in Indo-Ganges plain including the northern districts of Rajasthan, central districts of Uttar Pradesh and the districts of north and south Bihar plain. The other sub-region of significant areal strength runs from north to south in the centre of the Peninsular India. There are only 6 per cent of the districts having insecurity of food and are found in the districts of Meghalaya, Bihar, Madhya Pradesh, Tamil Nadu and Maharashtra. The severe food insecurity covering about 15 per cent of the districts is mainly concentrated in two pockets. One is located in northern India running from east to west covering the districts of Rajasthan and Madhya Pradesh, while the other is found in central Peninsular India covering the districts of Karnataka and Maharashtra. The other sub-regions are found in

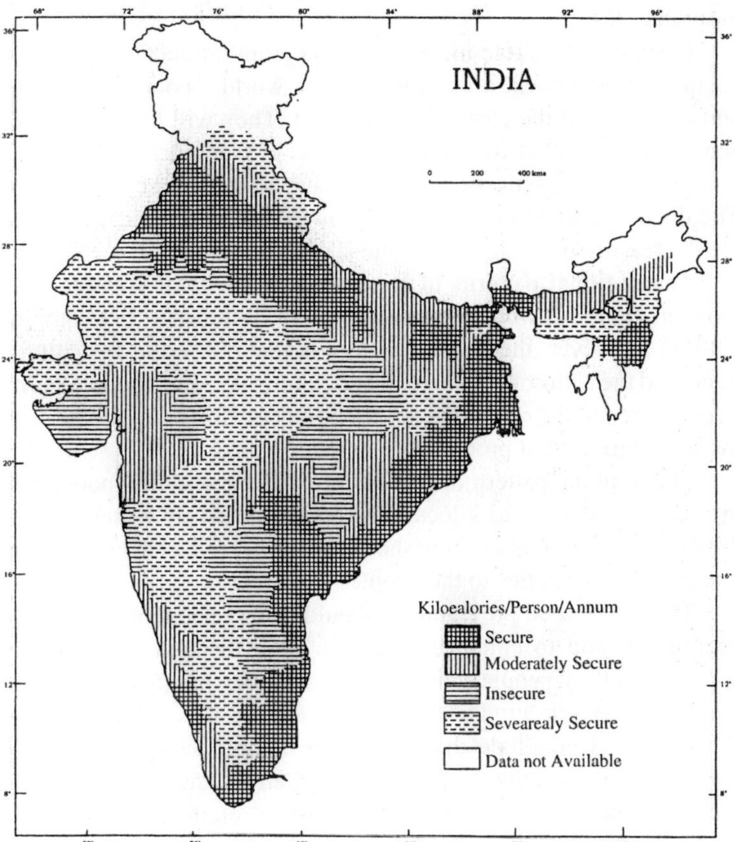

Fig. 25.8 : Food Security of Agricultural Population in Kilocalories.

Uttaranchal, hilly districts of Himachal Pradesh, Nagaland and southern Assam. In all these sub-regions, the harsh environment conditions detrimental to agriculture are the basic causes of severe food insecurity.

Conclusions

The concept of food security is dynamic, complex and multi-dimensional. Prior to 1980s it was considered at global, regional and national levels, but now the conceptual model of food and nutritional security is advanced and the focus is shifted towards household and individual levels.

In spite of four fold increase in food grain production leading to adequate bufferstock, around half of the world's poor live in India and do not have the clear access to food. They will have economic access to food either from increased level or from increased prices.

The adequate buferstock should leave no room for complacency. The declining trend in population growth should be accelerated and the decreasing trend in growth rate should be reversed.

The regional patterns of food security for total population in monetary value as well as in kilocalories have shown a very complex pattern. However, the general pattern is that the fertile agricultural tracts and the areas of commercial agriculture have more food security as compared to the rest of the country. Hence efforts must be made to increase agricultural productivity in the regions.

The regional pattern of food security for agricultural population in monetary value and kilocalories is comparatively better. Hence, the dependency on agriculture should be further reduced by providing alternate opportunities to the people.

The future food programmes should be targeted to increase food production and to enhance access to food through participatory approach of local people with emphasis on resources effeciency, social equity and conservation of the environment.

Finally, I conclude that global level food security cannot be achieved unless deliberate action is taken to ensure an enabling political, social and economic environment; eradicate poverty and inequality; pursue sustainable food, agricultural and rural development policies; ensure that food, agricultural trade and overall trade policies foster food security; meet emergency food requirements in ways that encourage recovery, rehabilitation and development; and promote investment in agriculture and rural development.

REFERENCES

Ankroyd, W.R., (1956) *The Nutritive Value of Indian Diet and Planning of a Satisfactory Diet*, Madras.

Barun, J. *et al.* (1992) *Improving Food Security of the Poor. Concept Policy and Programme,* International Food Policy Research Institute, Washington.

Bhattacharya, B. (2001) Food Security, NAP and NATO, *Yojana*, Vol.45, No. 1.

Chakravarty, A.K. (1970) Food Sufficiency Patterns in India; *Geographical Review*, Vol. LX, New York.

Chopra, *et al.*(eds) (1998) *Crop Productivity and Sustainability—Shaping the Future*, Oxford and IBH Pub. Co. New Delhi

Chung, K. *et al.* (1997) *Identifying the Food Insecure: The Application of Mixed Method Approaches in India*; International Food Policy Research Institute, Washington.

Dantwala, M.L. (1985) Garibi Hatao : Strategy Options, *Economic and Political Weekly*, Vol.20, No. 11, March 16.

FAO (1999) *The Status Food Insecurity in the World*, Rome.

FAD (1999) *Food Outlook*, No. 4, Rome.

Frankenberger, T.R and McCarton, (1998) The Household Livelihood Security Concept, *Food Nutrition and Agriculture*, F.A.O., Rome.

Fresco, L. (2000) Scientific and Ethical Challenges in Agriculture to meet Human Need, *Food Nutrition and Agriculture*, F.A.O., 27.

George, P.S. (1994) Food Security in South Asia : Performance and Prospects *Economic and Political Weekly*, Vol.29, No.15 April 30.

George, P.S. (1996) Public Distribution System, Food Subsidy and Production Insecurities, *Economic and Political Weekly*, Vol.31, No.39, September 28.

George, P.S. (1999) Some Reflection on Food Security in India, *Indian Journal of Agricultural Economics*, Vol. 54, No.2.

Gopalan, C. (2000) Medium Term Strategy for Meeting the Nutritional Needs of the Population, *Population and Human Development—Meeting Some Critical Needs in the New Century*, Commissioned Papers, Planning Commission.

Johnson, G. (1976) Increased Stability of Grain Supplies in Developing Countries, Optional Carry Overs and Insurance, *World Development*, Vol. 4, No. 12.

Kendall, M.G. (1939) The Geographical Distribution of Crop Productivity in England, *Journal of Royal Statistical Society*, Vol.38.

Lupien, J.R. and Menza,V. (1999) *Assessing Prospects for Improving Food Security and Nutrition*, FAO.

Minhas, B.S. (1976) Towards National Food Security, *Indian Journal of Agricultural Economics*, Vol. 31, No. 4.

Mohammad, N. (1987) *Impact of Caste on Primary Occupation: A Geographical Analysis*, Concept, New Delhi.

Mohammad, N. (1978) *Agricultural Landuse in India*, Inter India Pub., New Delhi

Mohammad, N.(ed) (1992) *New Dimensions in Agricultural Geography*, Vol. I - VIII, Concept, New Delhi.

Murlidharan, S. (2001) Upholding the Right to Food, *Frontline*, Vol. 18(18), New Delhi.

Radhakrishna, R. (1991) Food and Nutrition: Challenge for Policy, *Journal of the Indian Society of Agricultural Statistics*, Vol. 53, No.3, December.

Rath, N. (1985) Garibi Hatao: Can IRDP Do it? *Economic and Political Weekly*, Vol.20, No.6, February 9.

Raza, M. (1992) Regional Dimensions of Agricultural Development in Mohammad, N. (ed.) *New Dimensions of Agriculture*. Vol, IV, Concept, New Delhi.

Reutlinger, S. (1977) *Food Insecurity: Magnitudes and Remedies*, Washington, U.S.A.

Sarris, A. and Taylor L., (1976) Central Stock, Food Aid and *Food Security, Food, Nutrition and Agriculture*, Vol.4, No. 2.

Sen, A.K. (1981) *Poverty and Famines : An essay in entitlement and Deprivation*, Clarendan Press, Oxford.

Shafi, M. and Aziz, A. (eds) (1989) *Food Systems of the World*, Rawat Pub., Jaipur.

Shafi, M. (1984) *Agricultural Productivity and Regional Imbalances: A Study of Uttar Pradesh*, Concept, New Delhi.

Singh, B. (1947) *Population and Food Planning in India*, Bombay.

Singh, S. (2001) Human Development and Food Security, *Yojana*, Vol. 45, No.8.

Srivastava, H.C. (ed) (1993) *Biotechnological Applications for Food Security in Developing Countries*, Oxford and IBH Pub. Co., New Delhi.

Suryanarayana, M.H, (1996) Food Security and Calorie Adequacy Across States, *Journal of Indian School of Political Economy*, Vo1.7, No.2.

Swanunathan, M.S. (1993) Food Security through Sustainable Agriculture, in *Biotechnological Applications for Food Security in Developing Countries*, Oxford and IBH Pub. Co., New Delhi.

Valdes, A.(ed) (1981) *Food Security for Developing Countries*, Westview Press, Colorado.

26

Landuse Pattern and Sustainable Agricultural Development in Nagaland

M. Istiaq and *Isatoli Kibami*

Introduction

Nagaland, the sixteenth state of the India Union is bounded by Assam in the west. Myanmar (Burma) in cast, Arunachal Pradesh and parts of Assam in the north and Manipur in the south. It is situated between 25°6′ N to 27° 4′ N latitude and 93° 15′ E longitude, covering an area of about 16579 sq. km.

The total population of the state is 19,88,639 as reported by the 2001 census. The density of population is 120 persons per sq. km. The state accounts for about 19 per cent of the total population and about 0.5 per cent of the geographical area of the country. The state has 67.11 per cent of population as literate. The high literacy rate in the state is quite striking and encouraging. It indicates the tremendous awakening of educational consciousness among the people of the state.

The state consists of eight administrative districts—Mon, Tuensang, Mokokchung, Zunheboto, Wokha, Dimapur, Kohima and Phek. There are 13 major tribes and various sub-tribes inhabit the region. Each tribe has its own culture, dialect and dress.

The Naga villages are generally built on a commanding position, which are quite often, happen to be on the top of a hill. The selection of the site is largely influenced by the consideration of defensive strategy. The state comprised of hilly and rugged topography with average height of peaks between 900 and 1200 metres. Rainfall is heavy and it ranges between 175 and 250cm.

The state with a high rainfall experiences some of the highest soil crosion rate in the country. According to ICAR (Indian Council of Agriculture Research, Medzhiphema) the estimates of soil erosion are about 40 metric tonnes per year per hectare. Due to heavy rainfall there is luxuriant growth of evergreen forests, mixed with herbs and shrubs. Some biologists consider Nagaland as a great treasury of fauna and flora.

The forests of Nagaland have good quality timbers like Nahar, Hallock and Hallong, Pinc, Sal, Bamboo and Cane are other important trees of economic value. However, the human interaction with the physical environment and the age-old practice of jhuming have done incalculable harm to the forest wealth and ecological settings.

Socio-Economic Setting

Nagaland has basically an agriculture economy. Most of its workforce is engaged in primary occupation. The industries have come up but are few in numbers and small in size. Participation of workers in different economic activities is increasing over the decades, but agriculture still absorbs about 70 per cent of the main workers.

A remarkable feature of the economy is that there is no absentee landlord and there is no class of landless peasants. The village society is so organized that the basic requirements of food, clothing and shelter are guaranteed to all members. The population on the whole remains gainfully employed in the production activities and there is no surplus labour. During the cropping season, it becomes difficult to get hired labour and if available, very high wage has to be paid. There is however a system of providing casual labour by forming what are called 'field companies' of men and women of the same age group. Every member of the company gets the benefits of company's labour. The major source of revenue for the government until recently has been the house-tax. Now it is suplemented by the sale of forest and forest products and tax on urban commodities.

The state has a mountainous topography and is agrarian region. The primitive subsistence still dominates the economy of the state with about 70 per cent of its rural population subsisting on jhum cycle. The two methods of cultivation practiced are jhum cultivation and terrace cultivation. The state being hilly, jhuming has been the order of agriculture setting. Terrace though prevails is not widespread and is practiced only by few tribal groups like Angami. Chakhesang and Zeliang tribes. The remaining tribal groups—Ao, Chang, Khiemnugun, Konyak, Lotha, Phom, Pochury, Rengma, Sangtam, Sema and Yimchunger practice jhum cultivation. The area under jhum cultivation is about 87.339 hectares and under terrace cultivation is about 62,091 hectares.

Over the years the shift in diversification in agriculture sector is noticed in most of the tribal groups. It is to be noted that there is an increase in terrace cultivation, plantation, horticulture and in other agriculture sectors. The development in fisheries and scriculture is also making progress in the economic landscape.

Rice is the dominant crop and also the staple diet of the people. Of the gross cropped area under the food grains rice accounts for about 84.4 per cent. Oilseeds are´also important crops in the state. They include rapeseeds, mustard etc. Coffee, cardamom and tea are also grown as plantation in Nagaland. Other important crops grown are arums, yums, millets, maize, potato and sugar cane. Vegetables and fruits such as melon, oranges, onions, cucumber, spinach leaf, mustard, chilies, carrots, tomatoes, brinjals, etc. are commonly grown by the farmers in the state.

Landuse Pattern

Out of the total geographical are 16579 sq. km. forest occupies 14,164 sq. km accounting for about 85 per cent. The bulk of Nagaland's total forest area has been degraded due to the jhum and logging. Village controls as high as 92 cent of the forest while the government controls only 8 per cent of the total forest area. It is to be noted that twons, villages, roads, rivers, etc. occupies only 15 per cent of the total geographical area of the state.

As the land belongs to the people and not owned by the state, there is hardly any system of externally regulating land utilization. Individual and occasionally changes from on individual to another but generally remains within the village. The operational holding of the region in terms of size and number can be identified in the table 26.1.

Table 26.1 : Land Use in Nagaland

Land use	*Area in sq. km*	*Pecentage*
Forest under Government control	1133	6.83
Forest under of village, clans and families under tribal system, current jhumland	13031	78.59
Area under towns, village, roads, rivers, etc.	2415	14.56

Source : State Forest Report 1999 : FSI : 2000.

Table 26.2 : Area and Size of Operational Holding, Nagaland

(In hectares)

Size of land holding	*Area*
Less than 1.0	4750
1-3	41668
3-5	53027
5-7.5	74065
10-30	404851
Above 30	1817008

Source : Basic Statistics of the North East Region, 1987, NEC, Shilong 1988.

Jhuming

Jhuming cultivation are also called 'slash and burn' agriculture. It is widely practiced in the state. It involves rotation of fields rather than rotation of crops. Each village has its own traditionally determined land for cultivation. As a result of population pressure the jhum cycle became shorter, i.e. reduced to 7 years. In some areas the cycle is even shorter resulting in poorer yields and causing harm to the environment. The extensive jhuming cultivation is detrimental with the ecosystem and hence likely to generate micro-climatiç and other environmental change. It is to be noted that in a jhum field, farmers cultivate over 20 different agricultural crops and vegetables along with paddy. The jhum invariably does not belong to individual. It is the property of entire community (the people living in the village). The operation and the period of jhum cycle as obeserved by Hussain (1998) is as follows :

(a) Search for jhum land.
(b) Cutting and burning of trees and clearing of shrubs/grasses.

(c) Allotment of land.
(d) Digging and hoeing of fields.
(e) Sowing of seeds and plantation of trees/vegetables.
(f) Weeding and harvesting of vegetables.
(g) Harvest of main crops.
(h) Fallowing.
(i) Collection of ratoon crops.

Terrace Cultivation

Terrace cultivation is relatively a modern method practiced by preparing terraced fields. The complete hillside is cut to prepare the terrace fields. Normally the terrace are so graduated that water flows down continuously from one terrace to another below it, and so on. Bamboo pipes are used to regulate the flow of water. The excavating of terracces require colossal efforts, and one marvels at the amount of human energy used in cutting them to shape, but these terraced fields are prepared, are much casier to maintain than the jhum plots. They also have the advantage of being closer to the village site. The are under terrace cultivation is increasing gradully. In 1962-63, the area under terrace cultivation was reported to be 21,363 hectares, which increased to 62,091 hectares in 1995-96.

Case Study

To the study the land use pattern a fieldwork was conducted in 260 households belonging to 13 major tribal groups, of the state. It was noticed that the shifting cultivation and plantation constitute a major share in agriculture. It is to be noted that in some cases plantation outnumbered the shifting cultivation. Similarly, share of the terrace cultivation among the Angami and Zeliang is 58 per cent and 83 per cent respectively. There five tribal goups, i.e. Ao, Angami, Chakhesang, Konyak and Zeliang, in which the share of plantation and terrace cultivation together is 54 per cent, 70 per cent, 96 per cent, 56 per cent, and 70 per cent respectively. The six maor tribal groups : Sangtam, Yimchunger, Phom, Khicmnugun, Chang and Rengma still practice shifting cultivation at the higher order of 90 per cent, 83 per cent, 80 per cent, 79 per cent, 71 per cent and 70 per cent, respectively. On the basis of the fieldwork data one may conclude that about 50 per cent of the major tribal groups of Nagaland have shown a tendency to change from their traditional agriculture (i.e. shifting) to the plantation and terrace cultivation. Plantation and horticulture are becoming more popular and gradually gaining momentum. Since, these require more incentives in the form of capital investment, new varieties of good plants and seeds they should be available to the farmer up to their expectations for fulfilment of agriculture development.

Table 26.5 : Land use Pattern among the major Tribal groups in Nagaland

Tribal Group	*Terrace Cultivation*	*Shifting Cultivation(%)*	*Plantation(%)*
Ao	2	46	52
Angami	58	30	12
Chakhesang	46	4	50
Chang	11	71	18
Khiemnugun	8	79	13
Konyak	23	44	33
Lotha	18	58	24
Rengma	12	70	18
Phom	2	80	18
Sangtam	-	90	10
Sema	17	59	24
Yimchunger	2	83	15
Zeliang	83	5	12

Source : Data collected through the fieldwork during winter 2001.

Sustainable Agriculture Development

Agriculture and forestry are dominating the economy of Nagaland. Over 70 per cent of its rural population subsisting on jhum cultivation. Jhum cycle was productive and sustainable when the duration was over 30 years. However, increasing population has led to shortened jhum cycle and increased and degradation. A possible alternative is terrace cultivation, although not prevalent on a large scale. But this too, has its limitation because most parts of Nagaland are hilly with steep gradients.

In oder to feed the growth of population, diversification in agricultural sector is needed. A decade has passed and the need to make jhuming sustainable and viable was felt by the Naga communities. Farmers, therefore, cultivate over 20 different agricultural crops and vegetables along with paddy. The idea to introduce on more item, i.e. tree sampling was devised by the planners. This tree sampling would grow to commercial items, which would reap rich dividends to the cash strapped farmer in the state, by the time they go to jhum plot. Besides, there has been a growth in other agricultural sector such as plantation, horticulture, fishery, sericulture, etc. to meet the population demand and to sustain themselves.

The climate of Nagaland and soil are best suited to the cultivation of fruits, vegetables, spices, ornamental and medicinal crops. Fruits that extensively grown in the region are pineapple, oranges and banana. Other fruits grown in the specific areas are litchi, mango, guava, jackfruits, peach, plum and papaya. Ginger and tapioca are grown in abundance, *Jarnapani Horticulture Farm* at Pfutsero and Mokokchung, *Integrated Extensive Training Centre* at Medzhiphema, *Agriculture Research and Extensive Centre* at Wokha and Zunheboto are major and upcoming small agro-based industrial areas of the state.

The measures undertaken to fight land degradation are terracing for permanent cultivation, afforestation and tree plantation, watershed management, fallow management and encouragement of the traditional practice of keeping nitrogen fixing trees in the jhum fields and other soil retaining plants and also introducing various horticulture and cash crops by agriculture and other departments.

The Nagaland Environment Protection and Economic Development Agency (NEPEDA) funded and managed by *Canadian International Development Agency* (CIDA) and *International Development Research Centre* (IDRC) have been working to improve the status and welfare of the farmers. Such organizations also help to improve the economy of the farmers through agro-forestry practice to sustainable resource management. In the goverment sector, departments like Forest, Soil and Water Conservation and Agriculture Wasteland Development are taking measures for agriculture development.

As has been mentioned that majority of the land belongs to the people and very little is owned by the state, there is hardly anyway of regulating the land utilization with the care to minimize the land degradation. The private owners do not care much for the protection of the land, as their main motive is to get the maximum yield. The stress now is that, as the people own the land, the desire to make the land sustainable and projective should come from the farmer, as it would head to success if people intitiate the projects and facilitate various sorts of land use pattern. Their cooperation with NEPEDA and other government agencies will go a long way for sustainable agricultural development of Nagaland.

REFERENCES

Census of India, (1991), *State Profile* 1991, India.

Census of India, (2001), *Nagaland*, Sesion 13, Provisional Population Totals, Paper-1 of 2001.

Hussain, M. (ed), (1998), *Encyclopaedia of North-East India,* Vol. VI, Nagaland, Rima Publishing House, New Delhi.

North-East Council, *Basic Statsitics of North-Eastern Region India*, Shillong, 1980, 1985, 1987, 1988.

State Forest Repost, 1998, FSI, 2000

Statistical Handbook of Nagaland 2001, Directorate of Economics & Statistics, Government of Nagaland, Kohima.

27

Agricultural Transformation and Rural Sustainability in a Himalayan Watershed

Case Study of Kullu Valley

R.B. Singh and *D.K. Mishra*

Introduction

The biophysical environment forms a fundamental basis of the existence of human society. It refers to the land, soil, and water bodies, flora and fauna that surround human habitat and from which human beings derive a living. Basic human subsistence and requirements are met from the different natural resources. Each and every component of the natural environment is significant for our livelihood. Furthermore, the environmental components interlink in a complex regional ecosystem embodying many subsystems whose indiscriminate disruption would upset the delicate ecological balance, with enormous consequences on human kind. It is a matter of great concern that the quality of the natural environment is deteriorating in many parts of the country at a time when the human population is growing at an accelerated rate (Sharma and Minhas, 1993). And also there is a high degree of diversity in land productivity in different regions of the country (Hanumantha Rao, 1994).

Study Area and Methodology

The study area consists of 4 villages of upper Kullu valley in the northern most part of Himachal Pradesh. The total area covers about 120 km^2. And the elevation varies between 2000-4000 m from mean sea level. The slope gradient is steep and the region has a thin layer of mountain soil. The river Beas passes through the region in north to south direction. Climatically the region lies in the temperate zone extending to the tundra zone, owing to its high elevation, where snowfall is a common phenomenon. The natural vegetation is dominated by coniferous and mixed forest, with large tundra and alpine meadow areas. The proportion of inhabitable land is low in the region and being a tourist area, the seasonal concentration of tourists is high. The research methodology involves comprehensive studies using a historical perspective and the application of a geographical framework for data representation and understanding both in space and time. Agricultural transformation has been evaluated and the rural watershed sustainability framework has been assessed.

Agricultural Transformation

The unprecedented growth in population is exerting high pressure on the limited land available for cultivation. Agricultural practices have undergone drastic changes since the mid-1960s through the introduction of agro-development programmes and the application of various newly developed techniques. People are giving importance to orchards because of their high profitability. Thus, some agricultural land has been encroached by orchards. In addition, people for orchards and agricultural purposes are now using wastelands. The bulk of agricultural land in the region lies on terraces, along the steep hillsides. The gradient of some agricultural land exceeds 35° (Singh and Mishra, 1995). The cropping intensity, agricultural productivity and additional income generated through agrohorticultural activities, increasing apple orchards are the main factor reducing the cultivated area. Although the orchard plantations were started in the valley during the British Period, its commercialization began only during the 1960s and 1970s. The major factors responsible for changing the biophysical environment are deforestation and change in the cropping pattern. The mass felling of trees during the 1950s and 1960s and the clearing of large tracts of forests have made the ecosystems more vulnerable by reducing the bio-diversity of the valley. The high increase of consumption of water has considerably influenced the hydrological set up of the entire valley (Singh, 1998).

Indigenous Farming Systems

Nearly all indigenous farming systems in the valley are founded on the principles of organic farming, which a large number of experts consider to be the most efficient way of producing food on an environmentally sustainable basis, i.e. without any major lasting threat to the natural ecosystem. Organic farming exceeds artificial agro-chemicals. It is sustained by natural recycling of natural soil-replenishing nutrients, e.g. biomass, compost, manure and other forms of organic matter. It relies on biophysical methods to control pests and on crop rotations and the inter-mixture of species to sustain agroecological stability, food security and balanced utilization of soil nutrients. Oldest indigenous farming systems were based on the shifting cultivation and land rotation. Under a relatively low population pressure prevailing in the past, this system enjoys the advantages of minimising soil erosion, preserving agro-biodiversity, maintaining ecological stability, optimizing utilization of the different soil nutrients, and enhancing food security and a balanced diet, under the influence of the growing pressure on land, which is associated mainly with population growth, the cultivation system has been modified into more settled system (Pandey and Singh, 1984). There is a need for more research in identifying the most productive species, breeding, improved varieties, and testing and management methods. Research into mixed cropping and other traditional farming practices is also yielding interesting results. Increasing number of poor people is also the cause of environmental deterioration, as they are cutting the slopes to produce their crops, feed their livestock and family members (Sarmiento, Monasterio, and Montella, 1993).

Challenge of Sustainability under Changing Farming Conditions

The farming systems are not static. They are changing as a result of changes in the environment, both natural and socio-economic. Some of these changes have rendered the traditional farming unsustainable and somewhat outmoded. A few examples of the changes in the agricultural systems and their impact on sustainability are presented hereunder (Table 27.l).

Table 27.1 : Major changes and effect on sustainability in Upper Kullu Valley

S.No.	*Changes*	*Effect on sustainability*
1.	Introduction of Apple Orchards	Positive and Negative
2.	Population Explosion	Negative
3.	Commercialisation of agriculture	Largely Negative for less Resourceful Farmers
4.	Agricultural Chemicals	Largely Negative Unless Strictly Controlled
5.	Fertilizer Use	Negative and Positive
6.	Population Influx	Negative and Positive

Source : Assessed from various observations and field survey.

The manner in which changes affect sustainability can be illustrated with two or more of these examples. For instance, introduction of apple orchards can be contributing to increasing stability of production. But it is also true that the introduction of the orchards has often been promoted at the expense of the indigenous food crops, some of which are so neglected that they are not much being grown and a considerable degree of biodiversity has been lost. Population explosion has considerably increased pressures on land, resulting in intensification of farming. Use of agricultural chemicals has different impacts on the environment, where reasonable amount of chemicals is appropriately used. The effects are largely beneficial, where, for example, no fertilizers are used and farming is intensified, the nutrients are depleted and yields drop as soil become degraded. But where excess amounts of farm chemicals are applied, the environment may become polluted and unsustainability is the ultimate result (Berkes and Gardner, 1997).

Micro Issues of Agricultural Transformation

Till the mid-80s the local people were depending heavily on the agriculture produce in the valley. People were self-sufficient in food grains. They never bought grains from the market; in fact buying grains from the market was considered a great shame to the family. The drift from agriculture to horticulture in the mid-80s greatly decreased the agricultural fields. These fields have been converted into apple, plum, pears and cherry orchards. As the local people had land in different far distance villages, they had built temporary houses there, known as "*doghries*", meaning the second house or a place meant for temporary shelter. People used these places during the sowing and harvesting period. The "*doghries*" which used to have inhabitants at least on four occasions a year in the past, now present a deserted look. The small numbers of agriculturists still around now go by bus, taxi and car to cultivate and harvest the crops and come back home by dusk. Whether these "*doghries*" will ever get an inmate is a big question mark, but the current prosperity has certainly changed the lifestyle of the local population in less than 15 years.

Replacement of Nutritious Traditional Food Crops

Before Independence "*Saryara*" was an important staple crop in the region, but after Independence it has lost importance because of the innovations of the new high yielding varieties of crops. Though since the beginning of the century the fluctuations in all the 4 villages were a very common feature, but the major decrease has been obvious after 1960-69. "*Saryara*", is one of the traditional food crops, which are very nutritious, but there has been constant decrease in the sown area under this crop in all the villages. The decrease has been very sharp from 1979-89, with maximum decrease in case of Buruwa village where it has decreased from 35 ha to 7 ha. In Palchan, it has decreased from 18 ha to 3 ha and in Vashisht, it experienced a decrease from 14 ha to 3 ha. It is also reflected in the total area under "*saryara*" production where it has decreased from 74 ha to 13 ha. By 1994, the total area under this crop has reduced to only 3 ha with Palchan, Buruwa, and Vashisht each having 1 ha (Table 27.2)

"*Kodra*" is another important crop, which is rapidly loosing its importance due to ever increasing production of rice. The decrease in the net sown area under "*Kodra*" has not been constant over time in all the villages. In Palchan, the area under this crop has remained around 2-4 ha except during 1959-69, when it has decreased to 20 ha. In Buruwa village the area under "*Kodra*" till 1959 was quite high, around 60-70 ha, which had a sharp decline in 1969 when it reduced to 6 ha, which further declined to 3 ha in 1979. In case of Vashisht village these decline starts from 1979 onwards, i.e. 14 ha in 1969 to 2 ha in 1979 to 1 ha in 1989. In Manali, also the decline in this nutrient crop area has started from 1979 onwards from 19 ha in 1969 to 6 ha

Table 27.2 : Change in the Sown Area of Saryara Crop in Upper Kullu Valley (1919-94)

(Area in ha)

Year	*Villages*				
	Palchan	*Buruwa*	*Vashisht*	*Manali*	*Total Area*
1919	17.88	41.46	21.54	19.91	100.79
1929	17.07	31.70	17.88	19.51	86.16
1939	15.85	34.55	21.54	24.79	96.73
1949	15.04	31.30	22.76	16.66	85.76
1959	13.41	51.62	18.29	18.69	102.01
1969	20.32	32.00	15.00	16.00	83.32
1979	18.00	35.00	14.00	7.00	74.00
1989	3.00	7.00	3.00	NA	13.00
1994	1.00	1.00	1.00	NA	3.00

Source: Lal Kitab (Village Statistical Record), Kullu Tehsil, Kullu, 1919-94.

in 1979 to 1 ha in 1989. This decrease in net sown area becomes more interesting when we look at the total area under "*Kodra*" in the valley where it has decreased from 114.75 ha in 1919 to 59 ha in 1969 to 3 ha in 1994 (Table 27.3).

Table 27.3 : Changes in the Sown Area of Kodra Crop in Upper Kullu Valley (1919-94)

(Area in ha)

Year	*Palchan*	*Buruwa*	*Vashisht*	*Manali*	*Total Area*
1919	4 00	74.39	18.69	17 07	114.15
1929	2,00	63.00	16.66	16.26	97.92
1939	2.00	59.75	21.13	14.22	97.1
1949	6.00	70.73	20.32	15.85	112.9
1959	4.00	67.47	15.85	9.75	97.07
1969	2.00	6.00	14.00	19.00	59.00
1979	7.00	3.00	2.00	9.00	21.00
1989	2.00	3.00	1.00	1.00	70.0
1994	N.A.	3.00	N.A	N.A.	3.00

N.A = Not Applicable.
Source : Lal Kitab (Village Statistical Record), Kullu Tehsil, Kullu, 1919-94.

This general decrease in the land area under traditional and nutritious food crops can be attributed to the following causes:

1. Most of the agricultural land is being converted into orchards.
2. Agricultural land is also being taken up for various constructions purposes.

3. Most of the traditional crops are being replaced by cash crops.
4. Growing influx of tourists and increasing commercialisation.

Introducing cash crops for economic upliftment

Introduction of cash crops like maize in the valley has replaced almost all the traditional nutritious food crops like *Kodra, Saryara* and *Kathu*. This can be only attributed to the following reasons :

1. Maize as a cash crop fetches a good amount to the farmers during the tourist season, and
2. It is also high yielding in nature.

Therefore, we find a general increase in the area under maize cultivation in the upper Kullu valley, with the total area increasing from 41.45 ha in 1919 to 110 ha in 1979 to 268 ha in 1994. At the village level, Buruwa village has the maximum area (153 ha) under maize cultivation (Table 27.4).

Table 27.4 : Change in the Area under Maize Cultivation in Upper Kullu Valley (1919-94)

(Area in ha)

	Villages				
Year	*Palchan*	*Buruwa*	*Vashisht*	*Manali*	*Total Area*
1919	N.A.	21.54	11.78	8.13	41.45
1929	4.06	47.15	13.00	16.26	80.47
1939	5.28	51.21	16.26	15.04	87.79
1949	8.94	64.63	18 29	19.10	110.96
1959	15.44	32.11	13.41	20.73	81.69
1969	18.00	37.00	6.00	8.13	69.13
1979	11.00	79.00	18.00	2.00	110.00
1989	37.00	117.00	9.00	50.00	213.00
1994	41.00	153.00	30.00	44.00	268 00

N.A.= Not Applicable.
Source : Lal Kitab (Village Statistical Record), Kullu Tehsil, Kullu, 1919-94.

Previously, potato was an important crop in the upper region of the valley, but due to better production in Lahul and Spiti, other crops are replacing the production of potato. Therefore the general trend of production of potato is showing a negative trend among all the villages.

The variation over a last 10 years is maximum in Palchan and Buruwa, where Palchan has shown a negative trend but in Buruwa we find a positive trend (Table 27.5).

Table 27.5: Change in the Area under Potato and Vegetables Cultivation in Upper Kullu Valley (1919-94)

(Area in ha)

Year	*Total Area under Potato*	*Total Area under Vegetables*
1919	2.00	10.14
1929	4.00	6.07
1939	7.00	6.67
1949	20.34	2.01
1959	65.43	3.38
1969	66.00	25.40
1979	58.00	50.00
1989	36.00	11.00
1994	21.00	3.00

Source : "Lal Kitab" (Village Statistical Record), Kullu Tehsil, Kullu, 1919-94.

Since, the region is a tourist dominated, the production of various types of vegetables should be a common feature of agriculture. But the pattern is not visible. It is because most of the productive fields have been devoted to apple horticulture and on remaining part of the land farmers produce subsistence crop for their own livelihood. Fluctuations overtime have been observed in the production of vegetables, Vashisht has shown a remarkable increase (3 ha) while other villages have stopped the commercial production of vegetables after 1990. The total net sown area under vegetable was 10.14 ha in 1919, which increased to 50 ha in 1979, but due to the orcharding culture in 1994 it declined to only 3 ha in the valley (Table 27.5).

Status of Horticulture

The per unit area production of fruit and income are higher than ordinary cereals grown all over the region. In 1993-94, during late March or early April snowfall caused severe damage to the production of apple. The whole valley suffered from loss of apples worth of million of rupees due to late snowfall. While crossing the area one can observe the growth of apple trees over the cultivated areas.

Undoubtedly, increasing number of apple orchards is the main factor in reducing the cultivated area. Therefore, on the basis of intensive fieldwork and analysis (Singh, 1998) of data the following features can be observed:

1. The pattern of agro-orcharding is very common among all the villages. Apple trees are planted in the cultivated areas. Till the trees were small in size; other crops were also grown in the same field. When they become larger and occupied all the space between trees, cultivation of other crops was not possible. In this way cultivated area is decreasing and orchard area is increasing.
2. Though the orchard plantation was started during the British period, commercialisation started during late 60s and 70s Manali was the main centre for apple production and even today it occupies a prestigious place in the Kullu valley.
3. Palchan village, situated in the extreme north of the valley, experienced very low orchard growth due to severe winter and heavy snowfall. Now the apple trees are unable to produce good quality as well as quantity of the apples under such conditions. There was only 1 ha land under apple orchards up to 1985 but between 1985-90 it has increased to about 12 ha.
4. In Buruwa and Vashisht, orchards started during 70s and extensive orcharding was experienced. In Buruwa and Manali, apple orchards increased rapidly during 80s and 90s.

Unsustainability in Agriculture

The intensification of agricultural production as a result of increasing population pressure, intensification of farming, overgrazing and conversion of land to several uses have resulted in several undesirable changes in the environment with the adverse effect on agricultural production. Changes in the form of intensification of land use for food production, shortened fallow periods, reduced capacity for soil regeneration, and decline in soil fertility, reduced and declining yields, food shortages, use of marginal sloping lands, overgrazing, loss of vegetative cover, etc. can be seen in the valley resulting in unsustainability in agriculture (Tisdell, 1988).

Interface between Agriculture and Allied Sectors

For appropriate and realistic design of a sustainable farming system it is necessary to ensure that threats to it from other sectors are significantly minimised. For example, poor road construction and maintenance of NH-21 could result in flooding, eutrophication and erosion, all of which can seriously damage agricultural land and stream fisheries. A related example is a policy issue, such as structural adjustment aimed at increasing export earning or reducing debt burdens. This may result in large forest areas being cleared for commercial crops, which will expose the soil to erosion. Similarly, removal of subsidies may result in farmers not using excess pesticides and insecticides, which in turn, will result in minimising environmental degradation. Therefore, sustainable agricultural systems and adhering to the design alone will not ensure sustainability unless policies, strategies, technologies, systems and components of resource management and input use in other sectors are designed to ensure sustainability in development. All these call for not only integrated land use planning but also adoption of a holistic approach in development planning, in policy formulation and in selecting strategies and the execution of development programmes (Smith, 1993; Singh, 1998).

Conclusions

In designing sustainable agricultural production system, it is necessary to give due consideration to the characteristics of various resources used in production, the way they are managed in the production process and the indigenous technologies and local practices which render the resultant production system unsustainable. Unsustainability results when impacts of practices and technologies used are economically not feasible and sometimes also culturally unacceptable. In addition to the selection of methods for manipulation of resources and use of practices and technologies, which ensure sustainability, there is a need to ensure that other sectoral activities do not render the farming systems unsustainable. To generate new agricultural practices and technology continuous and systematic research is required incorporating people's response. Agricultural implements and farm inputs needed by farmers should be locally available at reasonable price. Presence of adequate educational, training and support facilities for farmers are urgently needed. Mechanism for planning and directing agricultural programmes should be integral components of overall economic development.

REFERENCES

Berkes, F. and Gardner, J.S. (eds), (1997). *Sustainability of Mountain Environments in India and Canada*, University of Manitoba, Winnipeg.

Hanumantha Rao, C.H., (1994). Agricultural Growth and Rural Poverty in India: Emerging Trends and Perspectives, *Indian Economic Review*, Vol. XXVII, No. 2, pp.129-40.

Pandey, U. and Singh, J.S., (1984). *Energetic of Hill Agro-Eco Systems: A case study from Central Himalaya*, Agricultural Systems, 13, pp. 83-95.

Sarmiento, L., Monasterio, M. and Montella, M., (1993). *Ecological Bases, Sustainability and Current Trends in Traditional Agriculture in Venezuelan High Andes*, Mountain Research and Development, 13, pp. 167-76.

Sharma, P.D. and Minhas, RS., (1993). *Land use and Biophysical Environment of Kinnaur District, Himachal Pradesh*, in Mountain Research and Development, 13, pp. 41-60.

Singh, RB. and Mishra, D.K., (1995). Dynamics of Land Use and Cropping Patterns: Historical Prognostication in Upper Beas Basin in RB. Singh and Martin, J. Haigh (ed.) *Sustainable Reconstruction of High Land and Head Water Regions*, Oxford and IBH Pub., New Delhi, pp. 649 -661.

Singh, RB., (1998). Land use/cover changes, extreme events and eco-hydrological resources in the Himalayan region, *Hydrological Processes*, 12, 2043-2055

Singh, R.B. (ed), (1998). *Sustainable Development of Mountain Environment in India and Canada* (CIDA-SICI Project Experience), Oxford and IBH Pub., New Delhi.

Singh, R.B., (1998). *Understanding High Mountain Land Use towards Sustainable Environmental Management in Jammu & Kashmir*, Oxford and IBH Pub., New Delhi, pp 409-422.

Smith, V. Kerry, (1993). "Non-Market Valuation of Environmental Resources. An Interpretation Appraisal", *Land Economics*, Vol. 69, No. 1, Feb., pp. 1-26.

Tisdell, C., (1988). "Sustainable Development: Different Perspectives of Ecologists and Economists, and Relevance to LDCs", *World Development*, Vol. 16, No. 3, March, pp. 373-84.

28

Food Security in India
Challenges and Opportunities

Kamalvir Tyagi

India has lived with food shortages for a very long period. During British rule, there was regular occurrence of famines. It was said that the British rule began with famine and ended with famine (referring to Bengal famine in 1943). Even in independent India, the economy continued to face food security and it led the government to undertake 'fire fighting' measures from time to time like suspending normal activities of markets and trade through nationalization of wholesale trade, organisation of food zones (for purpose of procuring food grains from surplus zones and distributing them to deficit zones); introduction of rationing or Public Distribution System (P.D.S.); relying on concessional foreign food (PL-480) and taking stringent administrative and legal measures such as Defence of India Rules (D.I.R.) and Essential Commodities Act (E.C.A.) etc.

Trends in Food Production

As far as food grains output is concerned, the total production increased from 50.8 million tonnes in 1950-51 to 155 million tonnes in the Seventh Plan (annual average) and further to 187 million tonnes in the Eighth Plan (annual average) Food grains production stood at 199.5 million tonnes in 2000-01 and this rose further to 212.0 million tonnes in 2001-02 (highest in the planning period). However, the year 2002-03 experienced a draught with the result that food grains production is likely to be only 183.2 million tonnes in this year (a fall of 13.6 per cent). This would push the food grains production to the lowest level recorded since 1996-97'. The trends in food grains production 1950-51 to 2001-02 in India are presented in the following Table 28.1.

Table 28.1 : Trends in Food grains Production

(In million tonnes)

Year	*Foodgrains*	*Wheat*	*Rice*
1950-51	50.8	6.4	20.6
I Plan 1951-56 (average)	63.2	7.9	25.0
IInd Plan 1956-61 (average)	74.0	9.7	30.3
IIIrd Plan 1961-66 (average)	81.0	11.1	35.1
Annual Plans 1966-69 (average)	87.8	15.5	35.9
IVth Plan 1969-74 (average)	103.0	25.4	41.8
Vth Plan 1974-79 (average)	118.1	29.8	47.3
VIth Plan 1980-85 (average)	138.1	41.2	54.5
VIIth Plan 1985-90 (average)	155.0	48.3	65.1
VIIIth Plan 1992-97 (average)	189.0	62.9	87.7
2000-01	199.5	69.7	93.1
2001-02	212.0	71.8	93.1

Source : Plan documents and Economic Surveys.

There is increase in food grains production in successive plans. However, the period after the Third Plan (generally referred to as the period of Green Revolution) is particularly marked by rapid strides in food grains production (particularly wheat and, in recent years, rice as well). This has enabled the economy to overcome shortages and build up large stocks of food grains to counter any scarcity conditions. There is broad feeling among experts that the food security system in the form of P.D.S. has been able to eliminate famines and also bring a substantial measure of relief to the consumers during years of acute scarcity. Because of the increases recorded in food grains production over the years and mounting food grains stocks with the government, the food problem in the sense of inadequacy and shortage of food grains no longer exists. However, food problem from the point of view of food security continues to exist as millions of poor suffer from persistent hunger and malnutrition and many others are at the risk of doing so in future.

Food Policy—Key Assumptions

The present food policy is based on the following three key assumptions :

First Assumption—Given its limited export potential, India could not count on food imports as a viable strategy making food security synonymous with food self sufficiency. In the post independence period, the country continued to face balance of payments difficulties. When India became independent, it had a 'sterling balance' of Rs. 1733 crore. This was the result of a sizeable surplus on balance of trade with the U.K. during the Second World War period when U.K. had made large-scale purchasing from India to meet its war requirements. The foreign exchange position of the country was therefore satisfactory. However, the immediate post independence period was characterized by the release of 'pent up demand' for imports (suppressed during the war) and shortages of food and raw materials. Thus, import bill increased substantially while exports remained stagnant. The defcit had to be made good from the sterling balances. During the first plan period, the overall deficit in the balance of trade was Rs. 541.9 crore. However, the deficit on current account was only Rs. 42.3 crore.

The period[2] comprising I, II, III and IVth Plans and first two years of the Vth Plan saw heavy deficits in balance of payments and an extremely tight payments position. Though the government resorted to severe import controls and foreign exchange regulations, the current account deficit stood at 1.8 per cent of G.D.P. Foreign exchange reserves were at a low level, generally less than necessary to meet three month's imports. The period from 1976-77 to 1979-80 saw a vast improvement in balance of payments. India had a small current account surplus of 0.6 per cent of the G.D.P. during this period and also possessed foreign exchange reserves equivalent to about seven months' import. This comfortable foreign exchange position was due to the following factors :

(1) rapid increase in private remittances from oil exporting countries where large number of skilled workers from India had gone for work.

(2) increase in export growth which rose 31 per cent in 1975-76 and 23 per cent in 1976-77.

(3) check in the growth of oil imports. In fact, imports of crude oil and petroleum products did not grow at all in volume terms during the four year succeeding 1973.

The period from 1980-81 to 1990-91 witnessed severe balance of payments difficulties. While the trade deficit continued to soar, the net invisibles showed declining tendencies. In 1990-91 the deficit touched Rs. 16934 crore. accompanied by lower remittances and higher interest payments. The current account deficit rose to Rs. 17,369 crore. Thus, the country continued to face balance of payments difficulties and they culminated in crisis situation in 1990-91.

Second Assumption—Market and market based instruments can not be trusted to deliver food to the poor at reasonable prices. For a long period of time, Indian agriculture was mostly in the nature of 'subsistence farming'. A considerable part of the total produce was sold by the farmers to the village traders and moneylenders often at prices considerably lower than the market prices. In fact, there was a large chain of middlemen in the agricultural marketing system like village traders, wholesale traders, moneylenders etc. As a result, the share of farmers in the price of agricultural produce was reduced substantially. The middlemen took full advantage of the ignorance and illiteracy of the farmers and used unfair means to cheat them. Many malpractices prevailed in the market. In addition, there was absence of proper warehousing facilities, inadequate transport system. Thus, the market for food grains suffered from many defects.

Third Assumption—Private traders were likely to exploit buyers through the exercise of monopoly power in the local markets. Failure of crops always encouraged farmers and the wholesale dealers in agricultural products to indulge in hoarding with the expectation that the prices of these commodities would rise. Hoarding, when production goes down in the country, aggravates scarcity conditions and pushes up the price level. Hoarding has been a common phenomenon in India throughout the period of economic planning.

Agricultural Price Support Policy

The first assumption referred to earlier led the government to adopt price support policies.

Orgainisation of Food Zones

To introduce an element of stability in agricultural prices, food zones were organized in 1964. The country was divided into eight wheat zones. Rice zone were formed in South India. On the failure of this experiment, each state was made a separate one. Movement of food grains within a zone was free but restrictions were imposed on movements from one zone to the other. The government took upon itself the task of procuring food grains from the surplus and distributing them to the deficit states through the public distribution system.

Fixation of Minimum Support Prices and Procurement Prices

The Agricultural Price Commission was set up in 1965. It was renamed commission for Agricultural Cost and Prices (CACP) in 1985. Ever since its inception, the commission has been announcing minimum support prices, procurement prices and issue prices for a number of agricultural commodities. In recent years, Government policy to announce large increases in minimum support price (M.S.P.) have caused procurement to rise much above the stipulated buffer norms.

Table 28.2 : Minimum Support/Procurement Price of Wheat and Paddy

(Rs./Quintal)

Crop. Year	*Wheat*	*Paddy*
1980 - 81	117	105
1990 - 91	225	205
1994 - 95	350	340
1995 - 96	360	360
1996 - 97	380	380
1997 - 98	475	415
1998 - 99	510	440
1999 - 2000	550	490
2000 - 01	580	510
2001 - 02	610	530

Source - Economic Survey-2001-02.

Public Distribution System In India

The second assumption that market cannot be trusted to deliver food to the needy at reasonable prices led to the building up of Public Distribution System (P.D.S.). The basic objective of the P.D.S. in India is to provide essential consumer goods at cheap and subsidized prices to the consumers so as to insulate them from the impact of rising prices of essential commodities and maintain the minimum nutritional status of our population. To run this system, the government seems to procure a part of the marketable surplus with producers at procurement prices. The grain (mainly wheat and rice) thus procured, is used for distribution to the consumers through a network of ration/ fair price shops and/or for building up buffer stocks. The most important items covered under P.D.S. in India have been rice, wheat, sugar and kerosene. Together these four items have accounted for eighty-six per cent of the total P.D.S. Sales. P.D.S. distributes commodities worth more than Rs. 30,000 crore annually to about 160 million families and is perhaps the largest distribution network of its kind in the world.[3]

Food grains Allocation and Offtake Under P.D.S.

The offtake under P.D.S. reached a peak. level of 19.0 million tonnes on 1991-92 but, in recent years, has tended to decline.

Table 28.3 : Food grains Allocation and Offtake Under P.D.S.

(million tonnes)

Year	*Wheat*		*Rice*	
	Allocation	*Offtake*	*Allocation*	*Offtake*
1991 - 92	10.56	8.83	11.36	10.17
1995 - 96	11.31	5.29	14.62	9.46
1999 - 2000	10.37	5.76	13.84	11.31
2000 - 01	12.29	3.98	16.26	7.74
2001 - 02	13.14	5.68	17.23	8.16

Source : Economic Survey-2002-03.

The combined offtake of wheat and rice which was 86.7 per cent of allocation in 1991-92 fell to 70.4 per cent in 1999-2000 and to only 45.6 per cent in 2001-02. This is mainly on account of the narrowing differential between the P.D.S. and open market prices, particularly for Above Poverty Line (A.P.L.) families.

Mountain of Foodgrain Stocks

After experiencing food shortages during the three decades of 1960s, 1970s and 1980s, India was faced with the problem of excess food stocks in 1990s.

Table 28.4 : Central Food grains Stock and Minimum Buffer Stock Norms

(million tonnes)

Period	*Total (Wheat and Rice)*	
	Min. Norms	*Actual Stock*
January 1999	16.8	24.4
January 2000	16.8	31.4
January 2001	16.8	45.7
January 2002	16.8	58.0
July 2000	24 3	63.0

Source : Economic Survey-2002-03.

As is clear from the Table 28.4 the actual stocks of food grains with the Food Corporation of India (F.C.I.) have far exceeded the stipulated buffer stock norms in recent period. The food grains stocks in the Central pool increased rapidly and had reached an-all time high of 63 million tonnes by July 2002.

But that is a thing of the past now. A monsoon failure last year witnessed a massive 14 per cent fall in aggregate food grains production from 212 million tonnes in 2001-02 to 182.5 million tonnes in 2002-03. In actual terms, food grains production has fallen by a hefty 29.5 million tonnes in a single year. Worse, the huge food stocks in the central pool registered a sharp decline to 30.5 million tonnes—down 53 per cent from the July 2002 level. The government may not be too concerned now as the news of a good monsoon in the current year has already improved the prospect of higher production— food grains production is projected to grow by 14 per cent to 208 million tonnes[4] in 2003-04.

Unsustainability of Excess Food Stocks

It is now widely accepted that the government cannot go on holding unlimited stocks of food grains as this is fiscally unsustainable. The mounting food subsidy bill of the Central Government is presented in Table 28.5.

Table 28.5 : Food Subsidy of the Central Government

(million tonnes)

Year	*Amount (Rs. Cr.)*	*Percentage of Total Govt. Expenditure*
1990 - 91	2450.00	2.33
1995 - 96	4960.00	2.78
1999 - 2000	9434.00	3.08
2000 - 01	12060.00	3.67
2001 - 02	17612.00	4.87
2002 - 03	21200.00	5.17

* Budget Estimate
Source : Govt. of India, Tenth Five Year Plan 2002-07, Vol. II.
Govt. of India, Economic Survey, 2002-03.

The food subsidy touched the high level of Rs. 17612 or in 2001-02 and is projected at Rs. 21200 crore in 2002-03 accounting for as much as five per cent of Central Government Expenditure. The Expenditure Reforms Commission (E.R.C.) has recommended that a buffer stock of 10 million tonnes, comprising 4 million tonnes of wheat and 6 million tonnes of rice, is sufficient from the point of view of national food security. The Tenth Five Year Plan (2002-07) has endorsed this level. A comparison of this level with the present levels of buffer stock shows that there is a considerable surplus of food stocks in the country.

Food Management: An Outlook

Till 1990s, the country's food management policy had evolved against the background of scarcity and import dependence. Policy focus therefore was to increase productivity, particularly of rice and wheat. By the mid-1990s, the strategy succeeded and the country generated sizeable surplus and, even became a net agricultural exporter. Today, all three assumptions underlying our food policy have become invalid. Therefore, we have little option but to underline the need to reorient country's food management policy.

First, we must recognize that food security is not synonymous with food self-sufficiency. Given India's comfortable foreign exchange reserves of about $ 100 billion, imports remain a viable option.

Secondly, biotechnologists describe genetically modified (G.M.)

food as the 20th century gift to the 21st and a powerful weapon for food security. It has the potential to dwarf the achievements of first Green Revolution and it is being dubbed as version 2.0 of the revolution. There are several positive factors—yield is better, the nutritional value improves and wastage declines sharply.

Third, while controls and government involvement in other parts of the economy have been gradually yielding to market based policies, our food policy remains antiquated. Entry into the food grain trade should be freed up with all agents permitted to operate anywhere. The F.C.I. should be confined to maintain emergency food reserves especially to feed inaccessible and remote areas. The country has witnessed a paradoxical situation in recent period. While the godown of the F.C.I. are over flowing with excess food grains stocks, starvation deaths have been reported from across the country, especially from remote and inaccessible areas.

Huge as it may seem on paper, all is not well with the P.D.S. in India. P.D.S. was criticized on a wide front : its failure to serve the population Below Poverty Line (B.P.L.), for its perceived urban bias, negligible coverage in states with a high density of rural poor and lack of transparent and accountable arrangements for delivery. Given that backdrop, the Government acted to streamline P.D.S. during the Ninth Plan Period as Targeted Public Distribution System (T.P.D.S.) where each poor family is entitled to 10 K.G. of food grains per month (20 K.G. w.e.f. April 2000) at especially subsidized prices. The identification of beneficiaries is done by the states, based on state-wise poverty estimates of the Planning Commission. The aim is to limit the benefits to really poor. However, T.P.D.S. suffered from the following problems—diversion of P.D.S commodities, lack of purchasing power to buy 20 K.G. at a time, low quality of food grains, administrative inefficiencies.[5]

Fourth, the food security implies not only the availability but more importantly the access to food. In a country like India where about one-third of population continues to live below poverty line, it is a crucial requirement. In addition to T.P.D.S., two other schemes Integrated Child Development Scheme (I.C.D.S.) and Mid-Day Meal programme of the Central Government—aim to alleviate household malnutrition. However, operational inefficiencies limit the impact of the schemes and more than half of the children below 1—5 years old in rural areas are undernourished with girl children suffering even more severe malnutrition.

Finally, to prevent leakages and corruption in food management and inject transparency and accountability, greater reliance need to be placed on decentralization and empowerment of poor people. The panchayats and the local level participatory organisation would serve as an apparatus for reaching food to the poor.

REFERENCES

1. Government of India, Economic Survey, 2003-03, p. 156.
2. Bimal Jalan, "Balance of Payments, 1956 to 1991" in Bimal Jalan (ed.), *The Indian Economy—Problem and Prospects* (New Delhi, 1992), pp. 163-64.
3. Government of India, Planning Commission, *Tenth Five Year Plan, 2002-07*, Vol. II, p. 367.
4. *Economic Times,* ET Intelligence Group, Oct. 27, 2003.
5. Government of India, Planning Commission, *Mid-term Appraisal of Ninth Five Year Plan 1997-2002*, pp.162-63.

29

Information Strategy for Landuse and Food Security in India

P.C. Agrawal

Introduction

Agriculture plays important role in Indian Economy. It provides gainful employment to nearly two-third of the population of the country and contributed about 30 per cent in the national income. Agriculture sector also earns valuable foreign exchange to support Indian Economy in various sectors. It also supplies raw material to various key industries and serves as a big market for non-agricultural products. The most important role of agriculture sector is to provide food security for entire population of the country.

Food security is the vital issue for developing countries like India. Food is basic requirement for human survival. To meet the food requirment, consumption basket include cereals, pulses, oilseeds, milk products, meat products, fish products etc. For food security the sustainable green revolution, white revolution and blue revolution is called for.

The food production at global (world) level is sufficient to meet the food requirement. The only difference is that some countries produce more than their requirement or just to sustain their requirement and rest of the countries produce less than their requirement.

Factors Affecting Food Securtity

1. Level of Inequality;

2. Socio-Economic factors;
3. Population-level and growth;
4. Land degradation and land use;
5. Variation in use of Inputs and types of Inputs :

    (i) Seed (HYV *Vs*, Non-HYV).
    (ii) Manure & fertilizer,
    (iii) Irrigation,
    (iv) Insecticides and Pesticides,
    (v) Machinery and equipments,
    (vi) Production technology,
    (vii) Credits;

6. Level of education;
7. Rainfall;
8. Natural disasters (floods, earthquakes).

Review

While crop yield forecasting is important for national food security, including early determination of import/exports plan and price, it is also important in providing timely information of optimum management of growing crops. An accurate yield forecast gives farmers the advantage of beneficial marketing (S.K. Sharma *et al.*, 2001). Due to adoption of technology in cereal crops, the non-cereal crops as well as non-food grains suffered in production in the total sown area. Growing population and declining oilseeds production is adversely affecting the per capita supply of oilseeds (Kusum Chopra, 1981). Availability of pulses is reducing because of area of pulse is being shifted to other lucrative crops like wheat, rice, sugar cane etc. (Kusum Chopra, 1981-82). The study revealed the consumption aspects and inequality in consumption of food items consumed during 1965-66 to 1973-74. It has been obseved the diminishing trend in the share of pulses consumption in Inida (Ved Prakash, 1987). A study was carried out relating to the changes in comsumption demand and output of pulses in Uttar Pradesh. To find the trend, the 3 yearly moving average techniques have been used and lorenz ratio technique has been used for finding consumption inequality (Ganga Devi Gupta,

1991). Analysis of both demand and supply aspects of oilseeds were done. The demand aspect include estimation of consumption elasticity, inequalities in consumption and projection upto 2001 A.D. (Agrawal P.C., 1990). A study revealed the role of agricultural Institutions in Indian Agriculture. The Government of India has set its vision on doubling the food production in the next 10 years. Agriculture sector is targeted to grow at 6.11 per cent per annum. Credit would continue to be a key element in production process and rural financial institutions have to play a major role in translating it into reality (B.V. Patel, 2000). Rapid growth of agriculture is essential not only to activate self-reliance at national level but also for household food security and equity in distribution and rapid reduction in poverty levels. (National Agriculture Policy, 2000). Information Technology will be the major player in this process. Good data and timely information will serve the farmers. The information is an important element in process of farm business management. These include planning, organizing, controlling and decision making (Economic Development in India, 2000).

Objective of Study

1. To analyse various factors affecting food security.
2. To suggest the estimation of demand projection.
3. To analyse the supply of food crops.
4. To suggest strategy in securing food security and sustainable land use.

Analyse various factors affecting food security

Land Utilization : In India total geographical are is 328.73 million ha. The reporting area for land utilization is 304.81 million ha. out of this forests account 22.5 per cent, land not available for cultivation (non agricultural use and barren land) account 13.7 per cent, other cultivated land (excluding fallow land) account 9.4 per cent, fellow land 7.6 per cent and net area sown is 46.8 per cent and net irrigated area is 38.7 per cent. The cropping intensity has been observed 132.7 per cent. Thus there is a need to increase net sown area, net irrigated area and cropping intensity and reduce fallow land.

Level of Inequality : As per 1990-91 Agricultural census, the total are operated is 165.5 million ha. out of this, 15 per cent area is operated by 60 per cent of marginal (<1 ha) land holders and 17 per cent area is operated by only 1.6 per cent large landholders (10 ha & above). This inequality is still higher among SC/ST social group.

In India as per 1993-94 estimate of poverty line was 35.97 per cent (This estimate has again been revised in 2000 based on small sample data which is about 25 per cent). The states having poverty line below 20 per cent were Goa, Punjab, Delhi, Chandigarh and Daman & Diu. The states having poverty line between 20 per cent and 30 per cent were Andhra Pradesh, Gujarat, Harayana, Himachal Pradesh, Jammu & Kashmir, Kerala, Mizoram, Rajasthan & Lakshadweep.

The States having poverty line between 30 per cent to 40 per cent were Arunachal Pradesh, Karnataka, Maharashtra, Manipur, Meghalaya, Nagaland, Tamil Nadu, Tripura, West Bengal, Andman & Nicobar and Pondicherry. The States having poverty line above 40 per cent were Assam, Bihar, M.P., Orissa, Sikkm, Uttar Pradesh, Dadar & Nagar Haveli.

The Higher Poverty implies less purchasing power which adversely affect the food security.

Land Degradation : Extent of soil degradation under different types have been observed as water erosion (45.3 per cent affected areas), wind erosion (4.1 per cent), chemical deterioration (4.2 per cent) water logging (3.5 per cent) and total affected area was 57.1 per cent out of total geographical area.

Population : During 1991 census the total population was 846.3 million which has growth at the rate of 2.14 per cent per annum. The rural population was 74.3 per cent. The projected population in 2001 has been 1012.4 million and it will be 1094.1 during 2006.

The food security level is very much affected with the population growth. Thus to attain the sustainable food security, this production should have higher growth than population growth.

Natural Disasters : Flood disasters appears suddenly with little warning, similarly earthquake and its effects on hydrological conditions and other factors like ground water, surface water, water quality and cyclone impact assessment is also very important. Natural disasters wherever occur affects food security adversely and it has to be tackled by government, Non-Government Organizations and society.

Rain fall : Rainfall is most important among the meteorological factors in determining the agricultural production. Rainfall is of crucial importance for all human and livestock and responsible for plant growth hence it determines the food supply for both human and livestock. Rainfall variation in timing and amount determines success or failure of crop production.

It is seen from the Table 29.7 that rainfall actuals as percentage of normal during 1996 are presented subdivision-wise. At all India level the rainfall actual as percentage of normal has been 103.

Level of Education : The knowledge of Agricultural practices and Impact of various inputs, crop prices, markets are very helpful in planning Agricultural crop so that farmers get remunerative prices of their products.

Application of Inputs

Seed : Seed contributes about 20 per cent of production. During 1999-2000 production of Breeder seed have been 47.17 thousand Qtl. production of foundation seed have been 710 thousand Qtl. and distribution of certified seed have been 9100 thusand Qtl.

Chemical Fertilizer : Use of chemical fertilizer adversely effect the soil fertility. In India its use has gone up to 16798 thousand tonnes. Similarly chemical pesticides also effect the quality of production and degradation of land. The table presents the use of seed, chemical fertilizer and pesticides. Thus for sustainable food production, the use of excess chemical fertilizer and pesticides should be minimized.

Machinery and Equipment : Traditional machinery and equipment effect adversely the quantity of production, time and cost. The advanced cost effective manchinery and equipment should be used to increase the food production.

Production Technology : There is a need to develop and use micro level production technology for given inputs and rainfed areas. These should be economic and easy to understand.

Estimation of food demand

It is very important to estimate the consumption pattern at micro level.

In India NSSO conduct consumption surveys. This surveys covers different items of consumption for different size groups of expenditure. The estimate determine the level of consumption and growth in consumption for different times, using consumption surveys of two periods.

The elasticity of demand is worked out using time series data and cross-section data.

Thus the following factors are needed in determining the consumption demand of different consumable items :

1. Level of consumption of various consumption items.
2. Growth in consumption in various consumable items.
3. Growth in per capita total expenditure.
4. Elasticity of demand for different times.
5. Projected population.

Demand = population* per capita cunsumption $(1+rB)^t$

This exercise is to be done for projected population in time t, projected per capita consumption at time t, growth rate is total expenditure (r), elasticity of demand (B) and time period for which the demand is estimated. A complicated mathematical exercise is to be done for each items separately for poor and non-poor at micro level. Macro level demand will bc determined by aggregation micro level.

Analyses of Supply of Food Crops

It is seen from the Table 29.1 that the production of rice in the world is 563.9 million tonnes (1995-97) against India's production of 121.4 million tonnes (1995-97). The percentage share of India is 21.5 per cent. The wheat production in the world is 579.7 million tones (1995-97) against India's production of 65.8 million tones (1995-97). The precentage share of India's 11.3 per cent. The maize production in the world is 563.7 million tonnes against India's production of 9.4 million tonnes. India's share is 1.7 per cent only.

It is observed from Table 29.2 that food grain production in India has reached 209.17 million tonnes in 2001-02. The per capita

Table 29.1 : Population and Agriculture Works

	Total Population	*Annual component growth rate (10)*	*Rural population (Millions)*
1991	846.3	2.14	74.3
2001*	1012.4	1.79	-
2006*	1094.1	1.55	-

* Population production for India and States 1996-2016 one based on report of the technical group on population projection constituted by Planning Commission—August 96.

Table 29.2 : Area, Production & Yield of food grains

Year	*A million ha.*	*P million tons*	*Y Kg/hector*
1970-71	124.32	108.42	874
1980-81	126.67	129.59	1023
1990-91	127.84	176.39	1380
1999-2000	123.16	201.56	1637
Growth rate			
(i) during eighties	-0.11	3.54	3.33
(ii) during nineties	-0.18	1.80	1.42

Source : Agriculture Statistics at Glance 2000.

Table 29.3 : Per capita net Availability and food grains in India (kg per year)

	Rice	*Wheat*	*O. Cereal*	*Total cereal*	*Total pulses*	*Food grains*
1971	70.3	37.8	44.3	152.4	18.7	171.1
1981	72.2	47.3	32.8	152.3	13.7	166.0
1991	80.9	60.0	29.2	171.0	15.2	186.2
1995	80.8	63.4	23.9	168.1	13.9	182.0

Source : Agriculture Statistics a Glance–2000.

Table 29.4 : India's share in the world

	1995-97		1997		
Corp	*World*	*India*	*% share of India*	*Yield of world (Kg/ha)*	*Yield of India (kg/ha)*
1. Paddy	563.9	121.4	21.5	3827	2915
2. Wheat	579.7	65.8	11.3	2686	2654
3. Maize	563.7	9.4	1.7	4182	1554

Source : Agriculture Statistics at a Glance 2000.

Table 29.5 : Extent of Soil Degradation

Degradation Type		Area Affected (millions ha.)	Percentage
1.	Water Erosion	148.9	45.3
2.	Wind Erosion	13.5	4.1
3.	Chemical deterioration	13.8	4.2
4.	Physical deterioration (water logging)	11.6	3.5
	Total affected area	187.8	57.1
	Total Geographical area of India	328.7	100.0

Source : Data book on mechanization and Agro-processing since Independence—CIAE, Bhopal (1997).

Table 29.6 : Percentage of People below Poverty Line (1993-94)

All India	35.97				
Below 20% poverty	Goa 14.92	Punjab 11.97	Delhi 14.89	Chandigarh 11.35	Daman&Diu 15.80
20% to 30%	A.P 22.19	Gujarat 24.21	Haryana 25.05	H.P. 28.44	J &K 25.17
	Kerala 25.43	Mizoram 25.66	Rajasthan 27.41	Lakshadweep 25.04	
30% to 40%	Aruna Pradesh 39.35	Kranataka 33.16	Maharashtra 36.86	Manipur 33.78	Meghalaya 37.92
	Nagaland 37.92	T.N. 35.03	Tripura 39.01	W.B. 35.66	Andman & N. 34.47
	Pondicherry 37.40				
40% and above	Assam 40.86	Bihar 54.96	M.P. 42.52	Orissa 48.56	Sikkim 41.43
	U.P. 40.85	D & N 50.84			

Source : Ninth Plan Document, Planning Commission.

Table 29.7 : Sub-Divisional Actual Rainfall as percentage of normal Rainfall during 1996

Sub-Division	*Presentage of Normal*
All India	103
1. Andaman & Nicobar Island	96
2. Arunachal Pradesh	113
3. Assam & Meghalaya	95
4. Nagaland, Mizorm, Manipur, Tripura	95

Sub-Division	*Presentage of Normal*
5. Sub-Himalayan, W.B. & Sikkim	105
6. Gabgatic West Bengal	98
7. Orissa	119
8. Bihar platieu	94
9. Bihar plain	102
10. Uttar Pradesh East	105
11. Plains of U.P. West	124
12. Hills of U.P. West	96
13. Harayana, Chandigarh & Delhi	131
14. Punjab	108
15. Himachal Pradesh	87
16. Jammu & Kashmir	148
17. Rajasthan East	138
18. Rajasthan West	126
19. Madhya Pradesh East	95
20. Madhya Pradesh West	83
21. Gujarat Region	98
22. Saurasthra, Kutch & Diu	93
23. Konkan & Goa	94
24. Madhya Maharastra	102
25. Marathwala	111
26. Vidarbha	73
27. Coastal Andhra Pradesh	127
28. Telangana	96
29. Rayalaseema	174
30. Tamil Nandu & Pondicherry	131
31. Coastal Karnataka	94
32. North Karanataka	93
33. South Karnataka	84
34. Kerala	90
35. Lakshadweep	101

Source : Agricultureal Statistics of a Glance, 2000.

Table 29.8 : Use of Chemical Inputs (1999-2000)

I	*Seed (in 000 qtl)*	
	1. Production of Breeder Seed	44.17
	2. Production of Foundation Seed	710.00
	3. Production of Certified Seed	9100.00
II	*Consumption of Chemical Fertilizer (1998-99)*	
	1. Nitrogenous (N) (000 tonnes)	133354
	2. Phosphotic (P) (000 tonnes)	4112
	3. Potassic (K) (000 tonnes)	1332
	Total (000 tonnes)	16798
III	*Consumption of Pesticides (1998-99) (000 tonnes)*	46.16

Table 29.9 : Land utilization in India 1996-97

	Mill ha.	%
I. Total Geographical area	328.73	-
II. Reporting area for land utilization (1 to 5)	304.81	-
1. Forests	60.75	22.5
2. Not available for cultivation	41.54	
(a) Non-Agricultural uses	22.45	7.4
(b) Barren & uncultivated land	19.09	6.3
3. Other cultivated land (excluding fallow land)	28.64	
(a) Permanent Postures & other Grazing land	11.04	3.6
(b) Miscellaneous trees crops & groves	3.50	1.2
(c) Cultivated waste land	13.94	4.4
4. Fallow Land	23.22	
(a) Other than current fallow	9.89	3.2
(b) Current fallow	13.33	4.4
5. Net Area sown (6-7)	142.82	46.8
6. Gross Cropped Area	189.54	
7. Area sown more than once	46.72	
8. Cropping intensity	132.70	
III. Net Irrigated Area	55.14	38.7
IV. Gross Irrigated Area	73.28	

Source : Agricultural Statistical a Glance, 2000.

Table 29.10 : Percentage of Number & Area of Operational holders by type of holding (1990-91)

(*In percentage*)

	Number	*Area*
1. Marginal (velow 1 ha.)	59.4	15.1
2. Small (1-2 ha.)	18.8	17.4
3. Semi-Medium (2-4 ha.)	13.1	23.2
4. Medium (4-10 ha.)	7.1	27.0
5. Large (10 ha & above)	1.6	17.3
Total	106.6 mil. 100%	165.5 1000%

Source : Agriculture census 1990-91.

Table 29.11 : Crop-wise Irrigated Area 1996-97

	Million ha	%
1. Rice	22.19	51.0
2. Wheat	22.31	86.2
3. Course cereals	3.24	10.2
4. Total pulses	2.85	12.7
5. Total oilseeds	6.93	26.3
Gross Irrigated Area	73.28	38.7

Growth rate of irrigated area - 2.58%

Source : Articulture Statistics at a Glance 2000.

Table 29.12 : Cropwise Estimation of Food Grain Production

	1999-2000	*2000-2001*	*2001-2002*
Food Grains	209.8	195.92	209.17
Food cereals	196.39	185.25	196.02
Food rice	89.68	84.87	91.05
Food wheat	76.37	68.76	73.06
Food course cereals	40.34	31.62	32.01

Source : Monthly Review of Indian Economy, CMIE March 2002.

Table 29.13 : Centre + State plan out large during 2000-2001

	Total	*10108*	*100.0%*
1.	Crop Husbandry	3670	36.7%
2.	Soil & water conservation	1021	10.2%
3.	Animal Husbandry	575	5.7%
4.	Dairy development	135	1.3%
5.	Fisheries	452	4.5%
6.	Food storage & warehousing	318	3.2%
7.	Coorperative	917	9.2%
8.	Forest & wild life	1898	18.9%

Source : Monthly Review of Indian Economy, CMIE, March 2000.

availability of food grain is about 2 Qtl per annum. Cereals production is 196.02 million tonnes. The Table 29.3 presents the per capita availability of rice, wheat, other cereals, total cereals, total pulses and food grains during 1971, 1981, 1991, Table 29.2 presents the area production and yield rate and growth rates.

About 2/3 of food grain production grown in irrigated area occupying 40 per cent of area sown and 1/3 of food grain is grown in rainfed area which account about 60 per cent of the area sown. The disparity is to be removed. The Government of India has given top priority for development of rainfed farming. The scheme integrated development of rainfed farming is most important scheme under which micro level watershed are being developed.

Strategy for Sustainable Food Security

1. *Factor endowment technique :* Develop crop specific areas which are extermely fertile due to factor endowment. These areas possesses suitable climate, natural resources and land to grow specific food crops. In these areas productivity is very high and cost of

production is low. Thus, crops at optimum level should be grown. The inter-state or inter-regional demand may be managed by adopting inter-state movement.

2. *Management of food security globally :* International organization like FAO may lead in managing food security globally. The excess food grain producing countries should contribute a fixed agreed proportion of excess food for the disposal/movement to deficit countries on the direction of international organization like FAO. The cost of food grain should be nominal so that deficit countries may effort to meet their requirement.

3. *Match cropping pattern with consumption demand :* Inter-state or inter-regional consumption demand is to be forecasted on the basis of consumption basket. The government or non-government organization under the agriculture experts show change the cropping pattern in order to meet the country consumption demand of food. Wherever cost of production is higher, the government should provide renuneration price to meet the cost of production with moderate profit to farmers.

4. *Management of proper distribution system :* In the country to attain food security, the government food grain stock exist but it does not reach to target group either due to procedural delay or lack of purchasing power. Unless inequality is reduced the higher food production will not provide food security.

5. *Minimise inequality of income and enchancing purchasing power :* In the country food exist but due to large inequality in income the poor people do not get sufficient food to meet their requirement due to lack of purchasing power. Unless inequality is reduced, the higher food production will not provide food security.

6. *Other important strategies for food security*

(a) Land and water development

- Agro-ecological zoning system
- Information system to manage water for agriculture
- Development of climatic database
- Irrigation Planning
- Education and training

- Fertilizer and water retaining products development.
- Crop information data
- Multi-criteria decision analysis.

(b) Plant protection services development.
- Developnent and use of integrated pest management (IPM) in order to sustain land fertility and ill effects on human.

(c) Seed and plant genetic resources development.
(d) Development of field crops, grass land and pasture crops, horticulture crops and industrial crops.
(e) Population control.

Conclusion

In spite of achieveing sufficiency in food grain production India has miles to go for securing food security to all sections of society. There is an urgent need to grow factor endowment crops, consumption—demand based crops, proper distribution system, enhancing purchasing power of the poor, minimising disparities in income and providing right information to farmers as well as consumers.

REFERENCES

Agrawal, P.C. (1990) *Oil Seeds in India—Perspective of 2001 A.D.,* Book published by IBM & Oxford Publisher, Delhi (1990).

Agriculture Statistics at a Glance 2000. Directorate of Economics and Statistics, Min. of Agriculture, Govt. of India, New Delhi.

Ganga Devi Gupta (1991) *Changes is consumption demand and output of pulses in Uttar Pradesh—An Econometric Analysis*, Ph.D. thesis approved by Meerut University (1991).

Kusum Chopra (1981) *Oilseed in India—A Regionwise analysis of Problems of Production of five major oilseeds*, No. 3 School of Social Sciences, J.N.U., New Delhi.

National Agricultural Policy (2000). Ministry of Agriculture, Govt. of India.

Ninth Plan Document, Planning Commission, Govt. of India, New Delhi.

Sharma, S.K., Naveen Kalra and R.C. Harit (2001) Crop Yield forecasting. Conventional and modern techniques. *Employment News* 23-29, 2001.

Singh, Padam & Agrawal, P.C. (1985) *Analysis of gram pulses & Tur Pulses in India*, Official Paper of Planning Commission, New Delhi.

Ved Prakash (1987) *Consumption pattern in India*; Ph.D. thesis approved by Meerut University.

30

Spatio-Temporal Analysis of Agricultural Marketed Surplus

A Case Study of Faridabad District (Haryana)

Rocket Ibrahim and *M.R. Goswamy*

Database

The study is based on primary data collected from the selected Agricultural Market Committees of the district. The officials of the Market Committees record daily arrival of the agricultural commodities, quantity of the produce, price at which it was sold and the address of the farmer (producer). The above information was taken from the *Auction Register* for the said *mandis* and products of Faridabad District for the year 1975-76 and 1995-96. The major commodities for which the data collected were wheat, barley, *dhan* (rice) and *sarson* for two time periods, i.e. from 1-4-1975 to 31-3-1976 and 1-4-1995 to 31-3-1996. The mandi-hinterlands are based on the list of notified villages provided by the Agricultural Marketing Board, Chandigarh, Haryana.

The Research Problem

The present study is an attempt to find out regional disparities in the agricultural marketed surplus of the major agricultural commodities produced in the year 1975-76 and 1995-96 in the hinterlands of different mandis of district Faridabad. A spatio-temporal analysis has been made to distinguish the regional disparities in the patterns of agricultural marketed surplus as existed in the two given time periods in the selected mandi hinterlands. Attempt has been made to investigate the following points :

To find out variations in the quantity of arrival of different commodities during above mentioned two time periods. How far the hinterlands of the mandi centres are circular in shape on the basis of intensity of agricultural marketed surplus? To what extent the agricultural marketed surplus decreases with the increasing distance from the mandi? To what extent the agricultural produce brought in the mandi and frequency of arrivals decline as we go away from the mandi centre? Whether the average size of arrivals is larger from distant villages as compared to the village located near the mandi area. Is there exists any relationship between transport connectivity-accessibility and quantity of agricultural marketed surplus?

Methodology

The data on daily arrival of agricultural marketed surplus of wheat, *dhan*, barley and *sarson* was collected from the *Auction Register* of different Agricultural Market Committees for all the producers (farmers) who brought their products to the market during the agriculture year 1995-96 was processed. In case of a selected mandi of Hodal the above mentioned information was collected for two time periods, i.e.1975-76 and 1995-96.This daily arrival of major commodities was processed, tabulated and mapped by taking villages as the unit of study in the hinterlands of the respective mandis of the district. The arrival data was compiled to get the total arrival of each village for the said periods. The arrival of various commodities per 100 hectares, average arrival size, total arrivals, and frequency of various commodities for various distance zones have been calculated. Choropleth maps were prepared for each commodity to show the intensity of agricultural marketed surplus. The hinterland of the mandi has been divided into different distance zones. The area of the first distance zone was taken within the a radius of 5 km of distance from the mandi, the seconds distance zone was from 5-10 km and the third zone was from 10-15 km, the fourth zone from 15-20 km and the fifth zone was beyond 20 km of distance from the mandi. The total arrival of wheat, *sarson*, *dhan* and barley was calculated zonewise to know the *intensity* of marketed surplus. The proportion of marketed surplus of each product was calculated to know the proportionate contribution of each zone for different commodities. The proportion of agricultural marketed surplus was also calculated for each zone to know the proportionate share of each commodity within each distance zone. The frequency of arrivals for each village for each commodity was also calculated.

Spatial Pattern of Agricultural Marketed Surplus

During the two time periods of 1975-76 and 1995-96, there has been considerable structural changes in the arrival patterns of the agricultural marketed surplus in the study area. Some of the commodities, which recorded very high proportion of arrivals in the mandi yards twenty years back, either disappeared or remained negligible in proportion to the major commodities. Sarson and barley, which were considered among the major produces of the district, their total arrival has gone down and become insignificant products of the district.

The Hodel, which is the second largest mandi of the district, barley was the most dominant commodity in the year 1975-76. The arrival of barley as proportion to five major commodities was more than 50 per cent and the proportion of wheat was 29.42 in the year 1975-76 in the first distance zone of 5 km from the mandi centre. After 20 years of gap the proportion of barley remained only 0.23 and that of wheat increased to 74.26.

The proportion of *sarson*, which was 16.46, declined to 5.60 and production of *mator,* which recorded 3.38 per cent in the year 1975-76, almost disappeared from the mandi by the year 1995-96. It is interesting to note that *dhan,* which was not known 20 years back has become the second most important commodity in the mandi. In the immediate vicinity of Hodal march, i.e. within 5 km of distance range the total arrival was recorded 19.11 per cent. In the second distance zone of 5 to 10 km of radius, the proportion of wheat, which was 30.91 the year 1975-76, remained 54.03 per cent by the end of second decade, much less than the first distance zone of 5 km radius. The proportion of barley declined from 49.66 to 0.26 and proportion of *sarson* decreased from 14.66 to 5.47. *Matar,* which reported 4.76 per cent, totally disappeared. The arrival of *dhan* became almost double of the first distance zone. The arrival pattern of commodities of third distance zone of 10 to 15 km from the mandi yard is more interesting. The proportion of the wheat, which was 6.40 in the decade 1975-76, increased to 33.56. The proportion barley was very high, i.e. 64.50 but came down to 0.19 only. The proportion of sarson declined from 18.77 to 2.15. This was the distance zone, which recorded maximum arrival of *matar,* which totally disappeared from the mandi, and the proportion of *dhan* reached 63.35. In the fourth distance zone of 15 to 20 km of radius, the proportion of wheat increased from 16.71 to 40.35 in 1995-96. The proportion of barley which was 59.67 per cent remained only 0.10. The proportion of *sarson* declined from 16.16 to 7.10 and the proportion of *matar, which was 7.34,* recorded no arrival in the mandi and *dhan* as the first dominant commodity with a record of 51.94 per cent. Beyond 20 km and more distance, the proportion of wheat increased from 11.48 to 23.63, barley from 45.84 to 0.35, *sarson* 30.08 to 1.99 and *matar* declined from 6.60 to 0 per cent. The proportion of *dhan* was almost two-third of the total commodities arrived in the mandi.

The arrival trends of the catchment area of Hodal mandi shows that the dry farming crops totally disappeared from the mandi which have been replaced by *dhan*/rice. The other important fact is that the proportion of wheat decreased as we go away from the mandi centre in different distance zones in both the time periods, i.e. 1975-96 and 1995-96 though the proportion of wheat increased in different distance zones in the two decades of time. The proportion of barley decreased drastically during the said period. *Matar* totally disappeared and *dhan was* not at all seen in the yard, jumped to first rank in the mandi. It clearly shows that the farmer has opted for the double-cropped combination of wheat-rice wherever irrigation facility is available particularly along the Yamuma belt, Agra canal and other distributaries. The dominant commodity wheat has clearly demonstrated that with the increasing distance from the Mandi, average arrival per hundred acres and the frequency of arrival declines. The trends are very similar to that of 1975-76, however, arrival per hundred acres is much higher in each zone as compared to the previous time period. For example half of the catchment area of Hodal shows less than 5 quintals of wheat per hundred acres in the year 1975-76. Whereas in the year 1991-96 most of the hinterland of Hodal recorded more than 100 quintal of the wheat per 100 acres. The area within

5 km of radius received more than 270 quintals of wheat per hectare. Even the distance zone of 10-15 km received almost 100 quintals of wheat per 100 acres.

The Palwal mandi recorded very high arrivals of *dhan* (rice) from each distance zone. The arrivals in the nearest zone of 5 km of radius was 398 quintals per 100 acres. The average arrival size was 27.82 quintals and almost 1/4th of the total marketed surplus was generated within this distance zone. The most important feature of the Mandi is that the arrival per area sharply decreased with increasing distance from the Mandi and the frequency of arrivals declined. Surprisingly there was hardly any change in the average arrival size. The arrival of the *dhan* was 225.29 quintals per hundred hectares in the second distance zone of 5 to 10 km of radius, whereas that of wheat it was 200.85 quintals. In the third distance zone arrival of *dhan* declined to 130.54 and 136.50. In the fourth zone of 15 to 20 km it further declined to 100.92 and 105.52 quintals respectively. The area within 10 km of radius from the mandi contributed 58 per cent of the total marketed surplus *dhan.* One-fourth of the total arrival of *dhan* was produced in the distance zone of 10-15 km. There was hardly 5 per cent arrival from the distance beyond 20 km. In case of wheat the inner most zone contributed hardly 9 per cent of its marketed surplus. The distance zones of 5 to 10 km and 10 to 15 km contributed 31 and 36 per cent respectively. The mandi receives hardly 8 per cent of its wheat from beyond 20 km of distance. The other commodities, though little significant, are jowar, bajra and arhar. Their arrival pattern do not duplicate the pattern of wheat and rice/*dhan.*

Conclusions

There has been considerable structural change in the arrival pattern of commodities in the mandies of the district. The products of dry farming disappeared from the mandies. *Dhan*/rice *was* not seen in the *mandi* till 1975-76, became the dominant commodity of the market. With the increasing distance from the *mandi* the arrival of frequency of the major agricultural commodities declined. With the increasing distance from the *mandi* centre, arrival of commodities per hundred acres also declined.

REFERENCES

Barbara Harriss, (1979) "Measuring Agricultural Market Performance", *Food Research Institute Studies,* Vol. XVII.

Caves, Rechard E., (1977) "Organization, Scale, and Performance of the Grain Trade", *Food Research Institute Studies,* Vol. XVI.

Goel,Veena, (2000) "A Study of Market Arrivals and Procurement Pattern of Wheat by Public and Private Agencies in Punjab", *Indian Journal of Regional Science*, Vol. XXXII, No. 2, Kolkata.

Harriss, Barbara, (1979) "Going Against the Grains," *Development and Change,* Vol. 10.

Harriss, Barbara, (1975) The Role of Punjab Wheat Markets As Growth Centres", *Journal of Geography,* Vol. 23.

Ibrahim, Rocket, (1984) *Market Centres and Regional Development,* Delhi, B.R. Publishing Corporation.

Johnson, G.D. (1994), "Does China Have a Grain Problem?" *China Economic Review* 4(I).

Jones , W.O., (1972) *Marketing Staple Food Crops in Tropical Africa, New* York, Ithaca.

Liang, Ernest P.L., (1981) "Market Accessibility and Agricultural Development in Pre-War China", *Economic Development and Cultural Change,* Vol. 30.

Raza, Moonis, (1992) *Transport Network, Commodity Flows and the Regional Structure of the Indian Economy,* New Delhi, Concept Publishing Company.

Smith, Carol A., (1972) "Market Articulation and Economic Stratification in Western Guatemala", *Food Research Institute Studies,* Vol. XI.

Smith, Vernon, (1975) "Marketing Agricultural Commodities in Pichincha Province, Ecuador", *Geographical Review,* Vol. 65.

Walter C. Neal, Harpal Singh and Jaipal Singh, (1965) " Kurali Market —A Report on the Economic Geography of Marketing in Northern Punjab", *Economic Development and Cultural Change,* Vol. XIII.

31

Sustainable Agriculture in 21st Century with Special Reference to Land Degradation

C. Sejwar

Introduction

The word 'Agriculture' came from a Latin term 'Agricultura'. 'Ager' means a 'Field' and culture means 'cultivate'. In a developing country like India with burgeoning population and limited land and water resources, it is imperative to boost farm productivity continuously in coming years without any undesirable impact on environment. The most tragic part is that land and water resources are finite. In fact, the land degradation and water pollution scenario is quite frightening. We should shift our approach from traditional agriculture to sustainable agriculture in 21st century as we have to produce more food and other agriculture commodities for a huge sized population under limited land and water resources and increased stresses along with conservation of resources.

Sustainable Agriculture in 21st Century

The fundamental challenges in 21st Century are to find ways for sustainable development that are environmentally sound, equitable and allow respect for individual and social rights. The concepts of sustainability lies in maintaining harmony between buoyancy and dynamism in agriculture growth for meeting the basic needs along with emphasis on protection and conservation of natural resources.

Sustainable Agriculture is the successful management of resources for agriculture to satisfy the changing human needs, while maintaining or enhancing the quality of environment and conserving the natural resources (CGIAR* in 1988). A sustainable system uses its inputs optimumly and judiciously. To optimise productivity and profitability with minimum adverse effect on environment.

Sustainability = Productivity + Conservation of resources.

India today is a leading nation in agriculture production with production of 89.48 million tonnes of rice and 75.57 million tonnes of wheat during 2000-2001. This rapid growth in the production of the two major cereals in the country has made India self-sufficient in food. India holds its position by virtue of not only quality but it is also endowed with rich diversity. We are also a leading nation in potato and cotton production. We have today surplus food stock in our godowns whose value is estimated to be Rs. 50,000 crore. In spite of our impressive achievements our productivity is very low in comparison to global average. Estimate indicates that India's geometrically increasing population will increases by 130 million by 2020 AD requiring production of 325 million tonnes of foodgrains. Lester Brown and Hal Kane (1994) of the Worldwatch Institute predicted that by 2025 AD, India may be compelled to import 45 million tonnes of foodgrains as its production will fall short of its need by this extent.

This may be achieved by producing 5-6 million tonnes additional food grains per year against 3.1 million tonnes per year over last 40 years. Input use efficiency is continuously decreasing resulting in poor output : Input ratio. In global terms Indian land resources are under immense pressure as it has only 2 per cent of World's Geographical area, 0.5 per cent forest cover and 0.5 per cent of total grazing area receiving 1 per cent of total rainfall. But with these resources, it has to support 16 per cent of global human population and 15 per cent of farm population. The resultant over straining of land resources and disproportionate human and animal population is at the root of widespread land degradation and non-sustainability of an ecosystem.

After near stagnation in 1999-2000 and negative growth of 0.2 per cent in 2000-01 agriculture sector is expected to grow at nearly 6 per cent during the current year, pushing up the GDP growth rate to

* CGIAR = Consultative Group on International Agriculture Research.

5.4 per cent from a lower level of 4 per cent in the preceding year as shown in Table 31.1. Much of growth in agriculture sector is attributable to the rise in food grains production to 209 million tonnes, up from 196 million tonnes in the preceding years as also increase in the production of non food crops including oilseeds, jute and cotton. This will help stimulate other sectors of the economy since an increase in incomes of farmers would generate fresh demand for goods and services during 2002.

Table 31.1 : GDP Agriculture and Food Grains Growth rate

Year	*GDP (At 1993-94) Prices*	*GDP, AGR & Allied Sector*	*Food Grain Production (Per cent)*
1998-99	6.5	6.2	5.9
1999-2000	6.1	1.3	3.0
2000-01	4.0	–0.2	–6.6
2001-02*	5.4	5.7	6.8

* Advance Estimates.
Source : Economic Survey 2001-2002, GOI, pp.186.

This year's food grain output is likely to be 209.20 million tonnes, representing an increase of around 13 million tonnes over the previous years production of 195.9 m. tonnes (2000-2001).

With the intensive use of agro-chemicals like fertilizers, pesticides, synthetic growth regulators etc. all these unilateral approaches of increasing production have generated several undesirable impacts, viz. loss of biodiversity, degradation of land and water resources, environmental pollution and climatic changes. Which are the "major challenges" to be met for a sustainable agriculture system. There is a need to shift towards optimum resource utilisation through better and efficient management system.

Sustainable agriculture avoids maximizing any single outcome variable. It attempts to find a balance among agronomic, environmental, economic and social aspects. Goals of sustainable agriculture in 21st century will be achieved much quicker and better if social scientists realise this challenge and act accordingly.

Hon'ble Shri Atal Bihari Vajpayee, Prime Minister of India, in his speech at the conference of Chief Ministers on W.T.O. and Agriculture and Food Management in New Delhi on May 21, 2001 has said that,

> "In 21st century, the challenge before us is to sustain food security and have some surplus for exports to take advantage of globalization of agriculture. Agriculture is one of the areas in which India has an inherent strength to dominate global markets we should resolve to make India a leading agriculture power in the world in 21st century."

Land Degradation

Land which is the most precious heritage and the physical base of bio-mass production of life supporting system is finite and inelastic. In this natural non-renewable endowment the share of our country is fixed at 328.73 m.ha.

Land the major input in agriculture is getting reduced both in terms of size and its productivity. Land degradation problem has become a major and serious issue not only in India but it is also being recognized globally. Land degradation, which is not a result of a single factor but a set of factors, damages the soil, water and vegetation and restricts the productive capacity of plants and crops. A close look at the present health of the soil and water resources reveal their wanton misuse and degraded environment. About 173.64 m.ha. covering slightly half of the country are threatened by various types of degradation like salinity, alkanity, water logging, ravenous and gullied lands, areas under ravages of shifting cultivation, desertification etc. Soil loss occurs in both water scarcity and water surplus zones in the form of wind and water erosion. There are specific problems of land degradation due to open-cast mining operations, using good productive land for brick kilns, coastal erosion and sea water ingress, excessive erosion and land slides in the crumbling hill areas.

Types of Land Degradation

(a) Water and Wind erosion.
(b) Physical degradation—water logging, soil crushing, compaction, desertification etc.
(c) Chemical degradation—Salinization, sodification, acidification, nutrient removal, decrease of organic matter.
(d) Biological degradation.

Extent of Land Degradation

The extent of land degradation varies from one state to another depending upon the topographical features, geological formulations, soil characteristics, rainfall and other climatic parameters, land use, measures of soil conservation management practices.

Table 31.2 : Extent of land degradation in India '

S.No.		*Description*	*Estimated Area in million hectare with 1981-82 Land Utilisation Statistics 1996-97*
1.		Geographical Area	328.73
2.		Area subjected to water and wind erosion	144.12
3.		Area degraded through special problems	29.52
	3.1	Water logged Area	8.53
	3.2	Alkali soils	3.88
	3.3	Acid soils	4.50
	3.4	Saline soil (includin Coastal Sand Area)	5.50
	3.5	Ravines and Gullies	3.97
	3.6	Area Subject of shifting cutivation	4.91
	3.7	Riverine and Torrents	2.73
4.		Total Problem Area	173.64
5.		Total Flood Prone Area	40.00
	5.1	Annual Average Area affected by floods	8.00
	5.2	Maximum area affected in the worst year	18.60
	5.3	Annual Average cropped area affected by floods	3.70
6.		Total drought prone area	260.00
7.		Culturable waste land	13.88

Source : Indian Agriculture in Brief 27th Editions Year 2000, p.14, Department of Agriculture & Cooperation, Govt. of India.

Wind and Water Erosion

In India 144.12 m.ha. area suffers from water and wind induced soil erosion. India is a reverine country, out of 328.73 m.ha. of the geographical area, 316 m.ha. is drained by major rivers and their tributaries, water induced erosion is thus acute in India and it is essential to arrest the menance of soil loss following the hierarchy of streams and rivers. The Himalayan mountains suffer from serious water erosion though water erosion is also rampant in the hills of western ghats and other areas of high intensity rainfall. Water erosion not only removes the productive surface layer of soil but also reduces the storage capacity of reservoirs sediments forms the largest chunk of pollutants. Ganga and Brahmaputra river systems carry annual load of 3000 million tonnes to the sea causing considerable damage in the form of loss of top soil, sediments also fills in lakes reservoirs and river causing displacement of water which can hinder flood control efforts and even usefulness of reservoirs.

Wind erosion is more prominent in hot arid region occupying 31.7 m.ha. More than 82 per cent of the arid zone is affected by severe wind erosion alone. Removal of vegetative cover and overgrazing enhance the intensity and extent of wind erosion and desertification. The sand movement causes calcine damage to the adjoining cultivated areas, roads, canals, buildings etc.

Waterlogging

Waterlogging is also a serious problem which has rendered 11.6 m.ha of lands degraded. Water logging caused by rise in water table poses a great threat to soil productivity and environmental ecology, specially in the irrigated areas. Roughly an area of 100,000 ha is estimated to be affected by water logging annually. Introduction of canal irrigation is the major reason for the once fertile lands to be affected by water logging. Degradation of land due to water logging can be due to seasonal waterlogging or permanent waterlogging.

Salinization and Sodification

Salt affected soils covering an area of more than 10 m.ha fall under two main categories :

(i) Saline soil having excess of soluble neutral salts like chlorides and sulphates of sodium, potassium and magnesium. The saline soil has lower values of pH and exchangeable sodium, better physical condition etc.

(ii) Alkaline soil with high concentration of carbonates and bicarbonates of sodium.

Soil salinity is more prominent in the arid and semi-arid areas though some coastal areas in the humid conditions also suffer due to the ingress of seawater. Continuous use of poor quality ground water for irrigation also leads to the development of soil salinity or sodicity, particularly in the slowly permeable soils. It is more serious in the Indo-Gangetic plain, black soil region, arid areas of Rajasthan and Gujarat and coastal areas. The sodic soils possess high pH and exchangeable sodium percentage values, preponderance of carbonates and bicarbonate salts of sodium deficient amount of organic matter, nitrogen available calcium and zinc. Presence of $CaCO_3$ (*kankar*) in the subsoil, impaired physical condition and poor moisture relations. Podosolisation process in green area results in acidity of the soil.

Nutrient Loss

The factor greatly responsible for enormous removal of the plant nutrients is soil erosion. The total loss of surface soil is estimated to be about 6000 million tonnes with major plant nutrients of nitrogen, phosphorus and potash varying from 5.37 to 8.40 million tonnes every year. This aggravates the problem of soil fertility depletion. Only 23 per cent of the applied chemical fertilizer is reported to be utilized by plants and remaining 77 per cent enters the soil and environment that not only pollutes the nature but also accounts for annual loss of huge funds (Ghosh, 1994). The production under intensive cropping system has resulted in large scale removal of nutrients from the soil resulting in a Negative balance of these nutrients.

Impact

The visible economic impact of land degradation is on agricultural productivity. The reduction in productivity of land resources has a telling effect on the socio-economic conditions. The immediate effects of land degradation includes reduced crop yields, increasing need for agricultural input and decreasing profits, reduction in the value of land, losses of water resources, increase in frequency and severity of floods, siltations of reservoirs, rivers etc. and adverse effect on power generation. All the above reasons ultimately results in shifting the land resources from agriculture to other uses including industry which bring environmental degradation. This is more common in Delhi where there is a growing number of brick kilns on the once productivc lands.

The greenhouse gases (GHGs) emitted through soil degradation are CO_2, CO, N_2O and NO.

Strategies for Prevention of Land Degradation

Since the problem of land degradation is very complex, it is important that we approach the issue in a systematic and logical manner. The following strategies are suggested :

(a) Proper records of land productivity status should be prepared with the help of soil scientists along with the latest technologies like remote sensing.

(b) For each soil properties there should be a clear demarcation of boundary in order to monitor our soil which is stressed by external degrading agencies.

(c) In areas of high water requirement like cropped area effective soil and water requirement like cropped area effective soil and water conservation techniques should be adopted with the guidance of engineers.

(d) Industrial activities in the nearby area which are hazardous to the environment should be regulated through laws.

(e) Sustainable and effective land use system should be followed.

(f) Waste and marginal lands should be amended properly so that they can be brought under crop cover. So as to reduce their further degradation.

(g) Implementation of agro-forest ecosystem wherever water erosion is a serious threat.

(h) Adoption of integrated watershed management system and integrated balanced nutrient management system.

(i) The use efficiency of different factors and production should be increased.

Sustainable Land Managements

Sustainable development and management of land resources have become a vital issue for the country in view of meeting the needs of food, fodder and fibre to the growing population. Agro-forestry is sustainable land management, which increases the overall yield of the land, combines the production of crops and forest plants or animals simultaneously on the same unit of land. Poor land management practices have resulted in rapid rates of soil salinisation.

Rehabilitation of Degraded Land

It can be achieved by various land reclamation techniques, viz., saline soils require leaching of the soluble salts. The waterlogged areas are rehabilitated by development of drainage network, conjunctive use of canal and ground water, adopting suitable cropping system etc. Through biological techniques rehabilitation of sand dunes, ravines, and mined land can be achieved up to a greater extent.

Watershed Management

Watershed management is a major land development programme in the country land resources development through watershed approach allows to manage soil and water resources without disturbing the harmony of natural habitats. Watershed is a natural hydrologic entity governed by the terrain topography from where runoff is drained to a point. Watershed is a good device for water harvesting. Drainage line treatment measures like earthen loose boulder structure, checkdams, drop spill ways, sediment detention structure etc. Proper watershed management can check further deterioration of ecosystem.

In the end watershed development is one of the best and most efficient way of reclaiming degraded lands.

Conclusion

To achieve food security in 21st century there is a need to generate an evergreen revolution based on congruence of enhanced productivity, sustainability, profitability and social equity. So more concerted efforts are needed in view of natural resource management by adopting a paradigm shift in our approach towards increasing agricultural productivity without any detrimental effect on environment, as in the new millennium we have to produce more food and other agricultural commodities for a huge population under limited land and water resources and increased stresses. Agriculture needs to be based on judicious use of renewable resources of energy. We should have some surplus for exports to take advantage of globalisation of agriculture.

Land degradation, a dynamic process influenced by a host of natural and human induced factors, periodic monitoring of the emerging changes is of paramount significance. Reclamation, development and sustainable management of land have to receive inputs not only of funds but also competent technical advice and institutional support. It is a matter of satisfaction that the government has recognized that watershed development is one of the best and most effective ways of reclaiming degraded lands.

REFERENCES

Annual Report 2000-2001, Department of Agriculture and Cooperation, Ministry of Agriculture, Government of India, New Delhi -110012.

Annual Report 2000-2001, All India Soil and Land use survey.

Anon (1996) *Soil management in relation to land degradation and environment*, Indian Society of Soil Science, IARI Campus, New Delhi.

Barrow, C.J. (1991) *Land Degradation : Development And Breakdown of Terrestrial Environments*, P-10 Cambridge University Press, New York.

Brown, Lester, Rand Hal, Kane. (1994). *Full House : Reassessing the Earth's Population Carrying Capacity*. Worldwatch Institute, New York.

Chisholm, M. (1975). *Rural Settlement and Land use*. Hutchinson, London.

Das, S.N. (1998). Watershed Development-Planning and Strategy: *Journal of Rural Development,* Vol. 17, No. 3; 463-69.

Ghosh, S.K. (1994). Impact of Land and Water Resource Degradation on Agricultural Production, p. 95, *National Resources Management for Sustainable Agriculture and Environment*, Angkor Publishers (P) Ltd., New Delhi.

Mishra, P.L. (1985) *Environmental Influence on Agriculture*. Ritu Prakashan, Hapur (UP).

Technical Bulletin-1/2001, Soil and Land Information for Watershed Development, All India Soil and Land use Survey. Department of Agriculture and Cooperation, Ministry of Agriculture, Government of India, New Delhi, 110012.

Virmani, S.M. (1997). *Hindu Survey of Indian Agriculture*, Published by the Hindu, New Delhi. pp. 161-63.

32

Population Dynamics and the Question of Food Security in South Asia

Mrityunjay Mohan Jha

Introduction

Throughout the developing world and specially in the South Asian countries the population growth has been regarded as the most vital parameters in the development process. In the recent years, food production has increased satisfactorily and, kept both Malthus (1798) and Ehlrich (1968) at bay. South Asia as a whole has managed to keep its food production ahead of population growth. The margin is not large. Despite the marginal advantage in food production there is wide nutritional gap which has led to hunger. A country like India, which has achieved the level of self-sufficiency in food production, is still facing the problem of accessibility to food in year of globalization and liberalization (Chopra *et al.*, 1992). Pakistan has almost achieved self-inefficiency but maldistribution and malnutrition is prevalent. Bangladesh has emerged as a country which has almost reached self-sufficiency in rice production (Goletti, 1994). Though poverty and malnutrition and its associated problems are posing serious challenges. Most of the countries in this region are struggling to check their high population growth. In the entire region, high population growth along with deep rooted poverty is biggest stumbling block in the socio-economic development process (Khadka, 1990).

While food security at the national level is a necessary condition, it is not sufficient to ensure automatically that all the people have enough to eat. With increasing pressure over the limited usable land

and non-farm employment very hard to find the problem of food insecurity get accentuated (Osmani,1989). Since poverty is taken as the root cause of food insecurity and malnutrition, one has to analyse these two.

Prof. Amartya Sen (1987) and John W. Mellor (1987) stressed the need for increasing the purchasing power to households to improve the household food security as well as national food security. The problem has to be attacked from both ends, food supply and fertility (Sen, 1989). There are examples of grain surplus and widespread hunger, this confirm that lack of purchasing power leads to under-nutrition; malnutrition and chronic food insecurity (Swaminathan, 1987).

The need for mustering food grain stocks and reaching them to the places where they are needed, is highly desirable (Acharya, 1983; Alam,1992). The examination of the population pressure on land and its impact on agricultural productivity is inevitable (Chaudhary, 1986). In another attempt (1988) he stressed the need for subsidized rationed shop to provide accessibility to food. Overpopulation in South Asia with high incidence or poverty and wide spread under-nutrition, are closely inter-linked. Therefore, the countries should remain cautious against the declining land/person ratio (FAO, 1996). A pro-active approach is needed to bring changes in the political policy, and institutional factors responsible for development failures, which can be debated.

Fred H. Sanderson (1984) insists on food and other kind of assistance for improving nutritional status. David Norse (1994) describes conduction of the environmental, economic, population and climate change threats to regional food security, but these do not come in a spatially or temporally consistent manner. For securing food, the adoption of a powerful employment oriented strategy of development with the support of food aid, is one of food security strategy (Ezekiel, 1988). At the same time the benefits and damaging consequences of food imports and food aid for the domestic economy cannot be ruled out (Fonseca, 1983). The relevance of food aid in solving the basic problem can also be questioned. Self-sufficiency in food is a crucial elements in the development of a country. And if this is not possible at the individual level, it should be an achievable objective at least at the national level.

This study attempts to analyse the population, food imbalance in

South Asia; to review the existing population, food-nutritional/ agricultural policies with special reference to India; and, to assess the challenges faced by South Asian countries.

Population—Dynamics and Food and Nutrition Status

There is a two-way interaction between population and food resources, the extent and complexity of which may vary in time and context. The size and composition of the existing population and its growth over time has its impact on various aspects of development, socio-economic and food security. Food security, obviously is the basis of social and economic security (Rodgers, 1989). At a given level of living, any increase in population generates increased demand for resource including 'food' which is the minimum requirement.

During the last few decades the world as a whole has experienced tremendous increase in population of human being, mainly in the poorer parts of the world (Table 32.1). A rapid decline in mortality

Table 32.1 : Population Sizes and Rates in World Region 1994

Region	*Population (Millions)*	*Growth Rate*	*Birth Rate*	*Death Rate*
Developing	442.6	1.9	28.0	9.0
N. America	157.4	2.5	33.0	8.0
C. American/Carribean	158.5	2.2	29.0	7.0
Sub-Saharan Africa	542.2	3.0	45.0	15.0
South America	311.1	1.9	26.0	7.0
West & Central Asia	218.9	2.6	33.0	7.0
Southern Asia	1274.6	2.1	32.0	11.0
China, Hongkong, Taiwan	1219.3	1.1	18.0	7.0
East/Southern Asia	560.6	1.8	26.0	8.0
Industrialised Development	1164.4	0.3	12.0	10.0
N. America	290.3	0.7	16.0	9.0
Japan, Australia, New Zeland	146.3	0.3	11.0	7.0
W. Europe	383.2	0.2	12.1	12.1
E. Europe	344.8	0.0	12.0	12.0
World Total	5606.0	1.6	25.0	9.0

* Includes Iran and Afganistan.

Source : Population Bulletin, Vol. 49, No.1, 1994.

and slower decline in fertility rates have been responsible for rapid population growth. In the developing countries of Africa and Asia, large population live in several areas which are poverty stricken. The Data relating to population and food growth clearly explains that in spite of the currently existing balance between the above mentioned, the magnitude of food insecurity persists.

Demographic Trends

Population size and rates in different Regions of the world (Table 32.1) clearly suggests that South Asian Region has to go few notches up in controlling its birth rate, which is significantly high. In all countries of South Asia, population has been increasing quite significantly since 1950. With decline in mortality rate due to improved medical facilities and almost unchecked fertility in some parts, it is going to exceed the East Asian Region by 2025. With Indian population already passed one billion mark on May 11th, 2000 and large population of Bangladesh and Pakistan (Table 32.2), this 'region' is going to take longer time to stabilise. With around 75 per cent of the total population residing in India, a serious implementation of population policy becomes imminent.

Table 32.2 : Proportion of Population in countries of South-Asia to Total Population of the Region 1950-2000 AD

Country/Region	*1950*	*1980*	*1990*	*2000*
Bangladesh	9.25	9.87	10.54	11.25
Bhutan	0.15	0.13	0.14	0.14
India	78.37	77.06	75.00	74.35
Maldives	0.02	0.02	0.02	0.02
Nepal	1.78	1.63	1.74	1.87
Pakistan	8.77	9.63	10.27	10.87
Sri Lanka	1.66	1.66	1.69	1.50
	100.0	100.0	100.0	100.0

Source : UN World Demographic Estimates and Projections, 1995-2025, New York, 1988.

Population of South Asia grew continuously because of the faster decline in death rate (Table 32.3). As per the earlier United Nations projections, in case of India it has come down from 25 in 1950-55 to

Table 32.3 : Birth Rate, Death Rate and Annual Rate of Population Growth in South Asia

	Birth Rates Per 1000/Year			*Death Rates Per 1000/Year*			*Population Growth Rates per cent/year*		
Country/Region	*50-55*	*85-90*	*95-100*	*50-55*	*85-90*	*95-100*	*50-55*	*85-90*	95-100
Bangladesh	40.2	41.7	34.8	24.2	15.6	12.3	1.61	2.64	2.27
Bhutan	43.5	37.0	32.1	27.2	16.7	13.9	1.63	2.06	1.84
India	44.1	28.1	23.4	25.0	10.9	8.9	2.02	1.74	1.46
Madlives	50.0	42.9	39.7	28.0	12.0	8.0	1.01	2.35	2.29
Nepal	45.5	40.0	35.4	27.0	14.2	13.8	1.18	2.62	2.34
Pakistan	49.5	40.4	31.9	28.5	13.8	10.6	2.12	2.26	2.15
Sri Lanka	38.5	24.2	18.5	11.5	6.1	5.9	2.58	1.49	1.11

Source : World Demographic Estimates and Projections, 1950-2025 United Nation, New York, 1988.

almost 10 in 1995-2000 (projected). Bangladesh and Pakistan are following India. Whereas Sri Lanka has the least death rate. Average Annual growth rates in the countries (Table 32.4) indicated that it is slowing down. But due to large population in the reproductive age group its momentum will continue. Declining average annual growth rate in a country like Bangladesh is a pointer towards the success of Family Planning Programme (NIPORT, 1997). South Asia as a whole and other less developed regions, fertility began to decline from 1965 and mainly from 1980 onwards. In this region Sri Lanka registered lowest growth rates and is expected to slow down further.

Table 32.4 : Population Growth in South Asia

	Total Millions		*Population Growth*	*Avg. Rate*	*Annual %*	*Aged 15-64 millions*
	1980	*1988*	*1980-90*	*1990-98*	*1980*	*1998*
B.Desh	86.7	125.6	3.7	1.9	44	71
India	687.3	979.7	3.5	2.0	394	596
Nepal	14.5	72.9	4.5	2.8	8	13
Pakistan	82.7	131.6	4.6	2.8	44	71
Sri Lanka	14.7	18.8	2.4	1.4	9	12

Indian population registered unprecedented growth in the last five decades. Many parts of India have shown fertility decline in recent period. Projection made by the Population Foundation of India (PFI) suggests that India will reach the replacement level fertility (1.6 TFR) by the year 2026 (Natrajan and Jayachandran, 2000). Some of the larger states like Uttar Pradesh will not be able to achieve this even by 2051. At that point of time UP's population will cross 40 crores and total fertility rate would be around 2.4. Considering this within the holistic framework of South Asian Region its implications on food security, environment and development are larger. The past trends in fertility and mortality in South Asia in general and India and Bangladesh in particular have shaped the typical population age structure, which has tremendous in-built potential. Even if the total fertility rate comes down below replacement level, yet, absolute number will keep on increasing. Therefore, even after reaching NRR by 2010 both in cases of India and Bangladesh (Table 32.5), it would take even longer time to stabilise.

Table 32.5 : Child malnutrition, maternity rate and adult literacy in South Asia

	Prevalence of Child malnutrition % of children under age 5	*Under 5 maternity rate per 1000*		*Adult literacy rate % of people 15 and above*	
	1992-97	*1980*	*1997*	*1997*	
				Male	*Female*
Bangladesh	56	211	104	50	73
India	53	177	58	33	61
Nepal	47	180	117	44	79
Pakistan	38	161	136	45	75
Sri Lanka	38	49	19	6	12

Table 32.5 (a) : Demographic Characteristics of South Asian countries

	CBR		*CDR*		*Projected year of reaching NRR*	*TFR*	
	1970	*1992*	*1970*	*1992*		*1970*	*1992*
Nepal	46	38	22	13	2030	6.4	5.5
B.Desh	48	31	21	11	2010	7.0	4.0
Bhutan	41	39	22	17	2035	5.9	5.9
India	41	29	18	10	2010	5.8	3.7
Pakistan	48	40	19	10	2030	7.0	5.6
Sri Lanka	29	21	8	6	2000	4.3	2.5

Food Production

The "begging bowl" image of South Asian countries has been erased, as far as 'food grain production' is concerned. With 'evergreen revolution' in past decades and 'gene' revolution round the corner, we are in a position to draw a balance sheet and chalk out a strategy for the future (Swaminathan, 1999). The food grain situation has improved considerably. In most cases growth of cereal production has been steadily increasing (Table 32.6)

Being an essentially agricultural region of the world, the entire economy and life of people having dependence on agriculture. The desired average annual growth rate should be around 4 per cent in South Asia, as set by World Food Conference. In contrast this falls below target (Table 32.7). Here projected rate is also not that encouraging.

Table 32.6 : Cereal Production in South Asian Countries

(In '000 metric tonnes)

	1982	*1986*	*1991*
Bangladesh	22340	24266	29655
Bhutan	166	164	105
India	136101	164955	195109
Nepal	3221	4000	5919
Pakistan	18076	20866	21198
Sri Lanka	2194	2640	2440
Maldives	Not Available		

Source : *United Nations Statistical Year Book* 1992, New York.

Table 32.7 : Average Annual Agricultural Growth Rate in South Asia

	1970-80	*1980-92*	*1992-2000**
Nepal	2.5	2.6	2.4
Bhutan	1.8	2.1	2.4
Bangladesh	2.6	2.3	1.8
India	2.3	2.1	1.7
Pakistan	3.1	3.1	2.7
Sri Lanka	1.6	1.4	1.1

*Maldives not included. ** Projected growth rate.
Source : World Development Report 1994.

This kind of trend does not give a positive signal and here arise the question of per capita availability of food for direct human consumption which can be termed as 'per capita food supplies'. So, how far South Asian region has reached towards per capita food supplies is a pertinent question. Region-wise per capita food supplies for direct human consumption (Table 32.8) prepared by Food and Agriculture Organization projects South Asia as number two from below. This has propelled incidence of chronic under-nutrition. There is tremendous pressure on the available arable land. In Bangladesh, almost all cultivable lands have been brought under-cultivation and there is no scope for further expansion (Islam,1991). Arable land hectares per capita is continuously declining from 0.10 to 0.07 in Bangladesh and 0.24 to 0.17 in India and 0.24 to 0.17 in Pakistan. This also gives an indication of declining farm sizes.

Question of Accessibility to Food

While India, Bangladesh and Pakistan continue to maintain the self-

Table 32.8 : Per Capita Food Supplies for Direct Human Consumption (Historical and Projected)

	1961-63	*1969-71*	*1979-81*	*1990-92*	*2010*
Developing Countries	1960	2130	2320	2520	2730
African Sub-Sahara	2100	2140	2080	2040	2170
N.East/N.Africa	2220	2380	2840	2960	3120
East Asia	1750	2050	2360	2670	3040
South Asia	2030	2060	2080	2300	2450
Latin America/Caribbean	2360	2510	2720	2740	2950
Developed Countries	3020	3180	3270	3330	3470
Former CDES	3130	3290	3350	3160	3380
Others	2980	3130	3230	3410	3510
World	2300	2440	2580	2710	2860

Source : United Nations FAO, NFS 96/Tech. 1, 1995.

sufficiency in food availability marginally, these countries are facing more difficult task of providing economic accessibility. Although data on below poverty line shows decline, yet absolute number is still very high (Table 32.9). Both India and Bangladesh have managed to bring it down to approximately 30 to 35 per cent. Their absolute figures remained almost unchanged. Nepal still has above 40 per cent of its population below poverty line. If we look into the distribution in the income/consumption at the country level, it almost homogenous (Table 32.10). Highest 30 per cent of the total population is consuming more than per cent, in India, Bangladesh and Sri Lanka. Whereas in Pakistan it is more than 70 per cent. This clearly depicts the inequality in the distribution of income. Lowest 30 per cent consumes only round 10 per cent in all countries. This makes a case for the provision of employment-oriented strategy. Gross inequality in consumption pattern has its implications. Country-wise data on percentage of infants with low birth-weights indicate pathetic situation—Bangladesh with 50 per cent; India 33 per cent; Nepal 26 per cent and both Pakistan and Sri Lanka at 25 per cent. These figures are quite higher than South-East Asian countries.

With one third of its population living below poverty line, the inaccessibility to safe drinking water further accentuates the problem of food and health insecurity. Around 85 per cent of Indian population do not have access to safe drinking water (Table 32.11) in 1995 and this figure was 54 per cent in 1982 (World Development Report, 1999-2000).

Table 32.9 : Population below Poverty Line in South Asia

	Year	*Rural*	*Urban*	*Total*	*Year*	*Rural*	*Urban*	*Total*
B.Desh	1991-92	46	23.3	42.7	1995-96	39.8	14.3	35.6
India	1992	43.5	33.7	40.9	1994	36.7	30.5	35.0
Nepal	1995-96	44.0	23.0	42.0	1990-91	–	–	–
Pakistan	1991	36.9	28.0	34.0		–	–	–
Sri Lanka	1985-86	45.5	26.8	40.6	1990-91	38.1	28.4	35.3

Source : World Development Report 1999/2000.

Table 32.10 : Distribution of Income or Consumption in South Asia (Percentage Share of Income or Consumption)

	Lowest 10%	*Lowest 20%*	*Highest 20%*	*Highest 10%*
B. Desh	4.1	9.4	37.9	23.7
India	4.1	9.2	39.3	25.0
Nepal	3.2	7.6	44.8	29.8
Pakistan	4.1	9.4	41.2	27.7
Sri Lanka	3.8	8.9	39.3	25.2

Source : World Development Report 1999/2000.

Table 32.11 : Health Conditions in South Asia

	Access to safe water % of Population		*Infant mortality rate per 100 level births*	
	1982	*1995*	*1980*	*1997*
B.Desh	40	84	132	75
India	54	85	115	71
Nepal	11	59	132	83
Pakistan	38	62	127	95
Sri Lanka	37	70	34	14

Source :World Development Report 1999/2000.

Infant mortality is another indicator reflecting the prevailing health condition and accessibility to food and nutrition. Almost all countries in this region have brought down their IMR, it is still significantly high. This, in turn, becomes one of the factor why rural population go for large family size. So, here it shows cause and effect relationship.

Challenges Ahead

Following issues are going to emerge as serious problems in the coming decades :

1. Meeting increased demand for food with declining safe water and arable land,
2. Removing inequality in income,
3. Increasing urbanization would influence/enhance consumerist culture,
4. Preserving bio-diversity, protecting land from further degradation such as salinity and alkalinity,
5. Providing non-farm employment and
6. Aging population.

Population and Agricultural Policies

To have food and nutritional security in South Asia one must think of reduction of population growth rate as the basic necessity. Considering the number of malnourished and undernourished persons in this region, its essential to think in terms of demographic sustainability *vis-a-vis* food security. Problems of food and nutritional security is multi-dimensional in nature. Its close relationship with population growth, poverty and socio-economic inequality cannot be ruled out. Depending on the situation prevailing in each country with regard to the population—natural resource equation, suitable population policy has to be implemented.

India has the distinction of being the first country in the world to launch a nation-wide family planning programme with full government support (Bhende *et al.*, 1985). Main goals were to control population growth by reducing fertility, decreased morbidity and mortality. The Indian government adopted 'national population policy' in 1976 with unnecessary coercive measures. The earlier programme focused on a total emphasis on sterilization as the main method of family planning, whereas, the latter came out with enactment of some far reaching consequences. All these policies lacked teeth as proper ground was not prepared for creating awareness among rural masses. Without proper rural infrastructure and opportunities for employment these drew flak low level of political commitment of these programmes along with illiteracy and poor health care it will not yield any positive result in some states.

The proposed 'National Population Policy, 2000' has set 14

National socio-demographic goals, which is to be achieved by 2010 (GoI, 2000). This cover wide ranging issues of literacy (focusing women), immunization, institutional deliveries and access to information plus convergence in implementation of related social sector programmes.

Throughout the world there is a broad commitment to reduce hunger and malnutrition. Increased food production is essential to achieve this objective. Reducing hunger involves food distribution, food consumption and human nutrition. To address these issues, in south Asian Region 'agriculture' was brought to the centre of planning. Initially, it aimed machete at increasing gain production.

In India, when the food stocks became in excess of the requirements under the buffer stock policy, a 'food for work' programme in 1978. Later on that was merged with 'Jawahar Rojgar Yojana'. There are several other programmes. With globalization and liberalization, there is a shift in agricultural subsidy policy and this will have its implications.

Need for an Integrated Population Food Development Policies

One must look beyond demography as to search for a truly effective population control strategy. While population policies focused mainly on family planning and sterilization, our agricultural policies focused on to increase food production. Non-food sectors have to take care of. Such as, education safe drinking water and infrastructure development. National policies should include effective participation of rural people in decision-making specially, women. All countries talk about reaching net reproduction rate (NRR) of one and health for all. For this, policy maker should adopt rural— centric and women centric development policies. And population education and awareness progrmmes have to be integral.

Fast spreading science and technology has proved to be significant gene revolution can still increase food production manifold as suggested by Prof. M.S. Swaminathan. Our yield per ha in many corps is one-third of other developed agrarian economies. Preserving our rich bio-diversity can be done through by using space technology to greater effect. The fast breeding information technology has to be taken to the villages. The space technology and 'Geographical

Table 32.12 : Estimates and Projections of the Incidence of Chronic under Nutrition in the Developing Countries

Region	*Year 3-year average*	*Total Population Millions*	*Undernourished % of total population*	*Persons Million*
Africa,	1969-71	268	36	96
Sub-Sharan	1979-81	375	39	140
	1990-92	500	41	204
	2010	874	35	302
Near	1969-71	178	25	44
East/North	1979-81	233	10	24
Africa	1990-92	317	10	32
East Asia	1969-71	1147	41	468
	1979-81	1393	27	371
	1990-92	1674	16	262
	2010	2070	5	105
Latin America	1969-71	279	18	51
and Caribbean	1979-81	354	13	46
	1990-92	443	14	61
	2010	593	8	49
Total	1969-71	2583	35	893
	1979-81	3228	27	878
	1990-92	4064	20	809
	2010	5668	13	730

Source : Word Development Report 1999/200.

Information System' can be very effective for 'future land use planning'; crop forecasting and also to assess natural resource management. India has to take leading role in the South Asian Region. Because changes or alteration in the natural resources and environmental degradation in its neighborhood has its impact on India with damaging consequences.

Conclusion

The rapidly increasing population with large reproductive base along with structural poverty is the root cause of food inaccessibility in the South Asian Region. On the population and food production front countries have improved. But issues like malnutrition among women and children, safe drinking water and unemployment still need serious attention. Due to large population base pressure on land and environmental resources is likely to increase. In a country like India, democratic decentralization which has come through 73rd and 74th

Constitutional Amendments, raising some ray of hope. Lack of employment opportunities in the rural areas may trigger migration towards urban centres, resulting increased urbanization with poor health and sanitary conditions. In Bangladesh, growing population is likely to force them to go out in the neighbouring country which will have its regional and political implications and ramification. Considering the recently introduced structural changes in this 'region' the role of 'state' needs to be evaluated in the changing food security system. The main challenge for the coming years will be to sustain the process of economic growth and reduction of poverty in the sub-continent 'Green Revolution' is a matter of past. 'Sustainable gene revolution' and 'evergreen revolution' is an ecological and economic necessity in India as well as in South Asia. Facing common problems, countries of this 'region' should cooperate at the highest level to eradicate poverty and combat chronic food insecurity. These countries will have to focus on the urgent humanitarian issues than other issues because after all human being is going to be our prime focus in the 21st century.

REFERENCES

Acharya, K.C.S. *(1983) Food Secruity System of India*, Concept Publishing Company, New Delhi, p. 3.

Alam, S. (1992), "Have supply responses increased for major crops in Bangaladesh". *Bangladesh Development Studies*, Vol. 20, No.1, March.

Bhatia, B.M. (1985) *Food Secruity in South Asia* Oxford & IBH Publishing Company, New Delhi-16.

Chaudhary, R.H. (1981), "Population pressure and agricultural productivity", *Bangladesh Development Studies*, Vol. 9, No.1, March.

Choudhary, N. (1988), "Accounting for subsidized food resources distributed in statutory rationing in Bangladesh", *Bangladesh Development Studies*, Vol.16, No. 4, December.

Durham, David, Fandrem, Jain, C. (1988) " The Food Surplus: a strapple illusion of economics, a cruel illusion for population", *Population and Environment*, Vol.10, No.2, New York, pp. 115-25.

Ezekiel, Hennan (1998) "An approach to a food and strategy": *World Development*, Vol.16, No.11, Washington D.C., p.1378.

Goletti, Francesco (1994) "The changing public role in a Rice Economy Approaching self-sufficiency: The case of Bangladesh", *IFPRI Research Report.*

Islam, Nurul (1995), "Population and Food in the early Twenty-first century: Meeting future demand of an increasing population", *IFPRI*, Washington D.C.

Mellor, John W. (1987) "Food and Development the critical Nexus" in *Economic Impact*, No.61.

Natrajan, K.S. and Jayachandran, V. (2000) Population growth in 21st century—India, Paper presented in the Millennium Conference on Population, Development and Environment Nexus. Organised by IASP, PFI and UNEPA, New Delhi, Feb. 14-16.

Norse, David (1994) "Multiple Threats to Regional Food Production: Environment, Economy, Population", *Food Policy*, Vol.19.No.2, p. 146.

Sanderson, Fred H (1984), "World Food Prospect to the year 2000", *Food Policy*, Vol.9, No.6, p. 373.

Sen, A.K. (1987) " Poverty and Entitlement" in J. Price Gottinger, Joanne Leslie and Casoline Hoisenglen (eds) *Food Policy, The World Bank*, The John Hopkins University Press, p. 200.

Sen, Rathindra P. (1989), *Demographic change and levels of living*, Daya Publishing House, Delhi PP.71

Swaminathan, M.S. (1987) "Building a National Food Security System'. *Indian Environmental Society*, New Delhi, p.124.

Swaminathan, M.S. (1999) "Science and Sustainable Food Security, 31st Jawaharlal Nehru Memorial Lecture, New Delhi.

UN, FAO,WFS 96/Tech 1 (1995), *Food Agriculture and Food Security—The Global Dimension*, p.1.

UNEPA (1991), *Population Resources and Environment: The Critical Challenges*, London.

United Nations System in India (1997), *India Towards Population and Development Goals*, Oxford University Press, Delhi.

World Bank, (1995), *World Development Report-1992.*

World Bank, (1999), *World Development Report-1999/2000*, Washington, D.C.

33

Monsoon Drought
A Natural Hazard in India

D.R. Sikka

Introduction

Monsoon drought in India is one of the worst natural hazards, which has struck India throughout its long history. Its devastating role in the socio-economic sector has been stressed in many studies. In the past drought on all India and regional scales had often resulted in famines with consequences to vast human mortality and destruction of cattle wealth. The dilemma of the socio-economic aspects of drought in the summer monsoon season (June to September) is highlighted by the fact that over 75 per cent of India's annual rainfall occurs during the monsoon season and if it fails, the probability of cessation of the drought till the next monsoon season is very slim. Thus drought in the monsoon season in India leads to parched fields, loss in agricultural productivity, unemployment in the poor rural households, stress on depleted water resources, loss in hydro-electric generation, fall in industrial production and above all a sense of desperation not only over rural India but also among the urban population. Monsoon drought is locked to the annual climate cycle over India and its occurrence is a manifestation of the year to year climatic fluctuations in a tropical/sub-tropical climatic regimes within which India is situated in its geographical setting. Agricultural cycle in India is controlled by the two-monsoon systems — the south-west or summer monsoon (June to September) and the north-east or the winter monsoon (October to December) with in between winter season (January-

February) and the pre-monsoon hot season (March to May). During winter season some rainfall/snowfall occurs in north-west India/Himalayan region. In the pre-monsoon season, the rainfall is scanty over most of India and is associated with thunderstorms. Summer monsoon is the presiding deity of India, which provides her water for its vast agricultural resources, recharges its ground water and rivulets and multipurpose water storage dams for irrigation and hydroelectric projects. The bounty of the monsoon rains, as witnessed by excess rainfall, herald's good prospects for the country and its occasional failure spells doom for many sectors of the economy. The more or less self-regulating monsoon rains in the annual climatic cycle make agriculture susceptible to small changes in the annual cycle with respect to its timing (for the sowing season to begin), its deficient behaviour in the rainy season with a prolonged temporary cessation of rains within the season, brings in vision of monsoon drought. A drought monsoon year leads to low crop yields with antecedent adverse consequences. Monsoon is somewhat fickle and is not always the same in showing its bounty. All crops might be lost if the rains fail at the crucial time. Forecasting of total quantum of monsoon rains for the season as a whole and its sub-seasonal variability on 3-5 day and longer time scales have remained some of the most challenging tasks and issues of considerable urgency before the atmospheric science community in India since the establishment of the Indian Meteorological Department (IMD) more than 125 years ago in 1875.

In this article we focus on the quantitative definitions of the monsoon drought, as used in India in Section-2. List of the infamous droughts of the past and simultaneity of occurrence of monsoon droughts over different regions of India are given in Section-3. Studies on Meteorological factors on the large scale are associated with the incidence of monsoon drought. Long range monsoon forecasting of drought are discussed in Section-4. Salient features of socio-economic consequences of the drought particularly in relation to agricultural productivity are covered in Section-5. Policy initiatives taken by the governments of the day at different stages of the long history of India with respect to drought relief and the recent shift in emphasis of the democratic India from drought—relief to drought proofing and drought management, with inputs from science and technology are presented in Section-6. Concluding remarks are offered in Section-7.

Definition of Monsoon Drought

Drought, in common parlance, denotes scarcity of water for meeting needs of agriculture and other human needs. In general term it implies a deficiency in precipitation of sufficient magnitude to interface with the needs of the socio-economic sector and setting up of a creeping long-term set back which has devastating effects on the very fibre of an otherwise vibrating society. Some authors rank drought above all other natural hazards in terms of proportion of people affected in recent decades. Whereas aridity or semi-aridity is a permanent incidence of low average annual rainfall, due to large scale climatic factors, drought is a temporarily sustained occurrence of deficient rainfall or less availability of water in natural systems (arid, semi-arid, moist, sub-humid and humid tropical and mid-latitude etc.) due to sub-seasonal or inter-annual fluctuations of the climatic regime. Drought, therefore, involves an abnormal failure of rains within a normally expected rainy season affecting a certain region in the global climatic regime. Drought, therefore is caused by abnormal climatic fluctuations, which under special ocean-atmospheric conditions may persist for a couple of years and in extreme conditions, like in the sub-Saharan Sahel area of Africa, may linger on up to two decades. Till about 50 years ago there were no scientific definition of the word "drought". Over the last 3-5 decades several definitions of drought, for example, hydrological, meteorological, agricultural are prevalent in the scientific literature (Singh, 1978, 1995; Kulshrestha, 1997, Sikka 1999 and others). Ramdas (1950) first adopted a definition of meteorological drought in the Indian context, which was 25 per cent deficient annual rainfall on regional basis in India. Subrahmanyam (1967) used the same definition for the study of incidence and spread of drought over India. IMD (1971) and Koteswaram (1976) used the definition of drought from meteorological angle depending upon spatial and temporal scales. Different percentages for defining droughts are used as in following table.

If the deficiency in rains on weekly/monthly/sub-division scales is between 26 to 50 per cent of the normal amount, the drought is moderate and if it is more than 50 per cent of the normal the drought is termed as severe.

Scale of Drought		*% departure of rainfall from long-term mean*
Spatial	*Temporal*	
1. Meteorological sub-division	Weekly and monthly	Less than – 25%
2. Meteorological sub-division	Seasonal	Less than – 25%
3. Group of homogeneous neighboring meteorological sub-divisions which have been traditionally drought prone	Weekly and Monthly	Less than – 25%
4. Group of homogeneous neighbouring meteorological sub-divisions which have been traditionally drought prone, such as north-west India and Peninsular India	Seasonal	Less than – 10%
5. India as a whole	Seasonal	Less than – 10%

For defining agricultural drought different indicators are prevalent

(i) Thornthwaite's (1948) aridity index (AI) is defined as:

$$\text{AI} = \frac{\text{Mean precipitation (P)}}{\text{Mean Potential Evapotranspiration (PE)}}$$

If the ratio is less than one it would indicate moisture deficiency in soil for vegetation P, PE are computed on monthly, seasonal and annual basis. If AI is less than 0.4, the situation is termed drought for agriculture.

(ii) Aridity Index (Ia) in percentage is given as :

$$\text{la} = \frac{(\text{AE-PE}) \times 100}{\text{PE}}$$

where AE is the mean evapotranspiration and PE is the mean potential evapotranspiration. If the aridity index (la) is negative, crops suffer from water deficiency.

(iii) Another Agricultural Drought Index (ADI) is expressed as:

$$\text{ADI} = \frac{\text{WD} \times 100}{\text{PE}}$$

where WD is the water deficiency and PE is the water need of crops calculated on an annual or seasonal basis. Drought intensity is defined in terms of the deviation of a particular aridity index from its standard deviation. For Agro-meteorological work, IMD monitors the drought index Ia, operationally on a weekly basis for over 250 stations in terms of (ii) above. AE is estimated by using a water balance technique and drought is considered mild, moderate and severe if the Index Ia is less than 25 per cent, between 26 to 50 per cent and above 50 per cent respectively. Different research workers in India have defined different indices to monitor the area under drought (Mooley and Parthasarathy, 1982, Parthasarthy *et al.*, 1992, Bhalme and Mooley 1981, Chaudhury *et al.*, 1989, and others). All authors have used the long-term monsoon rainfall series for calculating the area under drought or for the purpose of categorizing a monsoon season as a drought monsoon season. Parthasarathy *et al.* (1995) have prepared a long-term monsoon rainfall series for the years 1871-1994 and the mean monsoon rainfall of this series for All India, 5 Homogeneous Rainfall Regions of India and 29 non-hilly meteorological sub-division of India is based on the data from 1871 to 1990. Sikka (1999) has used this series for categorizing drought on All India and regional basis on the criteria that the drought is mild, moderate, severe or phenomenal if the rainfall of a monsoon season over the area under consideration is below 1.0 standard deviation, 1.25 standard deviation, 1.5 standard deviation and 2.0 standard deviation respectively.

Past Infamous Droughts in India

Ancient Period

References to drought and the associated food scarcity and famine conditions are available in the ancient literatures. The "*Vedic rishis*" sang hymns in praise of the rain God Indra who was worshipped as a 'beaker' of the persistent drought. *Sam Veda* even provided the methodology of performing *Yagna* to bring the bountiful rains for saving crops against being parched and destroyed by prolonged dry spell. The *Shashtra* literature also mentions the fear of the drought. There is mention in the writings of the Greek scholars, who accompanied Alexander in his invasion of India during 4th century

B.C., that the Indian philosophers indulge in philosophical discussions about the causes of drought. The ancient Indians thought that the drought is caused by the malevolent combination of the planets and heavenly bodies. They would forslow the lack of rains or otherwise in the ensuing rainy season at the beginning of the Hindu New Year by casting the position of different astronomical bodies. There was no other scientific method for any kind of weather or climate forecasting. *Arthashastra*, written by Kautilya (3rd century B.C.) mentions about the measurement of rains over the Mauryan Empire but we have no scientific authenticity as to the method followed for such measurements. *Arthashastra* also has reference to the large scale distribution of seasonal rains over different parts of India and divided the country into different rainfall regions geographically which broadly agree with modern meteorological records. However, after the collapse of the Mauryan Empire, the recording of rainfall in the country went into oblivion for nearly 2000 years and the measurements were re-started by the officers of the East India Company in different places in the beginning of the 19th century. The rain recording was established on firm scientific footings with quality controlled procedures after the establishment of the Indian Meteorological Department (IMD) in 1875. *Arthashastra* also mentions amongst the duties of the king to provide succour to the subjects and the farmers when they were suffering from drought. Irrigation, as a protection to save crops under dry spell, was well organized under the Mauryan Empire and there were officials in the empire who would provide equitable distribution of stored waters by operating sluices on the reservoirs. During Chandra Gupta Mauraya's rule there is a record of incidence of a severe country-wide drought in 298 B.C. *Rajtarangini*, an account of the ancient history of Kashmir, contains reference to a severe drought in Kashmir in 917-18 A.D. and the resulting famine "one could scarcely see the water of river *Vitasta* (Jhelam) as the river was covered with corpses soaked and swollen by the water in which they had long been lying.

Medieval Period

There are occasional references to the incidence of drought during the medieval *Sultanate* Period in the reign of Sultan Jalaluddin Khilji and Sultan Mohammad Tughlaq. Feroz Shah Tughlaq is credited with the reconstruction of the western Yamuna canal for bringing water for irrigation in parts of the present Haryana and for providing water to his palaces and gardens in Delhi. Eastern *Hindustan* was struck by persistent drought conditions for two successive years in 1554-55 to 1555-56. We have also an account of a severe famine caused by drought towards the end of the rule of the Mughal emperor Akbar, then royal food godowns were opened for the starving people. Heart rendering account of a severe drought/famine in 1630-31 A.D. in which large scale starvation and depopulation occurred in Gujarat and adjoining Madhya Pradesh and Maharashtra, are given by the European travellers during Shahjahan's period. Another terrible drought occurred during the reign of Aurangzeb. Historians report several sub-regional scale droughts and resultant famines or scarcity conditions during the Mughal period (Irfan Habib 2000, Indian Famine Commission Report 1880-1885).

Period since the 18th Century

The British colonial period began with the taking over of the *Diwani* of Bengal by the East India Company in 1765 and within 5 years of it, a devastating famine occurred over Bengal in 1770-71 in which nearly one-third (approximately 10 million) of the total population of the two provinces was lost due to starvation and drought related diseases. This shook the conscience of the East India Company. Rural India was bled to death by the systematic exploitation due to policies of officers of the East India Company and destruction of cottage industries which was the backbone of rural population in those times. India came under the British Crown in 1858. There were other regional and All India droughts in the first half of the 19th century. A drought/famine occurred in 1865-66 in Orissa and Bengal, which again killed several million people. The devastation caused by this drought waken up the British rulers and Orissa Famine Enquiry Commission was appointed in 1867 to report on the causes of the famine and related famine relief. As a result of this report as well as recommendations of the Sanitary Commission and to protection of British shipping from the destruction caused by the cyclonic storms (another natural hazard in Indian seas) the government of India established the IMD in 1875 with Dr. Henry Blanford as its first head with the designation of Meteorological

Reporter to the government of India. Prior to IMD provincial meteorological services were functioning for sometime in some provinces like Bengal, Bombay, United Provinces and the Punjab. Blanford had served for several years as the Meteorological Reporter to the Government of Bengal Presidency. Blanford worked very hard during 1875-1886 towards not only nurturing the newly formed IMD but also establishing the great scientific traditions of Meteorology of India under the IMD. He believed strongly in the paramount knowledge of the distribution of the rainfall in time and space in a country such as India, the wealth and welfare of which depend above all things on its agriculture. The creation of IMD spurred him to write the first exhaustive description of the Rainfall of India (Blanford, 1886). In this Memoir he emphasized several aspects of the Monsoon Meteorology of India such as the role of the Monsoon Trough along the Gangetic valley in controlling the rainfall of India and linked its shift to the foothills of Himalayas with the sub-seasonal weakening or break in monsoon rains over most of India. He mentioned that the distribution of seasonal monsoon rains over India in extreme years (drought or excess monsoon years) is such that 2 / 3 of the country is under drought (deficient) or *vice versa*. This observation has more or less stood the test of time (Sikka, 1999). During his tenure two major droughts occurred that of 1876 on the Peninsular India and of 1877 on All India scale. In his enquiry in the causes of the drought of 1877, he found above normal sea level pressure over Australia persisted for several months, which began prior to the monsoon season and lasted for an extended period till December 1877. Herein lay the seeds of the observations on Southern Oscillation (SO) which were exploited by Sir Gillert Walker after he took over as the Director-General of observatories of IMD in 1904, Walker had coined the term 'Southern Oscillation' to describe the out of phase pressure anomalies along the South Pacific Ocean and Indian Ocean. Southern Oscillation has been Walker's legacy to global climate research. It was later linked to the *El Nino* warming of the equatorial central and eastern Pacific by Bjerkenes (1969). The combined phenomenon of *El Nino*— Southern Oscillation (ENSO) is now regarded as the most important signal in global climate anomaly. This study remained a focus of a decade-long 1985-1995 special international programme known as Tropical Ocean Global Atmosphere Programme (TOGA). Sikka (1980) was among the first to link occurrence of warm phase of ENSO with the incidence of most of the monsoon drought years over India. The link between the Monsoon droughts and ENSO has been the subject of intense observational and modelling studies in the last two decades. It is now recognized that the links are very complex and the two systems (ENSO and Monsoon) are selectively interacting Webster and Yang (1992). Studies by Kriplani and Kulkarni (1996) and Krishnakumar *et al.* (1999) show that the ENSO-Monsoon connection has weakened since 1990s. However, this has happened too in an earlier period of 1930-1950 and hence weakening or strengthenig of ENSO- Monsoon connection is part of a complex climate system. Blanford (1884) has also linked the incidence of monsoon droughts with the preceding seasons (winter to spring) excessive snowfall over the Himalayas. The relationship between excessive snowfall over the Himalayas/Eurasia and the ensuing monsoon drought over India has been also a subject of intense observational and modelling studies since 1970s. Planetary and large-scale causes of the monsoon droughts were further studied in the early years of Monsoon Meteorology in India by Blanford's successor John Eliot and had reached commanding heights during the tenure of Walker from 1904 to 1924.

Information are available on the frequency of drought occurrence on regional and All India droughts in medieval period, colonial India under the rule of the East India Company (1765 – 1858) followed by the British Crown rule up to 1947 and democratic India since then. The records for the early years are mostly based on reports of food scarcity / famine conditions, and since 1875 based on quality controlled rain gauge measurements under the IMD with the definition of drought followed as per Sikka (1999). Decadal variability of monsoon droughts has been observed. It is also observed that about 9 to 10 droughts have occurred in each 50 years period since the beginning of the 19th century. The severity and areal extent of droughts on All India basis from 1871 – 2001 indicate that there is considerable decadal variability in the incidence of droughts and since 1871 the minimum number of drought occurred in the decades 1881-1890 and 1991-2000 and the maximum of 4 in the decade of 1981-1990. Sikka (1999) has also analyzed the droughts on sub-divisions of India and five Homogeneous Regions of India. According to his analysis the simultaneity of sub-divisional droughts with the All India Droughts varies from 40 to 70 per cent over different sub-divisions. With regard to the large scale eastern Indian sub-divisions are mostly out of phase with the All India droughts. According to information on the simultaneous occurrence of drought among pairs of 5 Homogeneous regions of India, it is observed that generally 3 Homogeneous regions are simultaneously affected by drought with the All India drought and there is only one monsoon drought of the year 1972 when all the 5 Homogeneous regions were simultaneously affected by drought conditions. It is observed that 30 to 70 per cent of the areal extent of India is affected by different monsoon droughts and the four worst droughts since 1875 have occurred in 1877, 1899, 1918 and 1972 in which the deficiency of monsoon rainfall on All India Scale exceeded twice the standard deviation and the rainfall below 25 per cent of its normal value. Fig 33.1 shows the pattern of seasonal rainfall in four years on sub-divisional basis. There have been two sequences since 1871 when two successive years witnessed monsoon droughts in 1904 and 1905, and 1905 and 1961 and one sequence of these successive years of monsoon droughts in 1985, 1986 and 1987. Hence, climatological records indicate that multi-year occurrence of monsoon failure on All India scale is rare. This is unlike the droughts in Sahel region of Africa where droughts could persist for one to two decades. Joseph (1976), Sikka (1980) and others have emphasized the multi-decadal epochal variability in the incidence of monsoon droughts and the period 1891 to 1930 witnessed more frequent droughts and the period 1931 to 1960 was relatively free from droughts followed by another period of 1961 to 1990 with enhanced incidence of All India monsoon droughts. Currently since 1991, we are under favourable conditions as no drought year on All India scale has occurred in this period, though there have been 4 years since 1988 (1991,1992, 2000 and 2001) when the deficiency of rainfall has reached between 8 to 9 per cent on All India basis. The year 2000 witnessed a drought over large parts of contiguous central Indian belt stretching from Orissa to Rajasthan. The drought of 2000 over the central Indian belt became a big issue, which was discussed in the Indian Parliament, Supreme Court of India and the Free Press in India as some starvation deaths were reported in Orissa (Sikka and Kulshrestha, 2002). Large scale climatology of droughts in India has been studied by several workers in India.

Meteorological Factors Associated with the Incidence of Monsoon Droughts over India : Features and Assessment of the Problem of Prediction of Droughts on All India Scale

The search for determining factors, which are associated with incidence of monsoon droughts over India began very early in India. *Varaha Mihira*, a 6th century A.D. natural philosopher echoes this enquiry in his famous work *Brihat Samhita* as a "Food is the elixir of life that food depends on rain. Therefore, it is worthwhile to investigate the rain with considerable effort." Till the beginning of scientific meteorology in India, the early efforts were merely based on speculative ideas. Meteorological measurement started on regular basis in the 19th century A.D. and data began to be analyzed. The attempts made between 1875 and 1930 were primarily based on seeking associations with surface pressure, temperature, snow cover etc. (features of lower atmosphere and land sub-systems). With the introduction of upper air measurements and availability of satellite data more factors have been included in studies, particularly since 1960. Sikka (1999) has reviewed in detail the work done by many investigators in linking factors of the coupled atmosphere-land-ocean system on the planetary and regional scales with the performance of the monsoon.

The factors discussed in this study, which are associated with drought condition over India are the following :

Planetary Scale Factors

- Negative phase of Southern Oscillation and Negative Southern Oscillation Index (SOI).
- Warm phase of EI Nino and associated SST anomaly in the equatorial central Pacific ocean.
- Excessive Eurasian snow cover in preceding winter and spring season.
- Zonal wind anomalies in the lower and middle stratospheric levels.

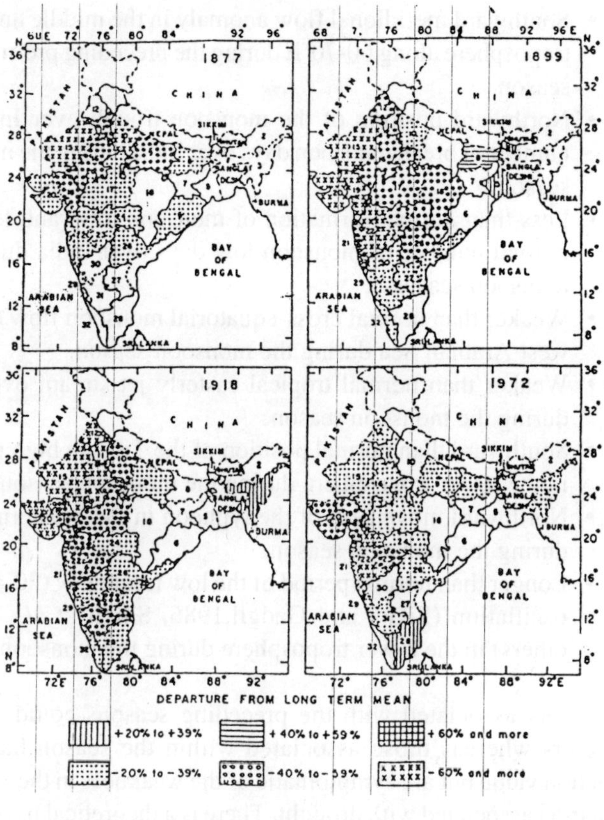

Fig. 33.1 : Pattern of Seasonal Rainfall in four years on Sub-divisional Basis

Regional Scale Factors

- Positive anomaly in surface pressure over Bombay in preceding winter and pre-monsoon seasons.
- Negative anomaly 500 hpa level sub-tropical ridge position over India (75°E) in April.
- Below normal surface temperature over central India and east coast of India preceding the pre-monsoon seasons in India.
- Negative SST normally in the preceding pre-monsoon season over the Arabian Sea.
- Excessive presence of Himalayan snow cover in the preceding winter and pre-monsoon seasons.

- Southward meridional flow anomaly in the middle and upper troposphere along 60-70°E during the preceding pre-monsoon season.
- Northward position of the monsoon trough over India and number of break monsoon days over India during the monsoon season.
- Less than normal formation of monsoon lows and less than normal number of monsoon low days over India during the monsoon season.
- Weaker than normal cross-equatorial monsoon flow over the west Arabian Sea during the monsoon season.
- Weaker than normal tropical easterly jet stream over India during the monsoon season.
- Southward than normal position of the Tibetan high pressure in the upper troposphere during the monsoon season.
- Northward movement of the typhoon in the Western Pacific during the monsoon season.
- Longer than normal period of the low frequency (30-40 days) oscillation (Sikka and Gadgil,1986, Singh *et al.*, 1992 & others) in the lower troposphere during the monsoon season.

Factors associated with the preceding seasons could serve as predictors whereas those associated within the season have little prediction value but are symptomatic of the weakness in the monsoon circulation associated with drought. There is a theoretical background (Charney & Shukla, 1981) that abnormal boundary conditions of the land-atmosphere-ocean system on the planetary and regional scales in the preceding seasons (Snow, SST etc.) dictate the performance of the ensuing monsoon. Sikka (2000) has found that there is component which is dictated by the evolution of the regional monsoon system such as the intra-seasonal oscillation in the regional monsoon system. Ajay Mohan and Goswami (2000) and others have stressed upon the relative impact of these two major aspects, which interact in a complex way to determine the total performance of the Indian monsoon. Webster *et al.* (1998) have also emphasized on complex interacting modes in the meridional direction (Hadley Cell) and in the east-west direction (planetary Walker Cell and the regional monsoon cell), which have profound influence on the performance of monsoon over India. What could be said at this stage is that the predictability of the monsoon is a 'tangled problem'. Even though preceding season boundary conditions may be used as predictors by global circulation models along with initial conditions, the season low-frequency oscillation of the regional monsoon can be predicted with the models with only limited skill. Palmer *et al.* (1992) have suggested an ensemble approach for using global circulation models with slightly different initial conditions to judiciously judge the performance of the monsoon from the scatter of predictions amongst the individual members of the assemble. The skill in the operational use of the global circulation modelling approach in predicting monsoon leaves much to be desired at the present stage of development but there is lingering hope that this approach in the near future would at least approach the empirical (statistical) approach presently in use by the IMD (Gowariker *et al.*, 1990). IMD's operational approach uses 16 antecedent seasons parameters—some global whereas others of regional character. Since 1988 the model is claimed to be successful in predicting the monsoon performance qualitatively in a satisfactory way (Rajeevan, 2001). However, judged in a quantitative terms there is considerable gap for the forecast to hit the targeted model error estimate (±4%). The prediction of seasonal monsoon performance on sub-divisional/ Homogeneous region scale is much harder but the efforts are underway amongst the Indian atmospheric research community.

Socio-Economic Consequences of Monsoon Drought

The social impacts of monsoon drought in India have been multi-dimensional since ages and can be said to have depended upon the level of development over the historical period. The capability of the society to face the drought hazard, pull through its adverse impacts and recover from its shocks economically and socially depends on the integration of the national economy and status of its infrastructural and resource development. Indian social and economic systems have gone through enormous change over long period of India's history from the *Vedic* period of pastoral settlements, passing through conflict ridden and political unrest of the middle ages and exploitation by the British colonial rule of nearly 2 centuries in the second half of the 18th century till the first half of the 20th century and finally emerging in the democratic India from 1947 onward.

The ancient people believed that drought is a natural calamity enforced by God for their ill deeds. Almost similar ideas remained fixed in national ethos through the middle ages till the beginning of the 20th century. Drought is now considered a natural hazard resulting from regional and global climatic fluctuations and hence is a disaster against which defenses can be built to minimize the societal damage. Traditionally the society suffered due to loss of agricultural output loss of employment in agriculture and related activities by rural and urban poor, weakening of resources by disposal of different assets, malnutrition disease and resultant high human mortality, loss of cattle wealth and a variety of other factors. To safeguard themselves against the calamity, the people would sell their land, cattle and other family assets and even undergo debt to provide food for the family at high cost. The shock was so great that people would migrate to safer places. As a result, villages were depopulated. Male members would move to urban centres for search of employment and would not return to their lands when rains would be better. There were limited institutional arrangements for the poor to tide over the crisis except benevolent charity of the rich or the owners of large land areas. There was hardly any science and technology input to combat the drought. Even in modern India when droughts struck between 1950 to 1970 at a more frequent interval the Indian economy had not yet diversified enough and the agricultural sector functioned on traditional lines, horrible socio-economic impacts resulted in 1951-52, 1965-69 and 1972-74 periods. India was said to live from ship to mouth as the government resorted to large food imports during 1951 to 1972. A study by Prasad and Singh (1994), has analyzed development and demographic features of the drought-prone states (Jammu & Kashmir, Haryana and West Bengal) based on the census data from 1961 to 1981. The study has found vast difference in the drought impacts in drought prone and non-drought-prone districts. Things would not have changed much better in the subsequent years as drought prone district have less infra-structural developments and the employment opportunities are also less. Drought in these districts leads to famine of work and hence low wages.

Several researchers (Mooley and Parthasarathy, 1980; Mooley and Pant, 1981; Singh, 1995; Kulshrestha, 1997 and others) have examined the socio-economic impacts of droughts in India in more details. These studies have highlighted the environmental and societal impacts. Environmental degradation is caused by droughts due to moisture stress on crops, natural vegetation and ecosystems. Societal harm is caused by drought in loss of agricultural productivity, reduction in farm income, fall in employment on agriculture-related activities and poor purchasing power, fall in wages, malnutrition, human mortality, increased morbidity, social unrest and conflicts, malpractices in food grain trade etc. There is also a gender bias as girl child, young and older woman suffer more under drought conditions. Even education of children suffers as a consequence of migration of families. The cost of socio-economic impacts of drought on the national, state and local levels is substantial. Amartya Sen (1981), after an exhaustive study of poverty and famines, has pronounced the thesis about the entitlement and deprivation of human society suffering from famines, which are mostly caused by droughts. Bhatia's (1967) exhaustive work on famines in India contains material in the socio-economic disruption caused by past droughts during the British colonial rule. The three reports of the Indian Famine Commission (1880-1885, 1898 and 1901) provide horrible accounts of the people who suffered from the droughts of 1877, 1896 and 1899. The suffering of the droughts of 1899 were so much that the drought has gone into the folk literature of north-western India as the unforgettable drought of '*Chappaniya*' (Year 1956 of the *Vikram Samvat* era). However, vulnerability of the socio-economic system is never static as it depends on the stage of infra-structural development, health of the economy and diversification of the economic system. Presently, India is not so much susceptible as in the years prior to 1970s. India is now an industrial and diversified economy and agriculture provides about 26 per cent only of GDP whereas in the past the contribution of agriculture to GDP was nearly 70 per cent and proverbial Indian budget used to be known as 'Gamble on the Monsoon'. One could now even say that Indian socio-economic system is at present much less vulnerable to the vagaries of monsoon than even 2 decades earlier.

If the data from 1950 to 2000 is studied it has been found that the adverse impact of drought on food grain production during this period is almost clear. The average year-to-year absolute variability in food grain production split in three phases (i) 1950-51 to 1967-68, (ii) 1968-69 to 1980-81, and (iii) 1981 to 2000-2001 prior to the green revolution, the beginning of the green revolution and the mature stage of green revolution respectively in India clearly shows a fall in production in a drought year and a rise in the year following the drought year. Relationship between monsoon and food grain production has been examined by Gadgil *et al.* (1999) and others. Year-to-year absolute variability in the food grain production is still very large, particularly in the Green Revolution period, which raises question about the unstable production system necessitating costly input like water, fertilizers, pesticides etc. Sikka and Kulshrestha (2002) have examined the question of persistent vulnerability of Indian food grain production to climate variability under the Green Revolution technologies, which have resulted in environmental degradation too. Opinions in some quarters that Indian agriculture is out of its dependence on the bounty of the monsoon, however doesn't appear to be correct as vulnerability of agriculture output is still about 5 to 7 per cent due to poor performance of the monsoon on All India basis and perhaps 20-30 per cent on sub-regional basis.

Progress in Policy Initiatives to Combat Droughts and Drought Proofing India

Famine/Drought Relief

In the ancient and medieval periods there was hardly any policy to combat drought except for the benevolence of the ruler and public charity. Of course, wherever possible irrigation was used to save crops. The means of irrigation were limited to channels dug in the river valleys, Persian wells in north-west India, traditional water harvesting in Rajasthan, village ponds in many regions and construction of large and small tanks over the peninsular India. Government policies to combat famines or food scarcity, mostly caused by monsoon droughts, began to evolve during the late 19th century which was strengthened during the British rule under the Crown. Prior to the Commission (1880-1885) there were two enquiry commissions which reported on the famine in United Provinces (now Uttar Pradesh) in 1860-61 and the Orissa Famine of 1865-67. Both the enquiry commissions recommended construction of canal for irrigations, strengthening of rainfall and agricultural statistics and improvement in communication and land revenue systems. The recommendations did not receive much attention. However, the devasting drought of 1876 (Peninsular India) and 1877 (All India) led to the appointment of the First Famine Commission in 1878 under General R. Sratchey and an exhaustive report of the Commission was prepared during 1880-1885. The Commission enquired into many matters connected with the Indian famines with a view as to "How far it is possible for government by its action to diminish the severity of famines, or to place the people in a better condition to endure them". "The Commission also recognized the paramount duty of government to offer protection to the people of India from the effects of uncertainty of the season. In the report the Commission identified drought-prone and relatively drought immune areas of India. The following are some of the major recommendations of the Commission :

- Maintenance of systematic record of rainfall of the country.
- Establishment of a department for agricultural statistics.
- Gratuitous relief in the form of food and money to be provided to rural poor through village communities and monitoring of famine relief.
- Suspension or remitting of taxes and offering loans to peasants with small holdings.
- Establishment of rural credit facilities so that traditional money lenders do not exploit the poor.
- Employment of able bodied persons on public works projects such as buildings of roads, tanks, canal irrigation facilities on large scale etc.
- Extension of railway network so that food could reach the drought affected regions from surplus regions.

The First Famine Commission prepared board framework for mitigating the immediate sufferings of the drought stricken people, establishment of an organizational framework to monitor the early warning signals of the creeping famine/drought and a long-term policy for extension of irrigation and railways, roads and other communication networks. Two other Famine Commissions followed. The Second one reported in 1898 and Third reported in 1901. These Commissions reviewed the working of the relief works during the famine/drought of 1896 and 1899, and made further recommendations particularly in respect of improvement in measures of irrigation, communications, public health and quantitative assessment of the crop loss due to drought (the *Annawari* system) for assessment of the magnitude of damage to crops. More or less same policy for drought/famine relief was followed till India's independence in 1947. However, for agricultural improvement the Imperial Council for Agricultural Research (ICAR) was established by 1930. The word Imperial was replaced by Indian on India's freedom.

The democratic Government of India was more receptive to the needs of the people and followed a policy of developmental planning on 5 years basis. The first Five Year Plan was launched in 1951. Developmental planning has followed comprehensive approaches for industrial growth, agricultural growth, extension of modern communication networks, improvement in higher scientific education and trainings, creating irrigation and power resources, technical training and expansion of science and technological facilities in the form of establishing new government departments and scientific laboratories/institutions and advancement of the existing scientific/technological infrastructures. The improvements brought about in science and technological infrastructures have played a great role in the management of the drought since 1970.

The Green Revolution Technology Introduction

A major government policy initiative was taken in 1967 to introduce high yielding varieties (HYV) in some food crops and establishment of facilities to provide quality seeds to the farmers of HYV. The ICAR research laboratories and state agricultural universities with their extension services have played a big role in the spread of Green Revolution across the country. The Government policy framework to support the Green Revolution also covered creation of large scale irrigation and power resource, availability of organic and inorganic fertilizers and pesticides etc. These inputs are essential for the success of the HYV. These policy imitative and work of the ICAR and allied institutions has made India self-sufficient for food requirement. This has been a major success and has helped India to come out of the nexus between drought and famine. Droughts do occur even now, as they would continue to occur in future as a result of climatic fluctuation, but food availability is assured such that starvation would not take place. Other measures like facilitating availability of food procurement through a centralized agency, stocking of buffer food stocks and release of food grains through the public distribution system on subsidized prices for the people below the poverty line have also helped in alleviating the sufferings of the drought affected people. Another approach initiated after the drought of 1972 and 1974 was the appointment of the Second Irrigation Commission (1972) and the National Commission on Agriculture (1976). Their reports were submitted to the Government in 1972 and 1976 respectively and several important policy decisions were taken which have helped in combating droughts. National Water Policy was adopted by the government of India (GOI 1987) in 1987 and Department of Agricultural Statistics has been strengthened.

A programme was launched to identify drought prone area (DPA) for taking special measures to fight the adverse impacts of droughts (Singh 1978). Some of the characteristics of DPA are :

- Continuous degradation of land and vanishing green cover.
- Depleting water resources and drinking water problem.
- Decreasing productivity of crops, livestock and human resources.
- Malnutrition and distress sale of family property and cattle wealth under drought condition.
- Migration of both human and cattle population in times of drought.

The Ministry of Human Resource Development has undertaken a sound programme to relieve the rural poor by providing them opportunities through diversification of developmental activities, encouraging the rural committees for water harvesting works and some other measures to strengthen their economic conditions. These efforts provide local resources to combat drought and have consequently led towards drought proofing of the DPA.

Drought Mitigation Measures

Short-term drought mitigation measures at present cover traditional relief works as modified over the passage of time, creation of employment opportunities such as employment guarantee scheme on the Maharashtra pattern, Rural Employment Guarantee Scheme after the drought of 1972-74 etc. Short-term relief is administrated through the Central Ministry of Agriculture. It is recognized that drought leads to loss of work employment in agricultural related activities. Therefore, rural work programme are at the core to enhancement of income for victims of this natural hazard. The Food for Work Programme now forms an important strategy to combat drought. Long-term measures, which support the short-term ones are focused on betterment of the resources of the farmers as well as farming communities, improvement in farm production system by adopting improved technologies, creation of alternative livelihood systems for the asset-less poor population, provision of social inputs like health care and educational facilities, and launching of Intensive Agricultural Development Programmes and Crop Insurance Scheme etc. In spite of all these efforts the dry land areas and the DPA suffer from lack of technical backup and management systems. Heavy subsidies and modern inputs flow into resource-rich areas at the cost of dry land areas. Poverty alleviation programmes have also helped preferentially the resource-rich areas. Recently the National Watershed Development Programme for rainfed agriculture and Integrated Waste Lands Development Programme have been launched for more focused attention to rainfed agriculture and development of waste land respectively. There is an endeavour to make these programmes "peoples programmes in which the government participates" and make watershed development a people's movement and demand driven. The success of these is seen in several areas and the year 2001 Megsasay award was given to a NGO leadership in Alwar (Rajasthan).

In addition to the Watershed Development Programmes in the Government sector, a number of externally aided watershed projects under the FAO, World Bank, European Economic Community, Danish International Development Agency, Swiss Development Cooperation, UNDP and some other agencies, are working in different parts of India. Several of the watershed development projects have received commendation by special review committees. The component of technology in these watershed projects are location specific. The missing links have also been identified and suitable policy initiatives would continue to flow to bridge the gaps in these programmes.

Drought Monitoring as Coordinate Approach and Crop Monitoring by the IMD, Department of Space and the Ministry of Agriculture

A new strategy has been developed in 1980s for combating the abnormal development of monsoon conditions by real-time monitoring by assessing conventional rainfall data and drought development by space monitoring using Land Satellite Systems and providing advice to the agricultural developments at central and state levels to adopt to creeping abnormalities in the evolution of monsoon. A Crop-Weather Group, consisting of representatives of IMD, Ministry of Water Resources, Ministry of Power and Ministry of Agriculture meet every week during the monsoon season to discuss the evolving situation. Agricultural scientists bring into motion their prepared strategies about modulating sowing operations, use of drought resistant varieties, adjusting other inputs like allocation of irrigation, fertilizers and pesticides etc. Similar groups work at the State and District levels too. The long-range monsoon forecast of the IMD issued at the end of May and the mid-term adjustment by end of July, promise options for long-term contingency planning and adoption of alternative agricultural strategies for the *Kharif* crops. Swaminathan (1987) has argued in favour of adoption of a drought code that clearly defines necessary steps to maximize production for the rain-fed areas. Similarly, he also favoured a good weather code for a normal monsoon to exploit it for the advantage of crop production. The National Centre for Medium-range Weather Forecasting, New Delhi, under the Department of Science & Technology provides special advisories to farmers through their network to combat the vagaries of weather development for the crop production.

Concluding Remarks

Drought is a hazard in India, which is caused by abnormal fluctuations in the global and regional climate. Research in meteorological, agricultural and hydrological droughts now provides a handle to monitor the creeping phenomenon on real time and adjust strategies by different Government agencies to minimize the societal impacts of drought. With the specially designed relief measures, employment generation schemes, other schemes for the DPA and for the rural poor and poverty alleviation or other programmes the adverse impacts of drought are mitigated in much better way at present than in the past. Food procurement policy and the public distribution system with more than adequate stocks in central reserves have generated food security in India and avoided hunger during drought conditions. In areas vulnerable to drought special programmes, like watershed development approach with the cooperation of the rural communities are yielding good results. The public response during drought conditions and work of the NGOs in drought related relief also help in providing relief. Different inputs from science and technology sectors are deployed very effectively to combat and manage drought. As a result of several policy initiatives by the Government in post independent India, a resilience has been built in the social-economic sector to bear the drought shock and come out of it much less scathed than in the past eras. Improved monsoon forecasting on short, medium and long range scales provides an effective instrument in the hands of public and Government sectors to tackle the drought conditions effectively. Thus science and technology, human ingenuity and work of NGOs have provided opportunities to manage drought effectively and make India drought-proof. Reports by the Department of Science and Technology and by Narasimhan (1998) on the science and technology in drought and management of drought in 1987 stand testimony to India's effort in breaking the nexus between drought and famine in the case of 1987 drought under a democratic setup. Early response and dynamic approach by the Central and State Governments were the keys for the great success in management of drought in 1987. However, the problem of logistics in some States still impede progress. The Free Press in India plays a role in bringing to the notice of the Parliament, the Executives and the Judiciary the local stress caused by drought conditions as it happened in the case of the recent drought of the year 2000. It must be admitted that on the broader canvas the Indian efforts to manage and combat the hazards of monsoon droughts have been effective and are worthy of praise. Even though the Government agencies and NGOs provide succor to the drought-affected, the aspiration of the people have also increased over the years under the democratic system. There is now a need to develop very effective district level resource persons for combating drought. Also social scientists, economists, atmospheric scientists and

technologists must put their respective expert resources in combating hazards of drought in a more quantitative and effective manner to answer public questions, which are raised at times of local drought related distresses. There is also a need to create broader public awareness by the print and electronic media about the infrastructures developed under democratic India to effectively monitor, manage and combat adverse impacts of drought by using modern tools of science and technology.

Acknowledgement

The author is thankful to Prof. K.S. Valdiya in persisting me to write this article. Thanks are also due to Prof. J. Shukla, President, Centre for Ocean Land—Atmosphere, Maryland, USA for creating interest in the author for drought studies. The help provided by Dr. Tapeshwar Singh in arranging type-settings and final proof reading is deeply appreciated

REFERENCES

Bhalme, H.N., and D.A. Mooley, (1981) Cyclic fluctuations in the flood/drought area and relationship with the double (Hale) sun spot cycle. *J. Appl. Meteor.*, **20**: 1041-1048.

Bhatia, B.M. (1967) *Famines in India: A study in some aspects of the economic history of India (1860-1965).* Asia Pub. House, New Delhi.

Blanford, H., 1(884) On the connection of the Himalayan snow to the dry winds and season of droughts in India. *Proc. Ray. Soc. London*, **37**: 3-22.

Blanford, H., (1886) Rainfall of India. *Mem. Ind. Meteor. Dept.* **3**: 658 pp.

Bjerknes, J., (1969) Atmospheric teleconnections from the equational Pacific. *Mon Wea. Rev.*, **97**; 163-172.

Charney, J.G., and J. Shukla, (1981) Predictability of monsoons. *Monsoon Dynamics*, eds. James Lighthill and R.P. Pearce, Cambridge Univ. Press, pp. 99-100.

Chowdhury, A., M.M. Bavedkar and P.S. Raut, (1989) Variability in drought incidence over India – a statistical approach. *Mausam*, **40**: 207-214.

Gadgil, S., Y.P. Abrol and P.R. Seshagiri. (1999) On growth and fluctuations of Indian grain production. *Current Science*, **76**: 548-555.

Gowariker, V., V. Thapliyal, R.P. Sarker, G.S. Mandal, N. Sen Roy and D.R. Sikka, (1990) A power regression models a new approach to long-range forecasting of monsoon rainfall in India. *Mausam*, **42**. 125-130.

I.M.D. (1971) Rainfall and Droughts in India. *Ind. Meteor. Dept. Pune.*

Indian Famine Commission Report (1880) Blue Book, London.

Indian Famine Commission Report (1898) Govt. of India, Calcutta, 371 pp.

Indian Famine Commission Report (1901) Govt. of India, Calcutta, 207 pp.
Joseph, P.V., (1976) Climate cycles in monsoon and cyclones 1871-1974. *Proc. Symp Tropical Monsson* (IITM, Pune), 378-389.
Koteswaram P., (1976) Climate studies of drought in Asiatic Monsoon-particulalry India. *Proc. Nat. Sci. Acad.*, **54**: 1-14.
Kriplani, R.H., and A. Kulkarni, (1997) Climatic impact of El Nino/La Nina on the Indian monsoon. *Weather*, **52**: 39-46.
Kulshrestha, S.M., (1997) Management of monsoonal droughts in India. COLA/CARE Technical Rep., 1, *Calverton, MD, Institute of Global Environment and Society*, USA, pp. 105.
Mooley, D.A., and G.B. Pant, (1981) Droughts in India over the last 200 years, their socio-economic impacts and remedial measures. In *Climate and History*, (eds.) T.M.L. Wigley, M.J. Ingram and G. Farmer (Cambridge University Press).
Mooley, D.A. and B. Parthasarathy, (1982) Fluctuation in the deficiency of summer monsoon over the India and their effect on economy. *Arc. Meteor. Geophys. Bioklim., Ser. B.*, **30**: 383-398.
Palmer, T.N., C. Brankovic, P. Viterbo and M.J. Moller, (1992) Modelling interannual variations of summer monsoons. *J. Climate*, **5**, 399-417.
Parthasarathy, B., A.A. Munot and D.R. Kothawale, (1995) Monthly and seasonal rainfall series for all-India, homogeneous regions and meteorological sub-divisions. *IITM Rsearch Rep. No. RR 065.*
Parthasarathy, B. Rapa Kumar and D.R. Kothwale (1992) Indian summer monsoon rainfall indices 1871-1990. *Meteor. Mag.***123**: 174-186.
Parthasarathy, B., K. Rupakumar, and A.A. Munot, (1992) Forecast of rainy season food grain production based on monsoon rainfall. *Ind. J. Agri. Sci.*, **62**: 1-8.
Parthasarathy, B., and S. Yang, (1995) Relationship between regional Indian summer monsoon rainfall and Eurasian snow cover. *Adv. Atmos. Sci.*, **12**: 143.
Prasad, K.D., and S.V. Singh, (1992) Possibility of predicting Indian summer monsoon rainfall on reduced spatial and temporal scales. *J. Climate*, **5**: 1357-1361.
Ramdas, L.A., (1950) Rainfall and agriculture. *Ind. J. Meteor. Geophys.* **1**, 262.
Sen, A., 1981. Poverty and famines. An eassy on entitlement and deprivation. Clarendon Press, Oxford.
Sikka, D.R., (1980) Some aspects of the large-scale fluctuations of summer monsoon rainfall over India in relation of fluctuations on planetary and regional scales circulation parameters. *Proc. Ind. Acad. Sci. (Earth Plan. Sci.)*, 89, 179-195.
Sikka, D.R. (1999) Monsoon Drought in India. Joint COLA/CARE Technical Report No. 2. IGES, Kalverton, MD, USA, p. 270.
Singh, N., (1995) Large-scale interannual variation of the summer monsoon over India and its empirical prediction. *Proc. Ind. Acad. Sci. (Earth Plan. Sci.)*, 104: 1-36.
Singh, N., S.S. Mulye and G.B. Pant, (1992) Some features of the arid area variations over India, 1871-1984, *Pure Appl. Geophys.*, **138**: 135-150.
Singh,T. (1978) *Drought prone areas in India.* People's Publishing House (PPH), New Delhi.

Subrahmanyan, V.P. (1967) *Incidence and spread of continental drought*, WMO.IHD. Report No., WMO, Geneva.

Swaminathan, M.S., (1987a) The Indian experience, *Monsoons*, eds. J.S. Fein and P.L. Stephens, Wiley and Sons, New York, 121-132.

Swaminathan, M.S., (1987b) Prediction and warning of systems and international, government and public response : A problem for the future. *Monsoons, eds.* J.S. Fein and P.L. Stephens, Wiley and Sons, New York, 607-620.

Thornthwaite, C.W., (1948) An approach toward a rational classification of climate. *Geog. Rev.*, **38**: 55-94.

Webster, P.J., and S. Yang, (1992) Monsoon and ENSO: Selectively interacting Systems, *Quart. J. Roy. Meteor. Soc.*, 118: 877-926.

Webster, P.J. and S.Yang (1998) Monsoon process, predictability and prospects of prediction. *J. Geophy. Res.*, 103 : 14451-14510.

34

Recent Temperature Trends in Taiwan and their Spatial and Temporal Variabilities

Dirk Schaefer and *Manfred Domroes*

Introduction

A large amount of studies give clear evidence of the fact that climate has rapidly changed over the past 100 years showing, on a global scale, an increase of surface air temperature at around 0.5°C (Houghton *et al.*, 1990). Due to its great societal impact, spatial and temporal downscaling is of special interest in order to describe and understand recent climate change. The results of climate diagnosis are the basis for building models.

Focusing on Taiwan, so far only some studies have been carried out about recent climate change. Therefore, the question remains open whether climate change in Taiwan corresponds with the global observations or not. A study on temperature change in Taiwan therefore contributes to bridge the gap of lacking information. The aims of this study are as follows:

(a) to analyse temporal and spatial variabilities of recent temperature change in Taiwan, and
(b) to discuss the development of temperature in Taiwan with global warming.

Data and Methods

Monthly temperature records were collected for 21 meteorological stations under the Central Weather Bureau in Taiwan with the longest available time series under record that covers a 101 years period of unbroken observations for five stations (1897-1997). The observation periods for the other stations are accordingly shorter (Table 34.1). Therefore, the secular period (1897-1997) was compared with the more recent period (1960-1997). The stations are scattered over Taiwan, including five stations in the mountains (Table 34.1, Map 34.1).

The reference data of the global temperature anomalies were taken from Jones (1999). In order to analyse the annual and seasonal features of temperature variabilities and trends, the means for the year and the seasons were computed. Seasons were defined as follows:

Winter : December-February
Spring : March-May
Summer : June-September
Autumn : October-November.

Before analysing the data, the station histories were studied and a statistical test (ABBE) was applied on the time series in order to test the homogeneity of the temperature data under study. Afterwards different statistical methods were applied to achieve the above defined aims, such as low-pass filtering, trend analysis, spectral analysis and principal component analysis (for details see Schaefer, 1996).

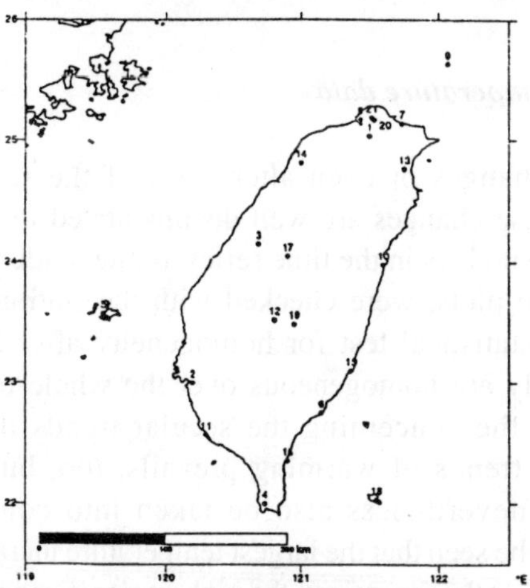

Fig. 34.1 : The location of the meteorological stations under study.

Table 34.1 : Stations under study with their location, elevation and observation period

No.	*WMO No.*	*Station*	*Latitude*	*Longitude*	*Elevation, m*	*Observation Period*
1	6921	Taipei	25.03	121.50	5	1897-1997
2	7410	Tainan	23.00	120.20	14	1897-1997
3	7490	Taichun	24.15	120.68	84	1897-1997
4	7590	Hengchun	22.00	120.73	22	1897-1997
5	7350	Penghu	23.57	119.55	11	1897-1997
6	7660	Taitung	22.75	121.15	9	1901-1997
7	6940	Keelung	25.13	121.73	27	1903-1997
8	6900	Tanshui	25.17	121.43	19	1903-1997
9	6950	Pengchiayu	25.63	122.07	102	1910-1997
10	6990	Hualien	23.98	121.60	16	1911-1997
11	7440	Kaohsiung	22.57	120.30	2	1931-1997
12	7530	Alishan	23.52	120.80	2,413	1933-1997
13	7080	Ilan	24.77	121.75	7	1936-1997
14	7571	Hsinchu	24.82	121.00	27	1938-1997
15	7610	Chengkung	23.10	121.37	33	1940-1997
16	7540	Tawu	22.35	120.90	8	1940-1997
17	7650	Jihyehtan	24.05	120.90	1,015	1942-1997
18	7620	Lanyu	22.03	121.55	324	1942-1997
19	7550	Yushan	23.48	120.95	3,845	1943-1997
20	6930	Chutzehu	25.17	121.53	607	1946-1997
21	6910	Anpu	25.18	121.52	838	1947-1997

Results

Homogeneity of temperature data

There are some changes or even alterations of the stations, but however, all of these changes are well documentated in the station histories. There are no lags in the time series; outliers, identified by a visual check of the plots, were checked with the station histories. According to the statistical test for homogeneity after ABBE, all stations under study are homogeneous over the whole observation period. Therefore the concerning the secular trends in seasonal respects, positive trends of warming prevails, too, but seasonal differences must nevertheless also be taken into consideration (Table 34.2). It can be seen that the largest temperature increase occurs in summer showing, at the same time, the highest significance. Ranking second, autumn shows also a large increase of temperature over the secular time series, though the trends at the stations under study are less significant than in summer. From the magnitude of warming, spring ranks third and winter forth with the lowest (but still positive) trend at a lesser level of significance. Not all the computed seasonal trends are linear (T/N-values >1.96), but all trends are significant positive because of the high values of Mann's Q (>1.96).

Such observations clearly explain that summer (and autumn) are the governing season(s) for the large annual temperature increase as shown above. These seasonal observations in Taiwan are, however, controversial to the global warming observations that clearly identify winter warming as the main reason of the annual temperature increase.

Short-term Trends of Temperature (1960-1997)

In order to study the more recent temperature development trends were also computed for the period 1960-1997. Comparing the computed trends for both the secular and the more recent period it can be seen whether there are differences in the trend behaviour over time or if the trends are stable.

Table 34.2 : Trend computations for annual temperature in Taiwan (1960-1997). T/N= Trend-to-Noise-Ratio; values >1.96 indicate a linear and significant trend. Mann's Q; values > 1.96 indicate a non-linear and significant trend

WMO No.	*Station*	*Trend*	*T/R*	*Mann's Q*
6921	Taipei	0.68	1.64	3.12
7410	Tainan	0.31	1.02	1.76
7490	Taichung	0.59	1.82	3.47
7590	Hengchun	-0.10	-0.35	-0.79
7350	Penghu	0.09	0.25	0.40
7660	Taitung	0.51	1.37	2.28
6940	Keelung	0.37	1.23	2.55
6900	Tanshui	-0.26	-0.79	-1.13
6950	Pengchiayu	0.21	0.75	1.35
6990	Hualien	0.67	1.71	3.40
7440	Kaohsiung	0.95	2.19	4.05
7530	Alishan	0.79	1.85	3.25
7080	Ilan	0.45	1.35	2.48
7571	Hsinchu	0.20	0.66	1.47
7610	Chengkung	0.38	1.11	1.85
7540	Tawu	0.00	0.01	0.20
7650	Jihyehtan	-0.13	0.45	-0.50
7620	Lanyu	0.36	1.29	2.28
7550	Yushan	0.51	1.38	2.29
6930	Chutzehu	0.27	0.94	1.62
6910	Anpu	0.27	0.90	1.79

Increasing trends can be seen for annual temperature at almost all stations under study for the more recent period. The trend values and values of significance are lower for the more recent period than the long-term trends. Kaohsiung shows the highest increase of 0.95°C over the last 38 years (Table 34.2). Concerning the spatial differences only small variabilities can be seen (Fig. 34.2). While increasing trends prevail over most parts of Taiwan, a slight decrease can only be seen for a small area in the northern part resulting from a slight decrease of 0.26°C for Tanshui.

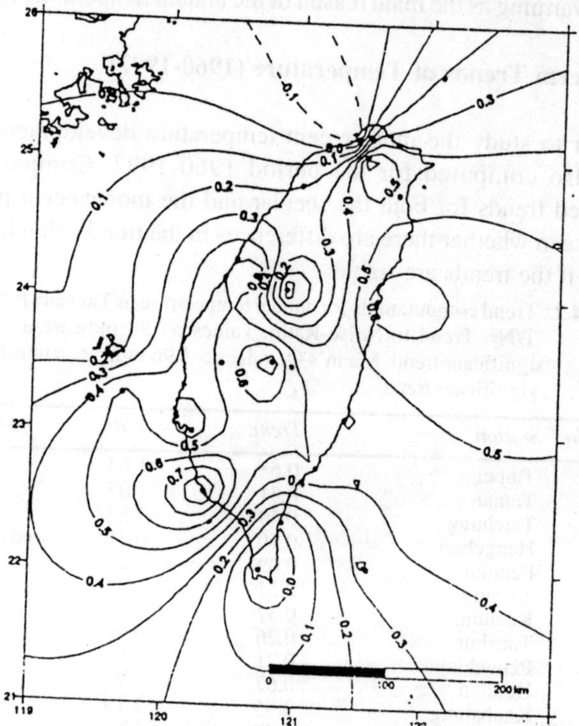

Fig. 34.2 : Spatial distribution of trends [°C] of annual temperature for the more recent period 1960-1997.

Concerning the seasonal temperature trends, positive values of warming prevail over Taiwan; yet seasonal differences must be mentioned, the largest temperature increase with the highest significance occurs in winter. Therefore, winter is the governing season for the annual temperature increase for the more recent periods (1960-1997). These seasonal trends in Taiwan correspond with the global warming observations identifying winter as the season with strongest increasing temperature trends.

Periods and Teleconnections

In order to detect any periods or quasi-periods in the secular time series 1897-1997 two different methods of power spectrum are applied, the Autocorrelation Spectral Analysis and the Maximum Entropy Spectral Analysis. The results of Maximum Entropy Spectral Analysis are more sophisticated but the testing of significance is more complicated (Monte-Carlo Tests). Therefore both methods can be combined in order to achieve best results in significance and spectral resolution of power spectrum.

Some periods were found which can be related to the well-known teleconnection of the El Niño/Southern Oscillation-phenomenon (ENSO), with periods of about 3.0-8.0 years. Besides the signals of the Quasi-Biannual-Oscillation (QBO) with periods of about 2.0-2.7 years can be found in the time series under study. Notably, the 11-year sunspot-cycle cannot be found in the temperature series of Taiwan.

All-Taiwan Temperature Series

A Principal Component Analysis (PCA) was used to construct an "All-Taiwan Temperature Series" for annual and seasonal temperatures. The time series of annual temperatures for all five stations were taken to build the data matrix (101*5): 101 cases which correspond to the temperature values, and five variables, corresponding to the five stations. All stations are well correlated ($r \geq 0.8$) and thus only one Principal Component (PC) is extracted that explains about 90 per cent of the total variance. The first PC describes the interannual temperature variability over Taiwan. The PC-values were constructed applying a multiple regression equation. Applying a back-transformation to the standardized PC1-values, the annual temperature values can be taken as an All-Taiwan Annual Temperature Series. The same procedure was applied for the seasons.

All five stations under study are highly correlated and thus the computed trends show comparable trends: Over the whole observation period a highly increasing and significant linear trend of 1.16°C can be identified for annual temperature (Table 34.3), whereas the global temperature shows an increase of 0.60°C for the same period. Obviously, temperature increase in Taiwan is double compared with global warming. Increasing trends can also be observed for the seasons, but only the trend of summer temperature is linear, the trends for the other seasons are, however, increasing and significant, but not linear (Table 34.3).

Table 34.3 : Computed trends of an All-Taiwan Temperature Series, in °C, their statistical significance (T/N) and Mann's Q-values for annual and seasonal means (1897-1997). T/N = Trend-to-Noise-Ratio; values > 1.96 indicate a linear and significant trend. Mann's Q; values > 1.96 indicate a non-linear and significant trend.

	Trend	*T/N*	*Mann's Q*
Year	1.16	2.57	8.05
Winter	0.78	1.08	3.47
Spring	1.19	1.75	5.11
Summer	1.14	2.63	8.42
Autumn	1.25	1.82	5.72

The annual and seasonal temperatures of the All-Taiwan Temperature Series are shown in graphs along with the 10-year Gaussian binominal low-pass filtered curves and the linear trends of temperature, resulting from the regression analysis trends over the secular time series (101 yr, 1897-1997) (Fig. 34.3). The annual and seasonal trends of the All-Taiwan Temperature Series show strong increases. Notably, the inter-annual variability of the All-Taiwan Temperature Series is high, especially in winter. A higher inter-annual variability means a higher statistical noise and consequently the significance of the trends are lower.

Acknowledgements

The authors would like to thank the DAAD (German Academic Exchange Service) and the NSC (National Science Council, Taiwan) for providing funds for the project. Monthly temperature data for the

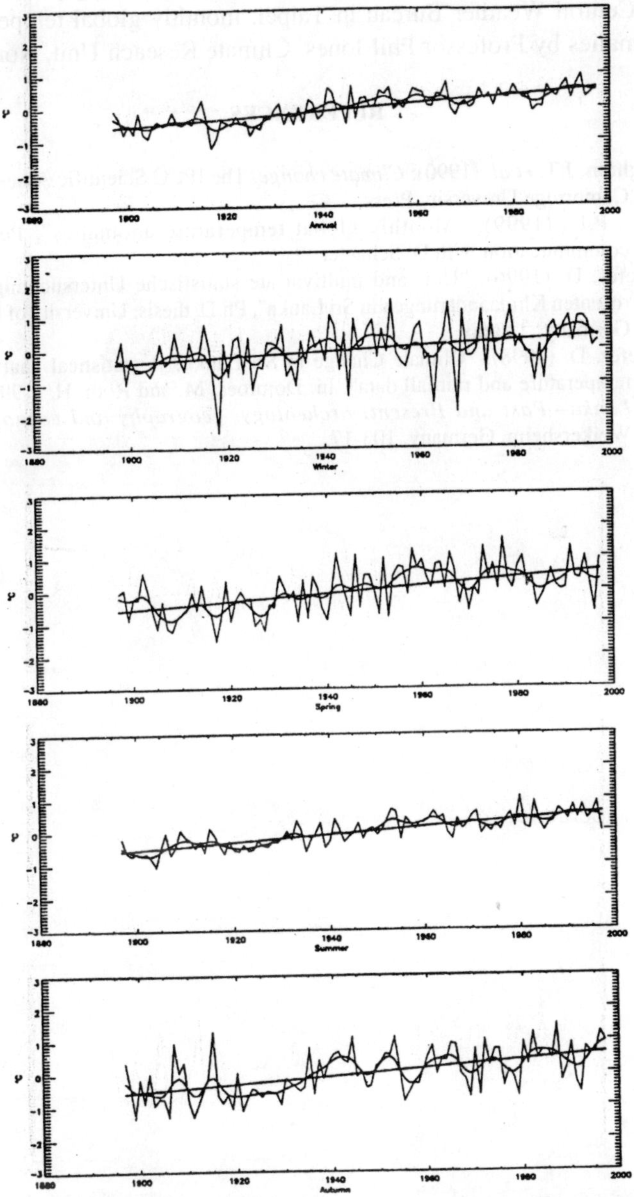

Fig. 34.3 : Annual and seasonal temperature of the All-Taiwan Temperature Series (thin line) along with the low-pass filtered values (smoothed line) and the trend line (1897-1997).

Taiwan meteorological stations under study were kindly provided by the Central Weather Bureau in Taipei, monthly global temperature anomalies by Professor Phil Jones, Climate Reseach Unit, Norwich.

REFERENCES

Houghton, J.T. *et al.* (1990): *Climate change*. The IPCC Scientific Assessment. Cambridge University Press.

Jones, P.D. (1999): "Monthly global temperature anomalies". Personal communication with D. Schaefer.

Schaefer, D. (1996): "Uni- and multivariate statistische Untersuchungen zu rezenten Klimaänderungen in Sri Lanka", Ph.D. thesis, University of Mainz, Germany, 226 pp.

Schaefer, D. (1998): "Climate Change in Sri Lanka? A statistical analysis of temperature and rainfall data". In: Domroes, M. and Roth, H. (1998): *Sri Lanka—Past and Present: archeology, geography and economics*", Weikersheim, Germany, 103-17.